页岩气开发理论与实践

（第二辑）

贾爱林　位云生　编著

石油工业出版社

内 容 提 要

页岩气开发是非常规气开发的重要组成部分。本书分为综合篇、开发地质篇、气藏工程篇、生产应用篇四部分内容，汇编了国内页岩气开发领域的最新研究成果，以及部分国外页岩气开发的最新进展，可为国内页岩气开发提供理论参考和方法借鉴。

本书可供从事页岩气等非常规天然气开发的科研人员使用，也可以作为高等院校相关专业师生的参考用书。

图书在版编目（CIP）数据

页岩气开发理论与实践．第二辑／贾爱林等编著
．—北京：石油工业出版社，2019.8
ISBN 978-7-5183-3525-1

Ⅰ．①页… Ⅱ．①贾… Ⅲ．①油页岩资源－油气田开发－文集 Ⅳ．①P618.130.8-53

中国版本图书馆 CIP 数据核字（2019）第 165429 号

出版发行：石油工业出版社
（北京安定门外安华里 2 区 1 号　100011）
网　址：www. petropub. com
编辑部：（010）64523708　图书营销中心：（010）64523633
经　销：全国新华书店
印　刷：北京中石油彩色印刷有限责任公司

2019 年 9 月第 1 版　2019 年 9 月第 1 次印刷
787×1092 毫米　开本：1/16　印张：18.25
字数：450 千字

定价：130. 00 元
（如出现印装质量问题，我社图书营销中心负责调换）

前言

中国页岩气资源丰富，技术可采资源量达31.6万亿立方米（据EIA预测），仅次于美国，同时中国也是世界上除北美地区之外，少数获得页岩气规模开发的国家之一，特别是近几年，中国页岩气发展迅猛，年产量从2014年的13亿立方米快速增长至2018年的108亿立方米，成为继美国、加拿大之后，世界第三大页岩气生产国。在国内非常规气藏开发生产中，页岩气已成为继致密气之后第二大非常规气藏类型。

中国页岩气的跨越式发展，得益于四川盆地及其周缘长宁—威远、昭通、涪陵三个国家级页岩气示范区海相页岩气开发技术的长足进步。“十二五”期间，在政府大力投资和补贴政策的激励下，通过系列技术攻关和开发试验，形成了埋深3500米以浅海相页岩气藏储层评价、优快钻井、体积压裂、产能评价和开发技术优化五大主体技术系列：高演化强改造复杂山地地球物理资料解释及储层综合评价技术、水平井地质工程一体化的优快钻井技术、以大排量滑溜水+低密度支撑剂+可溶桥塞+拉链式压裂模式为主的水平井体积压裂工艺技术、页岩气体积压裂水平井概率性产能评价技术、水平井及裂缝参数综合优化技术。这些技术的试验与应用，提高了单井产量，降低了开发成本，奠定了中国页岩气规模开发的技术基础。“十三五”前三年，中国埋深3500米以深的南方海相页岩气获得了多个井点突破，有望在“十三五”末及未来的页岩气规模开发中发挥主力军作用，同时为北方大面积的陆相及海陆过渡相深层页岩气资源开发技术的突破提供借鉴和示范作用。

《页岩气开发理论与实践》系列论文集，立足中国页岩气勘探开发所取得的成果，收集整理了近年来具有代表性的学术论文，按照综合篇、开发地质篇、气藏工程篇、生产应用篇进行分类编辑，接续出版，希望在总结页岩气勘探开发理论和技术成果的同时，在推广应用方面发挥积极作用，并对新理论、新技术的探索有所启发，从而对国内页岩气勘探开发技术的进一步发展起到推动作用。

目　录

一、综合篇

二、开发地质篇

三、气藏工程篇

四、生产应用篇

一、综合篇

井控页岩气可动地质储量和可采储量的评价方法

陈元千，齐亚东，傅礼兵，位云生

（中国石油勘探开发研究院）

摘　要: 页岩气是一种非常规的天然气资源。页岩气藏由超致密基质和天然裂缝系统组成。基质中的页岩气是自生、自储和自闭的吸附气，裂缝系统中的页岩气为自由气，其中吸附气是页岩气的主体部分，必须通过打水平井进行多段压裂，才能由压裂造成的裂缝面解吸出来。这种被水平井控制的解吸气和自由气量总称为井控页岩气可动地质储量和可采储量。评价页岩气藏基质中的吸附气资源量和裂缝系统中自由气的资源量，通常采用体积法。其有效应用受基质中原始吸附气含量、页岩视密度、裂缝系统的有效孔隙度和原始地层压力可靠性的影响。应当指出，中国于2014年颁布的《页岩气资源/可采储量评价规范》规定的体积法公式是错误的，而且也缺少动态法应用的具体内容。因此，根据页岩气以井为开发单元的特点，提出利用动态法评价井控页岩气的可动地质储量和可采储量方法，且页岩气井实例的应用结果表明，所建方法是实用有效的。

关键词: 页岩气；井控；可动地质储量；可采储量；评价方法

中国拥有比较丰富的页岩气资源，近几年来已逐步进入工业开发阶段。2016年页岩气的年产量已近$80\times10^8m^3$，约占全国天然气总产量的5.9%，并拥有近500口页岩水平气井。页岩气是一种非常规气，由超致密基质中的吸附气和裂缝系统的自由气组成，而吸附气是页岩气资源的主体。由于页岩基质渗透率为平方纳米级，比毫达西小3个数量级，如此低的基质渗透率，根本不存在吸附气的解吸和流动条件。因此，必须通过打水平井，并进行大型多段水力压裂，才能使吸附于裂缝面的气发生解吸，与储存自由气的天然裂缝系统相沟通，形成可用于页岩气可动地质储量和可采储量的井控评价单元。令人遗憾的是，在中国评价页岩吸附气和自由气资源量的DZ/T 0254—2014[1]中规定的体积法是错误的[2]。就体积法本身而言，影响页岩气资源量评价可靠性的因素很多，比如，页岩的原始吸附气含量、页岩的视密度、天然裂缝系统的有效孔隙度和原始地层压力等。

在拥有页岩气井大量动态资料的情况下，如何利用动态法评价井控页岩气可动地质储量和可采储量，陈元千等[3]已做出了有效尝试。对于垂直气井，陈元千[4]于1991年提出的垂直井压降曲线拟稳态方程，已被广泛应用于四川盆地大量定容裂缝性气藏探明地质储量的评价，并被引入SY/T 6098—2000和SY/T 6098—2010[5-6]。基于文献［7-8］的研究成果，笔者重新推导了评价水平气井井控页岩气可动地质储量的压降曲线拟稳态方程。陈元千[9]于2005年提出的快速直接评价油气井、油气藏和油气田可采储量的方法，于2010年被引入SY/T 5367—2010[10]。基于文献［11-12］的研究成果，经推导得到由文献［13］利用综合归纳法提出的直接快速评价井控页岩气可采储量的广义递减模型法。

1 评价方法的建立

1.1 评价井控页岩气可动地质储量的拟稳态法

压降曲线的拟稳态法在中国又称为弹性二相法。在稳定产量生产的条件下，压降曲线存在非稳态和拟稳态两大阶段。当气井以稳定产量开井生产时，井底流压随时间呈直线下降，当井底流压变化率为常数时，气井的生产动态处于拟稳态阶段。从理论上讲，由于地层气体黏度和气体偏差系数都是压力的函数，因此，应用拟压力表示压降曲线方程更为合适。但对页岩气井来说，由于页岩吸附气的解吸需要明显降低井口和井底流压，因此，当连通体内的压力低于 15MPa 时，应用压力平方代替拟压力可以得到比较满意的结果。目前，中国许多气井的稳定试井和不稳定试井基本上都采用压力平方法表示。尽管陈元千[4]已对垂直气井的压降曲线拟稳态方程进行了理论上的推导，但考虑到该方法在评价井控页岩气可动地质储量方面的重要性和页岩水平气井的特点，笔者又进行了一次简明的推导。当水平气井以稳定产量生产，而压力动态已达到拟稳态时，以压力平方法表示的气井产量公式[4, 8]为

$$q_{\mathrm{g}}=\frac{2.714\times10^{-5}KhT_{\mathrm{sc}}\left(\bar{P}_{\mathrm{R}}^{2}-P_{\mathrm{wf}}^{2}\right)}{P_{\mathrm{sc}}\mu_{\mathrm{gi}}Z_{\mathrm{i}}T\left(\ln\frac{r_{\mathrm{ed}}}{r_{\mathrm{w}}}-\frac{3}{4}+S\right)} \tag{1}$$

将式中水平井的椭圆形驱动半径转为产量与垂直井相等的等效圆形驱动半径[7, 8]：

$$r_{\mathrm{ed}}=r_{\mathrm{w}}^{(1-h/L)}\left(\frac{h}{2}\right)^{h/L}\left[\left(\frac{4a}{L}-1\right)^{2}-1\right] \tag{2}$$

将式（1）改为

$$P_{\mathrm{wf}}^{2}=\bar{P}_{\mathrm{R}}^{2}-\frac{3.684\times10^{4}q_{\mathrm{g}}\mu_{\mathrm{gi}}Z_{\mathrm{i}}TP_{\mathrm{sc}}}{KhT_{\mathrm{sc}}}\left(\ln\frac{r_{\mathrm{ed}}}{r_{\mathrm{w}}}-\frac{3}{4}+S\right) \tag{3}$$

对于由水平气井控制的页岩气连通体，当气井的产量从 t_0 时间进入拟稳态时的物质平衡方程式[14]为

$$q_{\mathrm{g}}\left(t-t_{0}\right)B_{\mathrm{g}}=G_{\mathrm{wc}}B_{\mathrm{gi}}C_{\mathrm{t}}^{*}\left(P_{\mathrm{i}}-\bar{P}_{\mathrm{R}}\right) \tag{4}$$

将式（4）等号两端同乘以 $\left(P_{\mathrm{i}}+\bar{P}_{\mathrm{R}}\right)$ 可得

$$P_{\mathrm{i}}^{2}=\bar{P}_{\mathrm{R}}^{2}+\frac{q_{\mathrm{g}}\left(t-t_{0}\right)B_{\mathrm{g}}\left(P_{\mathrm{i}}+\bar{P}_{\mathrm{R}}\right)}{G_{\mathrm{wc}}B_{\mathrm{gi}}C_{\mathrm{t}}^{*}} \tag{5}$$

其中，

$$C_t^* = C_{gi} + \frac{C_w S_{wi} + C_f}{S_{gi}} = C_{gi} + C_e \tag{6}$$

在拟稳态阶段早期，$\overline{P}_R \approx P_i$，$B_g \approx B_{gi}$，由式（5）可得

$$P_i^2 = \overline{P}_R^2 + \frac{2q_g P_i}{G_{wc} C_t^*} \tag{7}$$

将式（3）减式（7）得压降曲线拟稳态阶段的关系式：

$$P_{wf}^2 = P_i^2 - \frac{3.684 \times 10^4 q_g \mu_{gi} Z_i T P_{sc}}{K h T_{sc}} \left(\ln \frac{r_{ed}}{r_w} - \frac{3}{4} + S \right) - \frac{2q_g P_i (t - t_0)}{G_{wc} C_t^*} \tag{8}$$

将式（8）简写为

$$P_{wf}^2 = \alpha - \beta t \tag{9}$$

其中，

$$\alpha = P_i^2 - \frac{3.684 \times 10^4 q_g \mu_{gi} Z_i T P_{sc}}{K h T_{sc}} \left(\ln \frac{r_{ed}}{r_w} - \frac{3}{4} + S \right) + \beta t_0 \tag{10}$$

$$\beta = \frac{2q_g P_i}{G_{wc} C_t^*} \tag{11}$$

根据实测的井口压降曲线拟稳态阶段的直线关系，由式（9）进行线性回归，求得 α 和 β 的数值后，再将式（11）改写为评价井控页岩气井的可动地质储量关系式为

$$G_{wc} = \frac{2q_g P_i}{\beta C_t^*} \tag{12}$$

由于 $C_e \ll C_{gi}$，因此，式（6）可写为

$$C_t^* = C_{gi} = \frac{1}{P_i} - \frac{\partial Z}{Z \partial P} \tag{13}$$

由于 $\frac{\partial Z}{Z \partial P} \ll \frac{1}{P_i}$，故式（13）可以比较精确地简写为

$$C_t^* = \frac{1}{P_i} \tag{14}$$

将式（14）代入式（12）得评价井控可动地质储量的关系式为

$$G_{wc} = \frac{2q_g P_i^2}{\beta} \tag{15}$$

对于一口页岩水平气井，通过进行大型分段水力压裂后，从井口到水平井段，形成了一个由水平井主导的人工裂缝系统与天然裂缝系统相沟通的定容连通体。当水平气井以稳定产量生产，而压力动态达到拟稳态时，从井口到井底再到水平段内连通体内任意位置的

流动压力将进入等速同步下降的拟稳态阶段，此时的 $\frac{\mathrm{d}P_{\mathrm{whf}}^2}{\mathrm{d}t}=\frac{\mathrm{d}P_{\mathrm{wf}}^2}{\mathrm{d}t}=\frac{\mathrm{d}P_{\mathrm{xf}}^2}{\mathrm{d}t}=\beta=$ 常数。因此，可以利用井口流压的 β 值代替井底流压的 β 值，于是就可利用式（15）评价井控页岩气可动地质储量。

1.2 评价井控页可采储量的产量递减法

产量递减法是评价油气井、油气藏和油气田可采储量和剩余可采储量的重要方法，也是世界各石油公司和国际评估公司进行储量资产上市评估的主要方法。它的有效应用不受油气藏储集类型、驱动类型和开发方式的限制，只要产量进入递减阶段即可有效应用。Arps 于 1945 年提出的 3 种经典的产量递减法，只适用于投产即进入递减的模式，而且不能直接进行可采储量的预测。为此，笔者提出可以直接快速评价可采储量的方法。

设 t_0 为从投产计时进入递减阶段的开始时间，$G_{\mathrm{p}t_0}$ 为生产到 t_0 时间的总累计产量，q_{i} 为进入递减评价阶段的初始产量，则截止到 t 时间（$t>t_0$）总累计产量为

$$G_{\mathrm{p}t}=G_{\mathrm{p}t_0}+\int_{t_0}^{t}q_{\mathrm{g}}\mathrm{d}t \tag{16}$$

在产量递减阶段，文献［11］提出了产量与时间的广义关系式为

$$q_{\mathrm{g}}=\frac{q_{\mathrm{i}}}{\left[1+D_{\mathrm{i}}\left(1-m\right)\left(t-t_0\right)\right]^{\frac{1}{1-m}}} \tag{17}$$

将式（17）代入式（16）得文献［2］提出的累计产量与产量关系式为

$$G_{\mathrm{p}t}=G_{\mathrm{p}t_0}+\frac{q_{\mathrm{i}}}{mD_{\mathrm{i}}}\left[1-\left(\frac{q_{\mathrm{g}}}{q_{\mathrm{i}}}\right)^m\right] \tag{18}$$

将式（18）整理简化后得文献［13］利用综合归纳法得到的广义递减模型：

$$q_{\mathrm{g}}^m=A-BG_{\mathrm{p}t} \tag{19}$$

其中，

$$A=q_{\mathrm{i}}^m+\frac{mD_{\mathrm{i}}G_{\mathrm{p}t_0}}{q_{\mathrm{i}}^{1-m}} \tag{20}$$

$$B=\frac{mD_{\mathrm{i}}}{q_{\mathrm{i}}^{1-m}} \tag{21}$$

当 $q_{\mathrm{g}}=q_{\mathrm{EL}}$ 时，由式（19）得评价经济可采储量的广义关系式为

$$G_{\mathrm{RE}}=\frac{A-q_{\mathrm{EL}}^m}{B} \tag{22}$$

当 $q=0$ 时，由式（19）得评估技术可采储量的关系式为

$$G_{RT}=\frac{A}{B} \tag{23}$$

当 $m=2$ 时，由式（19）—式（23）可得线性递减的关系式为

$$q_g^2=A_L-B_LG_{pt} \tag{24}$$

$$A_L=q_i^2+2D_iG_{pt_0}q_i \tag{25}$$

$$B_L=2D_iq_i \tag{26}$$

$$G_{RE}=\frac{A_L-q_{EL}^2}{B_L} \tag{27}$$

$$G_{RT}=\frac{A_L}{B_L} \tag{28}$$

当 $m=1$ 时，由式（19）—式（23）可得指数递减的关系式为

$$q_g=A_E-B_EG_{pt} \tag{29}$$

$$A_E=q_i+D_iG_{pt_0} \tag{30}$$

$$B_E=D_i \tag{31}$$

$$G_{RE}=\frac{A_E-q_{EL}}{B_E} \tag{32}$$

$$G_{RT}=\frac{A_E}{B_E} \tag{33}$$

当 $0<m<1$ 时，双曲线递减的评价关系式即式（19）—式（23）。现对广义递减模型的递减率推导如下：

已知 Arps 于 1945 年定义的递减率为

$$D=-\frac{dq_g}{q_g dt} \tag{34}$$

为确定广义递减模型的递减率，将式（19）对时间求导得

$$-\frac{dq_g}{q_g dt}=\frac{B}{mq_g^{m-1}} \tag{35}$$

将式（34）代入式（35）得广义递减模型的递减率为

$$D=D_i\left(q_i/q_g\right)^{m-1} \tag{36}$$

当 $m=2$ 时，由式（36）得线性递减的递减率为

$$D_L = D_i q_i / q_g \tag{37}$$

当 $m = 1$ 时，由式（36）得指数递减的递减率为

$$D_E = D_i = 常数 \tag{38}$$

当 $0<m<1$ 时，由式（36）可得不同 m 值时递减率关系式，m 值越大，则递减率越大。

2 方法应用案例

笔者通过中国 3 口页岩水平气井的实际应用，说明所建方法的实用性和有效性。

2.1 拟稳态法的应用

SG-1 井的生产历史曲线如图 1（a）所示。可以看出，曲线最后一段产气量基本稳定，井口流压呈直线下降趋势，具有达到拟稳态的条件。按照式（9）绘制得到拟稳态阶段的压降曲线，曲线最后一段是一条很好的直线（图 1b）。经线性回归求得直线的截距为 246.1，斜率为 0.9688，相关系数为 0.9981。由页岩层埋深静水压力 1.5 倍估算原始地层压力，其值为 38.25MPa，评价段的稳定产量为 $12.64 \times 10^4 m^3/d$。将这些有关参数代入式（15），得该井的井控页岩气可动地质储量为

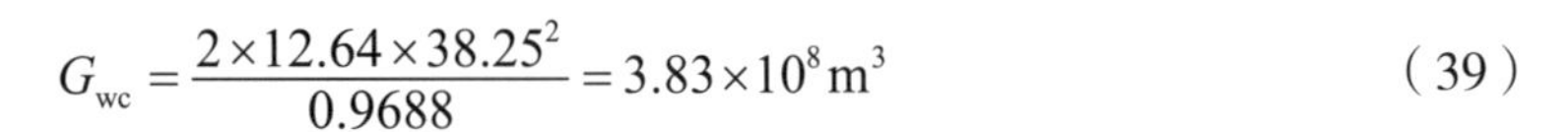

$$G_{wc} = \frac{2 \times 12.64 \times 38.25^2}{0.9688} = 3.83 \times 10^8 m^3 \tag{39}$$

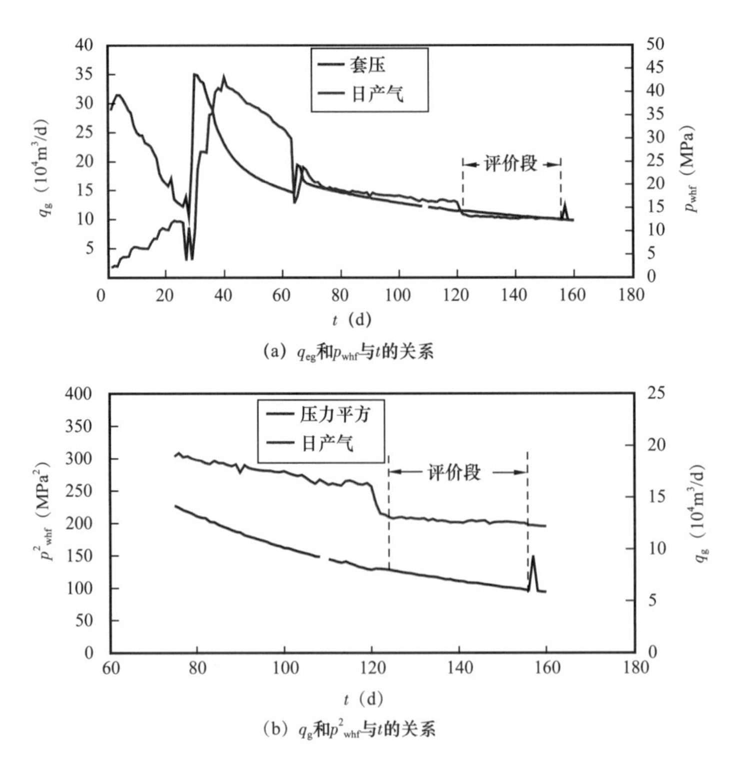

（a）q_{eg}和p_{whf}与t的关系

（b）q_g和p^2_{whf}与t的关系

图 1　SG-1 井的生产历史曲线及拟稳态阶段的压降曲线

2.2 线性递减法的应用

SG–2 井的 q_g 与 t 的关系曲线如图 2（a）所示。可以看出，该井投产 50 天后即进入线性递减阶段。在图 2（b）上绘制由式（24）表示的 q_g^2 与 G_{pt} 的直线关系，由分析可知，该井的递减属于线性递减率，m=2。经线性回归后，求得直线的截距为 1.2209，斜率为 0.0007，相关系数为 0.9486。将 A_L 和 B_L 值代入式（28），得该井的井控页岩气可采储量为

$$G_R = \frac{1.2209}{0.0007} = 1744 \times 10^4 \text{m}^3 \tag{40}$$

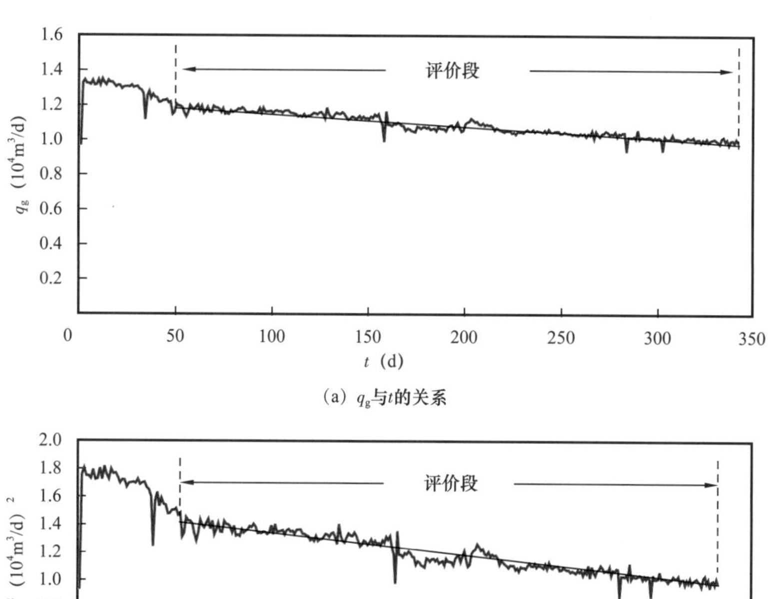

（a）q_g与t的关系

（b）$q^2{}_g$与G_{pt}的关系

图 2　SG–2 井的 q_g 与 t 及 $q^2{}_g$ 与 G_{pt} 的关系曲线

2.3 指数递减法的应用

SG–3 井的 q_g 与 t 的关系曲线如图 3（a）所示。可以看出，该井在排除钻井、完井、测井和压裂等施工液对产气量的波动影响后，即进入了正常的产量递减阶段。在图 3（b）上绘制由式（29）表示的 q_g 与 G_{pt} 的关系图可以看出，q_g 与 G_{pt} 呈很好的直线关系，因而符合指数递减，m = 1。经线性回归后，求得直线的截距 A_E = 32.138，斜率 B_E = 0.0048，相关系数 r = 0.9842。将 A_E 和 B_E 值代入式（33），得该井的井控页岩气可采储量为

$$G_{RT} = \frac{32.138}{0.0048} = 6695 \times 10^4 \text{m}^3 \tag{41}$$

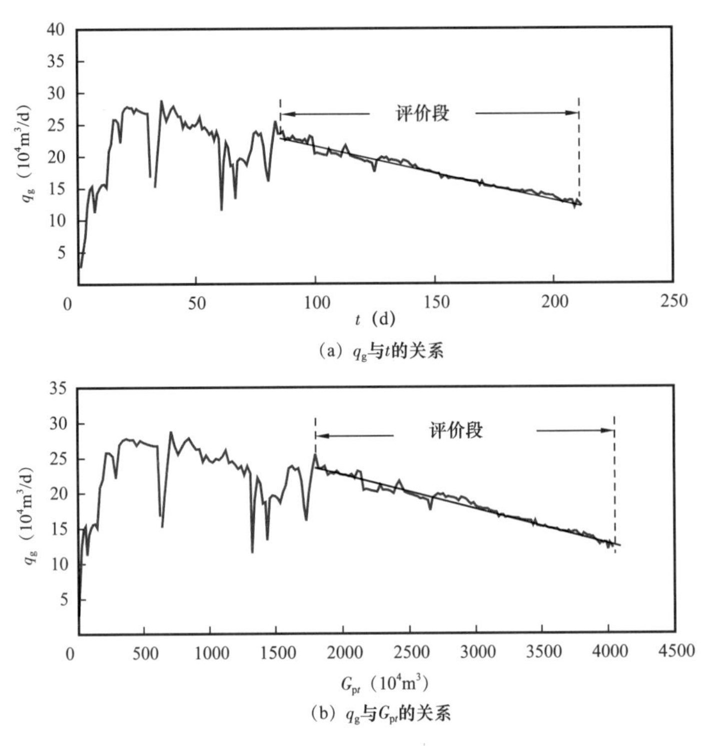

图 3　SG-3 井的 q_g 与 t 及 q_g 与 G_{pt} 的关系曲线

3　结论

页岩气是由超致密基质吸附气和天然裂缝系统的自由气组成，而且吸附气是页岩气资源的主体。根据天然裂缝的发育程度，可将页岩气藏分为裂缝性页岩气藏和基质性页岩气藏。前者是形成页岩气井高产的主要条件，后者是为页岩气井提供解吸气源的基础。评价页岩气资源量的方法主要是体积法，其可靠性取决于页岩基质的原始饱和吸附气量、页岩视密度、天然裂缝的有效孔隙度和页岩气藏的原始地层压力。令人遗憾的是，中国发布的 DZ/T 0254—2014 所规定的评价页岩气资源量的体积法是错误的。

页岩气的开发，必须通过打不同方式的水平井，并通过大型分段水力压裂，形成以水平井为主导的人工裂缝与天然裂缝相沟通的定容连通体，这就是笔者所提出的井控评价单元。笔者提出的评价井控页岩气可动地质储量和可采储量的方法，实例应用结果表明是实用、有效的。对于应用压降曲线拟稳态法评价的水平气井，当产量进入递减阶段后，应采用广义递减模型法评价井控页岩气的可采储量。对于已进入产量递减的水平气井，应当先根据气井的生产历史曲线，判断产量与时间是否呈直线关系。如果呈直线关系，则应先利用广义递减模型的线性递减法评价井控页岩气的可采储量；如果不呈直线关系，则再应用广义递减模型的指数递减法进行可采储量的评价，双曲线递减发生的概率较低。从递减的快慢而言，线性递减最快，指数递减其次，双曲线递减最慢。总之，m 值越大，产量递减越快。

页岩水平气井的生产，会受到钻井、完井、测井和水力压裂等施工液的严重影响，应当在投产初期尽快排出，否则，页岩水平气井就难以正常生产，而面临停喷停产的局面。除此之外，在页岩水平气井正常生产的时候，不要随意关井开井，避免引起气井的激动，降低气体携带凝析水的能力，导致气井过早的停喷停产。为了保持或提高页岩气井的产量，或降低产量的递减率，可能会采用换小油管、注气排液和井口安装压缩机进行负压生产等措施，对此，需做好仔细的经济评价工作。应当指出，在页岩气的开发过程中，始终存在着投入与产出的平衡问题，两者的严重失衡，将是对页岩气开发的最大挑战。

井控页岩气可动地质储量和可采储量的评价，是预测而不是算命，更不能一劳永逸。评价结果会因评价时间、评价方法和评价阶段使用的资料不同而改变。

符号注释：

P_i—原始地层压力，MPa；P_{sc}—地面标准压力（0.101），MPa；P_{wf}—井底流动压力，MPa；P_{whf}—井口流动压力，MPa；P_{xf}—水平井段任一位置的流动压力，MPa；q_g—拟稳态法的稳定产量，或产量递减法 t 时间的产量，$10^4 m^3/d$；q_i—产量递减法 t_0 时间的初始产量，$10^4 m^3/d$；q_{EL}—产量递减法的经济极限产量，$10^4 m^3/d$；K—有效渗透率，mD；h—有效厚度，m；L—水平井段长度，m；r_{ed}—水平井的等效圆形驱动半径，m；r_w—水平井段的井筒半径，m；μ_{gi}—P_i 压力下的气体黏度，mPa·s；Z_i—P_i 压力下的气体偏差系数；t—从投产计时的生产时间，d；t_0—从投产计时拟稳态的开始时间，或产量递减的开始时间，d；T—气层温度，K；T_{sc}—气体地面标准温度（293），K；α—水平井椭圆驱动面积的长轴半长，m；C_t^*—总压缩系数，MPa^{-1}；C_{gi}—P_i 压力下的气体压缩系数，MPa^{-1}；C_e—气层的有效压缩系数，MPa^{-1}；C_w—束缚水的压缩系数，MPa^{-1}；C_f—岩石的有效压缩系数，MPa^{-1}；S_{wi}—束缚水饱和度；D_i—广义递减模型法 t_0 时间的初始递减率，d^{-1}；D—广义递减模型的递减率，d^{-1}；D_L—线性递减模型的递减率，d^{-1}；D_E—指数递减模型的递减率，d^{-1}；m—广义递减模型的递减因子：$m = 2$（线性递减）；$m = 1$（指数递减）；$1>m>0$（双曲线递减），d^{-1}；G_{wc}—井控页岩气可动地质储量，$10^4 m^3$；G_{pt}—从投产计产到 t 时间的总累计产气量，$10^4 m^3$；G_{pt_0}—从投产计产到 t_0 时间的总累计产气量，$10^4 m^3$；G_{RE}—经济可采储量，$10^4 m^3$；G_{RT}—技术可采储量，$10^4 m^3$；α 和 β—拟稳态法直线的截距和斜率；A 和 B—广义递减模型直线的截距和斜率；A_E 和 B_E—指数递减法直线的截距和斜率；A_L 和 B_L—线性递减法直线的截距和斜率。

参 考 文 献

[1] 陈永武，王少波，韩征，等. DZ/T 0254—2014. 页岩气资源/储量计算与评价技术规范[S]. 北京：中国标准出版社，2014.

[2] 陈元千，周翠. 中国《页岩气资源/储量计算与评价技术规范》计算方法存在的问题与建议[J]. 油气地质与采收率，2015，22(1)：1-4.

[3] 陈元千，李剑，齐亚东，等. 页岩气藏地质储量、可采储量和井控可采储量的评价方法[J]. 新疆石油地质，2014，35(5)：547-551.

[4] 陈元千. 油气藏工程计算方法(续篇)[M]. 北京：石油工业出版社，1991：37-47.

[5] 国家石油与化学工业局. SY/T 6098—2000 天然气可采储量计算方法[S]. 北京：石油工业出版社，2001.

[6] 国家能源局 . SY/T 6098—2010. 天然气可采储量计算方法 [S]. 北京：石油工业出版社，2010.
[7] 陈元千 . 水平井产量公式的推导与对比 [J]. 新疆石油地质，2008，29（1）：68–71.
[8] 陈元千 . 确定水平井产能比、流动阻力比、驱动面积比和表皮因子的新方法 [J]. 中国海上油气，2008，17（2）：5–9.
[9] 陈元千 . 预测油气田可采储量和剩余可采储量的快速方法 [J]. 新疆石油地质，2005，26（5）：544–548.
[10] 国家能源局 . SY/T 5367—2010 石油可采储量计算方法 [S]. 北京：石油工业出版社，2010.
[11] 陈元千，周翠 . 线性递减类型的建立、对比与应用 [J]. 石油学报，2015，36（8）：983–987.
[12] 陈元千，吕恒宇，傅礼兵，等 . 注水开发油田加密调整效果的评价方法 [J]. 油气地质与采收率，2017，24（6）：60–64.
[13] 陈元千，唐玮 . 广义递减模型的建立及应用 [J]. 石油学报，2016，37（11）：1410–1413.
[14] 陈元千 . 油气藏工程实用方法 [M]. 北京：石油工业出版社，1999.

页岩气地质工程一体化建模及数值模拟：现状、挑战和机遇

鲜成钢

（斯伦贝谢公司中国地球科学与石油工程研究院）

摘　要：页岩气地质工程一体化核心内涵新模型，将钻井、完井改造、生产和开发四大工程系统有机联系，在多学科、多部门一体化共享模型基础上，从单井、平台、全气田多尺度，概括了针对四大工程系统的主要工程应用。从三维建模、压裂模拟和数值模拟等方面，回顾了国内外非常规油气藏建模及数值模拟的现状及进展。指出中国在地球物理资料应用、适时建模、气田尺度一体化共享模型、全气田数值模拟、对四大工程系统各种应用的发掘和支持等领域走在前列；北美在机理研究、基础数学模型建立、建模方法及数模新技术探索、井或平台尺度的精细化参数研究等领域更加详细和深入。从复杂天然裂缝系统定量表征、井筒完整性、平面（立体）加密井或立体布井、页岩气田全气田数值模拟、井筒动态和气态动态耦合5个领域分析了所面临的关键性现实挑战，探讨了主要研究方向和思路。

关键词：页岩气；地质工程一体化；三维建模；数值模拟；钻井；完井；压裂；裂缝系统定量表征；井筒完整性

页岩气地质工程一体化理念在国内已经被广泛接受[1-3]，该理念最早是在中国川南海相页岩气开发中被较系统明确提出和阐述的[1]。四川盆地及周缘海相页岩气区很难套用北美大规模、高密度、连片化布井的开发模式。与北美商业化开发的页岩气区相比，中国页岩气开发在地质、储层、工程、地表、地貌、环境生态、水资源、基础设施等方面存在更多的与基础理论、工程技术和经济因素等相关的挑战，这对中国页岩气开发提出了非常高的技术经济指标和要求，工程作业效率和开发效益必须并重，在获得产能重大突破的基础上降本增效。地质工程一体化是发挥综合技术优势，避开北美昂贵的学习曲线，实现中国页岩气开发跨越式发展的关键途径。它以油气藏认识为核心，在勘探开发进程中，在作业和工程实践中，通过一体化研究和一体化作业的及时互动，不断深化油气藏认识、持续优化工程应用，提高作业效率和开发效益[1-3]。

1　页岩气地质工程一体化核心内涵新模型

地质工程一体化有从单井尺度、平台或多井局部尺度拓展到整个气田开发宏观全局尺度的强烈需求。结合国内页岩气地质工程一体化实施经验[1-2]，进一步丰富和发展了地质工程一体化理念和核心内涵，提出了页岩气地质工程一体化核心内涵新模型（图1）。该

模型从全局的角度，把开发工程融入地质工程一体化研究和作业中，是以地质工程一体化研究为重要支撑，在多学科、多部门一体化共享模型的基础上，从总体上评价储层品质、完井品质和钻井品质[1]。在此基础上，结合各项工程实施部署及作业流程，通过适时建模等技术手段，从单井、平台、全气田多尺度，动态支持和优化钻井工程、完井及改造工程、生产工程和开发工程四大工程应用。

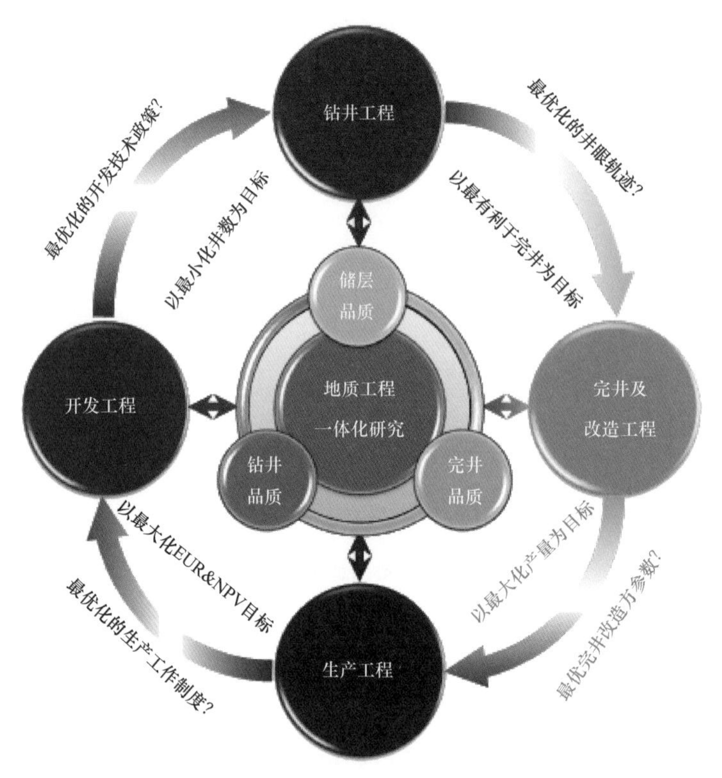

图1　页岩气地质工程一体化核心内涵新模型

在本文中，生产工程是指单井或平台如返排管理、油嘴制度、生产协调、人工举升、井下作业等生产管理和优化相关的工程措施，与英文Production Engineering相对应；开发工程指在气田或开发区块整体尺度，与布井、井距、开发部署、方案调整、开发技术政策等相关的工程应用和方案。钻井工程在保证井筒完整性的基础上，要以最有利于完井改造为设计和施工目标；完井及改造工程服务于最大化单井产量、实现产量突破和高产为优化目标；生产工程主要目的是最大化单井预测最终可采储量（Estimated Ultimate Recovery，EUR）和单井设计生产年限内的净现值（Net Present Value，NPV）；开发工程的终极目标是通过优化开发方案和开发技术政策，实现一定投资规模条件下全气田采收率最大化和全资产收益最优化。

地质工程一体化综合研究是支撑地质工程一体化作业的重要技术保障之一，根据不同的勘探开发对象，可以针对性地定义和定制地质工程一体化综合研究的主要内容和各种工程系统的主要应用。它需要建立多学科、多部门、不断更新的从一维到三维的一体化共享模型。该模型主要包括地球物理及地质模型、构造（断裂）及天然裂缝模型、地质力学模型、从单井到多井的水力压裂缝网模型及从单井到全油气田的数值模拟模型。在页岩气地质工程一体化核心内涵新模型的基础上，根据长宁国家级页岩气示范区开发建设的具体实践[2]，对页岩气地质工程一体化在工程中的主要应用进行了概括和总结（图 2）。

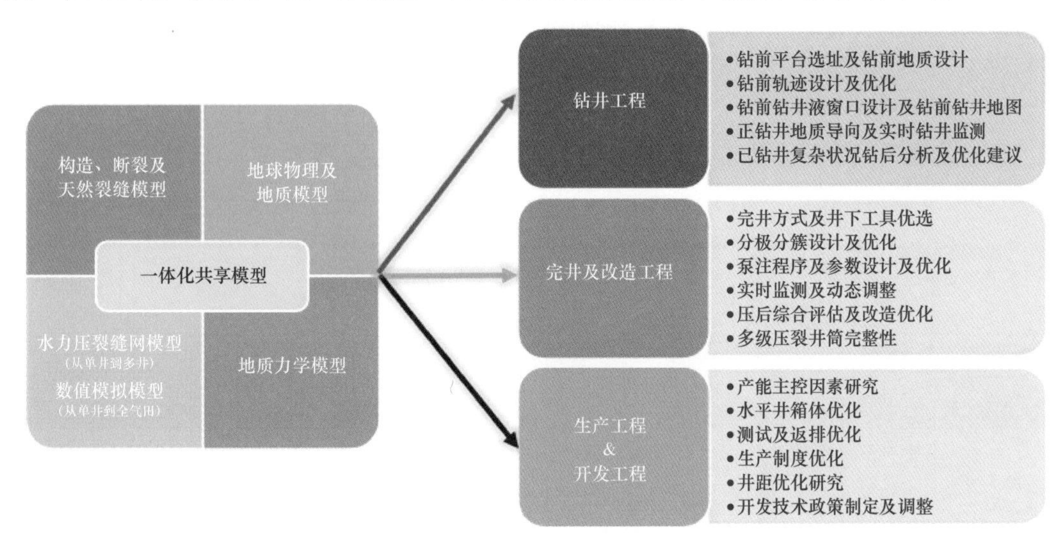

图 2　页岩气地质工程一体化在工程中的主要应用

2　非常规油气藏建模及数值模拟现状及进展

根据公开文献报道，从三维建模、压裂模拟和数值模拟 3 个方面，简要概括国内外非常规油气藏建模及数模的现状及进展。

2.1　北美非常规油气藏建模及数模现状及进展

2.1.1　三维建模（包括三维地质力学建模）

总体上，北美三维地学建模[4]（包括地质力学建模[5]）起步较早，但全油气田尺度的应用案例相对较少。强烈的非均质性是非常规油气藏的重要特征，这种非均质性在不同尺度又具有非常强的差异。如何有效地综合岩心、测井和地震这些具有不同尺度的数据，使非常规储层的非均质性在三维模型中得到合理和可靠的表征，是建模和数模工作重要的基础。在过去几年中，北美发展并实践了“非均质非常规油气藏多尺度一体化表征”方法[6-8]（图 3）。其中，地震非均质性分析（Seismic Heterogeneous Rock Analysis）概念及工作流程是其中一项较突出的技术[6-7]（图 3）。遗憾的是，这套针对性和实践性都很强的方法，除了在 Haynesville 的盆地级别研究中有应用之外，在公开发表的其他文献中没有看到更多应用实例，这套技术在国内一体化研究和建模工作中值得借鉴、应用和发展。

2.1.2　非常规储层一体化水力压裂建模

北美水力压裂建模及模拟技术发展很快，已经形成较系统和完善的非常规油气藏压裂缝网建模、一体化压裂建模及评估技术、与现代压裂工艺相匹配的压裂建模技术、全耦合建模及数模技术等，并广泛应用于储层改造设计、后评估及优化研究[9-27]。储层改造效果评价和改造参数优化很大程度上取决于水力裂缝模型的准确性或可靠性。在非常规储层所期望达到的缝网或体积改造中，天然裂缝系统起到了至关重要的作用[10]。北美目前的水力压裂模拟器已经可以处理三维天然裂缝系统，近期又进一步拓展到多期次（多套）复杂天然裂缝系统储层的水力压裂缝网建模及模拟[21]。这种技术对于中国普遍经受过多次强构造运动改造的页岩储层来说，具有较强的应用价值，但其对天然裂缝系统的建模提出了更高要求。应力阴影是水力压裂设计、建模和施工中需要考虑的重要内容。近年来，已经可以在简单一维水力裂缝到二维复杂水力裂缝缝网中考虑应力阴影的影响[10]，近期基于以二叠纪盆地为代表的立体布井实践，把这项工作拓展到了三维建模[21]。

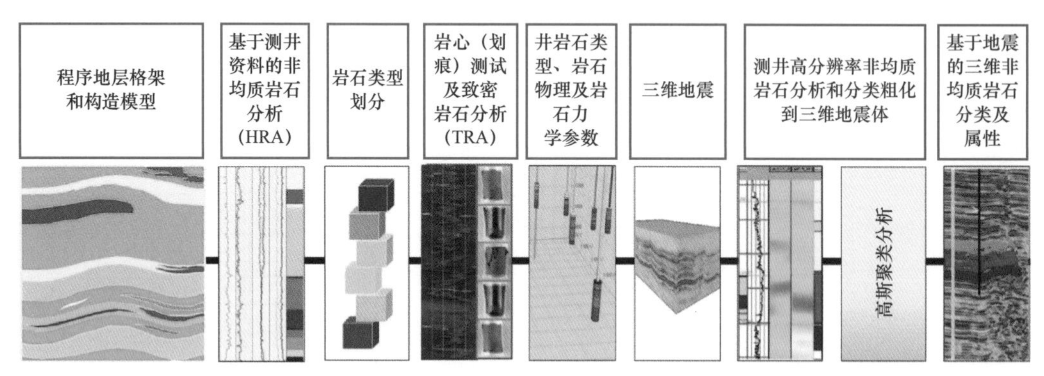

(a) 岩心尺度致密岩石分析（TRA）+测井尺度非均质分析　　(b) 地震非均质性分析+多尺度一体化表征

图3　非均质非常规油气藏多尺度一体化表征方法（根据文献［6–8］修改）

2.1.3　非常规油气藏数值模拟

数值模拟技术在非常规油气藏也得到了广泛应用和快速发展，随着新一代高性能数值模拟器的出现，以 Fan Li 等[28]为代表的结构化网格等效数模技术转向了非结构网格高分辨率数模技术，从而更好地模拟水力裂缝网络的生产动态。同时，北美发展了“水力压裂、地质力学和油气藏动态全耦合数模”技术（图4），考虑压裂过程中和生产过程中孔隙压力变化引起的地应力场改变，主要应用于水力冲击压裂（Frac Hit）、井距优化、加密钻井、重复压裂、高密度或立体复杂井网等研究[14–27]。从公开报道的文献看，由于这项技术应用的复杂性，目前主要在一个平台范围内的井组开展针对单井、双井或多井的研究工作。

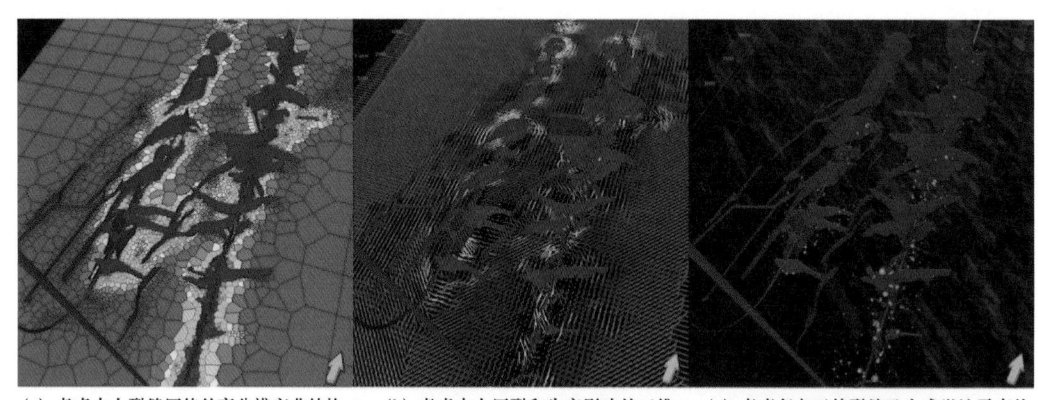

(a) 考虑水力裂缝网络的高分辨率非结构化网格数值模拟技术，生产过程中油藏压力及水力裂缝净压力变化　　(b) 考虑水力压裂和生产影响的三维地质力学模拟技术，在生产过程中水平应力大小及方向变化　　(c) 考虑复杂天然裂缝及合成微地震事件的非常规水力裂缝网络模拟技术，在生产过程中裂缝滑动产生微地震事件（合成）

图4　水力压裂、地质力学和油气藏动态全耦合数模技术（根据文献［17］修改）

2.1.4 云计算技术

从储层品质、完井品质、改造参数等方面，改造优化需要考虑的影响因素或参数很多[28-37]，而数学模型越来越复杂，运算量越来越大。北美开始采用云计算技术，结合多因素不确定性优化技术，摆脱单因素或多因素统计分析及有限模拟方案研究的局限性，更高效、更全面地掌握产能主控因素和主要作业优化参数。斯伦贝谢公司近期实施的一个Wolfcamp致密油改造优化研究项目，使用云计算技术运行了超过1200多个复杂缝网模拟方案，形成了基于主控成分的最优化"穹顶"或者最优化域"空间分布"，使改造优化认识的实用性和适用性得到了极大提高。

2.1.5 从地球物理到数值模拟一体化三维建模

在页岩气田开发建产阶段，北美很少有在全气田尺度涵盖从地球物理储层反演和预测、构造及地质建模、地质力学建模、水力压裂建模及数值模拟的全方位、全过程的一体化建模及数模案例。尤其是非常规油气藏全油气田数值模拟，迄今还没有看到北美公开报道的案例。Suarez等[4-8]较早在盆地级别开展从地球物理到地质力学一体化三维建模工作，但总体上北美对地球物理资料综合运用的案例相对较少。究其原因，可能是北美页岩油气主要产区地质条件较好，拥有大量的井资料，能够以井资料为主建立较可靠的三维构造及地质模型。近期Liang等[15, 25]发表的案例，很好地展示了如何以1100口直井测井资料为基础，建立高分辨率三维构造及地质模型、地质力学模型；基于此在典型平台开展"水力压裂、地质力学和油气藏动态全耦合"参数研究及敏感性分析。这个案例是北美公开发表、为数不多的气田尺度三维一体化建模实例，但没有表明其使用了三维地震数据。

2.2 国内页岩气建模及数模现状及进展

与北美相比，中国南方海相页岩气勘探开发的资料基础尤其是井资料基础非常薄弱，因此中国学者从一开始就非常重视地球物理资料的采集、应用和深度挖掘。在中国页岩气地质工程一体化建模、数值模拟研究及实践中，最初就涵盖了地球物理储层反演和预测、构造及地质建模、地质力学建模、水力压裂建模和数值模拟的全方位、全过程[1-2]，而近期的工作又将页岩气数值模拟从平台尺度拓展到了全气田尺度。

中国主要页岩气储层均经历了多期次强构造运动的强烈改造，发育非常复杂的断裂和天然裂缝系统，对其建模的精度和可靠性将直接影响地质力学建模和水力压裂缝网建模的精度和可靠性。因此，在中国页岩气地质工程一体化建模及数模研究中，断裂和天然裂缝系统建模是承上启下的关键环节。通过多年实践和探索，逐渐形成和发展了与复杂构造运动背景相匹配的多尺度断裂及天然裂缝系统建模方法和质量控制流程[37, 38]。复杂的构造及微构造、断裂和天然裂缝系统，形成了复杂多变的现今就地应力格局和分布，对三维地质力学建模提出了非常高的要求。在过去几年的地质工程一体化建模研究中，发展和完善了针对性的三维地质力学建模流程和方法[39-41]。由于中国页岩气地质工程一体化建模及数模面对的研究对象更为复杂，井控资料相对匮乏，在实践中采用了与钻完井作业进程相匹配的"适时建模"（Live Modeling）方法，不断利用各种新数据、新资料进行质控和更新各种模型，以提高模型精度，降低模型不确定性[37-40]。此外，在页岩气地质工程一体化建模及数模的基础上，运用气藏工程分析方法与单因素和多因素分析相结合，对产能主控因素开展较系统的分析和研究[42]。

宁201井区地质工程一体化建模及数模研究[43]代表了该领域国内发表的最新成果和进展。针对示范区建设过程中存在的“Ⅰ类储层钻遇率较低、井筒完整性较差和体积压裂效果仍需提高”等难题，在充分运用地震储层预测技术、测井储层评价技术等综合地质评价的基础上，发展完善了页岩气地质工程一体化建模技术。建立了涵盖构造、储层、天然裂缝、地质力学等各种要素的地质工程一体化模型，定量刻画了储层关键地质和工程参数在三维空间的展布规律，实现了页岩气藏的可视化，打造了“透明页岩气藏”[43]。

在宁201井区页岩气地质工程一体化研究中，开展了水平井多级压裂复杂缝网模拟、双井拉链式压裂缝网模拟、复杂缝网与气藏及地质力学动态耦合模拟等数值模拟工作，并在国内首次尝试了页岩气全气田尺度的数值模拟研究。根据公开发表的文献，国外尚未开展页岩气全气田数值模拟研究工作。尽管一些专家和学者根据北美非常规油气田开发实践，认为这种全气田数值模拟模型的必要性值得商榷，但如果数值模拟的最终目标是建立数字化页岩气田的核心动态模型，是为了实现整个气田甚至多个气田从地下到地面一体优化，那么它应当是一项非常必要的工作。

3 页岩气地质工程一体化建模及数值模拟面临的挑战

作为页岩气地质工程一体化研究的重要支撑，一体化建模及数值模拟仍然存在诸多挑战。对过去几年在页岩气地质工程一体化研究中开展的建模及数模工作进行分析，从与钻井、完井及改造、生产和开发四大工程系统紧密相关的应用角度，在5个领域总结和提出了关键性的现实挑战。

3.1 复杂天然裂缝系统定量表征

由于多期次、强构造运动的强烈改造，四川盆地及周缘的页岩气形成了复杂的断裂（天然裂缝）系统。这也是与北美著名页岩气储层相比，中国海相页岩气储层一个显著不同的地质特点。同时，从钻井取心和井筒电阻率成像测井上，可以观测到垂向上发育层面缝和水平薄弱面。由于缺乏水平井成像测井等测量手段，单纯依靠地球物理测井和直井取心，很难表征不同天然裂缝系统的产状和几何参数，而不同期次、不同尺度天然裂缝系统的地质力学参数更难直接测定。因此，天然裂缝系统定量表征和建模，仍然是一体化建模和数模中承上启下的最不确定环节，直接影响着地质力学模型和水力裂缝建模的可靠性和精度，同时也在一体化建模及数模过程中，制约地质模型、地质力学模型、水力裂缝模型和数模模型的尺度关联与分辨率匹配。因此，复杂天然裂缝系统的定量表征，从基础理论、关键参数测量到定量建模方法及流程，存在众多需要不断改进和提高的细节或环节，是今后研究攻关的重要技术领域。

3.2 井筒完整性

在四川盆地页岩气开发中，尤其是在威远地区，压裂中和压裂后发生的套损现象比较普遍。而国外其他高构造应力背景的页岩油气储层，比如阿根廷的 VacaMuerta，压裂中和压裂后发生的套损也比较严重。目前的套损研究，更多是集中在基于材料力学的静力学研究，或者是基于地质模型或施工响应分析等开展的定性分析，不能充分考虑下套管、固井这些作业过程对套管初始内应力分布的影响，不能充分考虑在多级压裂的动态受力条件下，套管、水泥环和储层尤其是断层或天然裂缝带的相互作用。

3.3 平面（立体）加密井或立体布井

五峰组—龙马溪组优质海相页岩段在部分地区厚度较大，通常为 30m 以上。而实践表明，基于目前的改造技术和改造强度，水力裂缝的垂向有效支撑通常是 10～20m，有的甚至可能小于 10m。对已开发建产区实施平面甚至立体加密布井提高资源动用率和采出程度、对未开发新区开展立体布井部署，存在强烈的现实需求。在垂向改造可能受目前压裂改造技术和工艺水平限制的情况下，要前瞻性探索五峰组—龙马溪组页岩立体开发的可能性和可行性（图 5）。

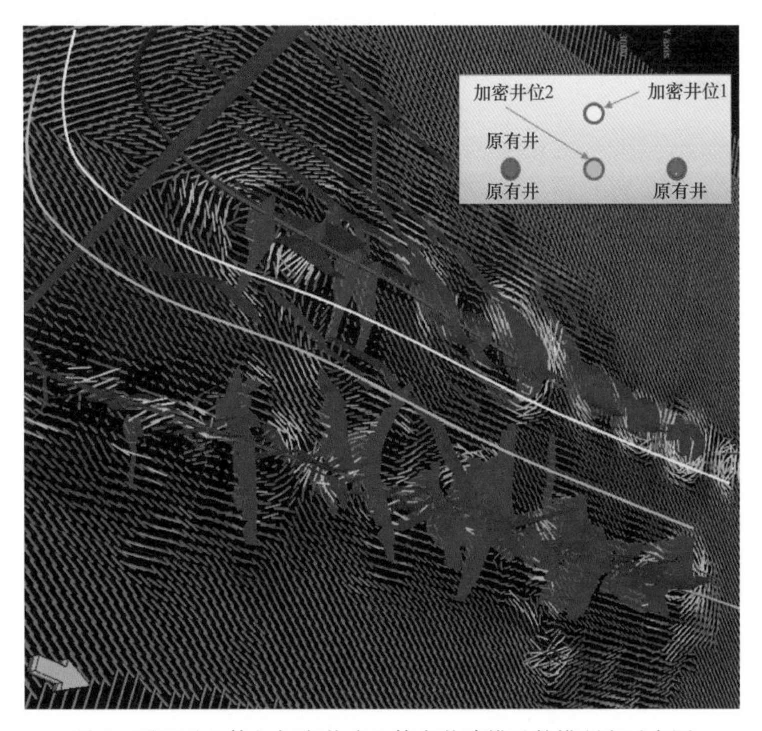

图 5　平面（立体）加密井或立体布井建模及数模研究示意图

3.4 页岩气田全气田数值模拟

坦率地说，页岩气田全气田或全资产项目数值模拟目前还处在“解决有无和解决可行性”的层面，因为它在理论和技术上还存在巨大挑战。这些挑战主要包括：由于涉及的系统众多，参数复杂多变，各种关键参数的获取比常规气藏困难，参数不确定性高；各种获取参数的实验室方法和矿场试验方法也有待进一步突破，如应力敏感性、水力裂缝有效性等；不同模型的不同尺度之间，精度、可靠性和计算效率之间，如何建立一致性工作流程并实现计算效率和计算精度平衡；在基础理论领域，业界对纳米孔隙的微观渗流规律、高压条件下纳米孔隙中甲烷分子的吸附态和相态等[44]还存在巨大分歧，相关的数学模型还无法取得业界共识，还不能满足大规模数值模拟计算和各种实际工程应用的要求。

3.5 井筒动态和气藏动态耦合

井筒水动力学对生产的影响及与气藏动态的相互作用，在一定情况下，可以对页岩气井的生产动态和产能带来不良影响，甚至影响到整个气田的开发效果[45-55]。由于井筒倾角的变化和局部“微狗腿”的影响，气水（压裂液）两相在井筒中可以形成非常复杂的流态（图 6）。根据实验室观察，气水两相在局部起伏结构可以形成多种流态，沿井筒的复杂流态和气液分布，可以产生较大的附加阻力甚至形成“液锁”，从而严重制约页岩气井产能[45-47, 49]。返排和生产过程中强烈的段塞流动可以造成井底巨大的瞬时压差变化，诱发支撑剂回吐和（或）高压差对近井地带水力裂缝导流能力和连通性的伤害，使页岩气井产能受到严重损害[47]。返排、生产尤其是生产中后期的举升系统选择和优化，必须充分考虑井筒中复杂流态的制约和影响[49-52]。井筒复杂流态形成复杂生产动态这个问题，在中国页岩气开发中可能会更加突出，因为国内页岩气储层微构造、微断层发育，水平井轨迹局部变化更频繁和剧烈，井筒内可以出现极其复杂的流态和气液分布。

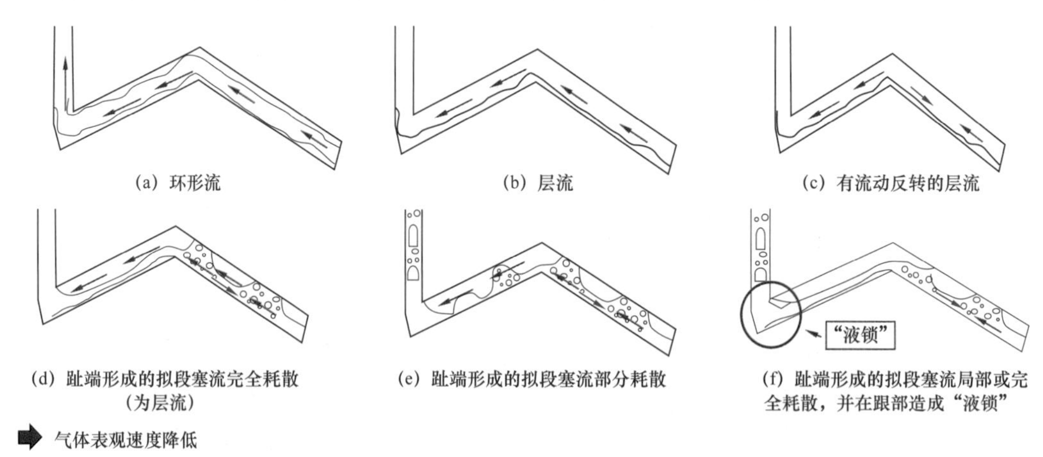

图 6 局部起伏结构形成的流态（基于实验观测，根据文献［49］修改）

4 页岩气地质工程一体化建模及数值模拟的发展机遇

前文论述的挑战，同时也是进一步完善和发展页岩气一体化建模及数模技术的重要机遇：

（1）在一体化建模中，要把复杂断裂及天然裂缝系统的定量化建模作为进一步攻关的重点。在深度挖掘地震资料在裂缝建模中应用的同时，强化与构造地质研究的互动和集成，通过对构造和应力演化史的物理模拟和数值模拟，从理论上掌握断裂和裂缝发育机制和模式、不同期次裂缝系统相互作用、不同尺度裂缝尺度理论比例关系及其宏观分布趋势，指导定量化建模。强化采集高品质地震数据的同时，要注重在水平井采集高分辨率井筒成像数据，以期为裂缝系统的平面分布及裂缝参数提供更直接的测量数据。

（2）开展高压高应力页岩气多级压裂水平井套损机理及对策研究，需要有针对性地设计相关材料力学实验，进行单井套损机理系统测井测量及综合评价，在此基础上，开展气田—平台—井筒多尺度全耦合动态建模及数值模拟，深入理解和掌握套损的动态机制（图7），从而为套管设计、固井及完井压裂优化、生产管理提供更加准确可靠的理论依据。

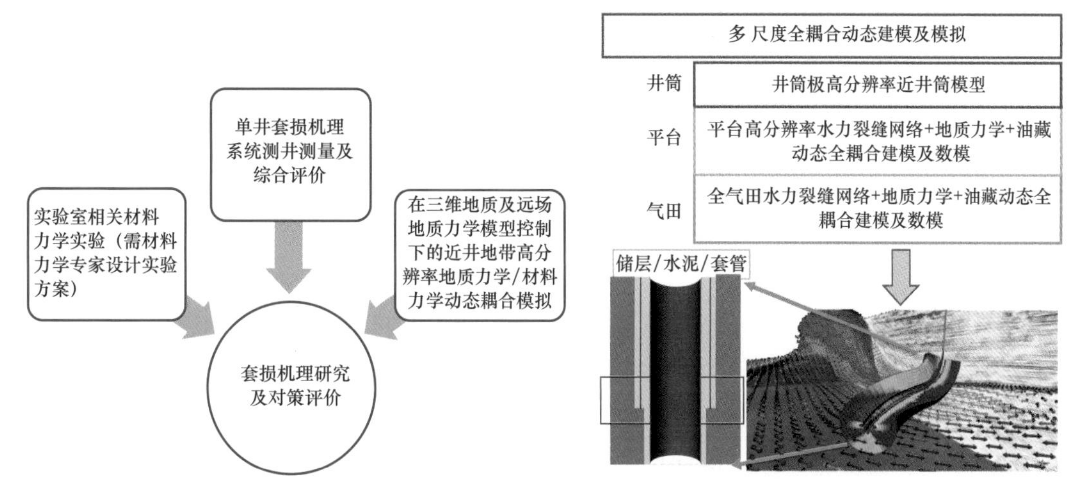

图7　建议的高压高应力页岩气多级压裂水平井套损机理及对策研究方法

（3）在平面、立体加密井或立体布井建模及数模研究中，除了考虑在平面或空间的井位优化，还要充分研究原有井生产如何改变孔隙压力和就地应力分布。在此基础上，需要系统性研究：加密井区选择、加密钻井时机选择、加密井完井与改造优化、冲击压裂（Frac Hit）[23]监测与评估、加密井及邻井在产井的井筒完整性、加密后加密井及邻井生产干扰与生产制度优化等。

（4）非常规油气藏的大规模数值模拟工作还任重道远，需要进一步从基础理论、关键参数矿场测量、建模优化等领域开展深入研究，从而将页岩气全气田数值模拟技术从理论和应用，提升到更适用更可靠和更高的层次，使其在智慧化页岩气田中充分发挥价值。

（5）北美 R.Jain 等[53]较早地提出在完井设计中，要充分考虑井筒复杂流态的影响，建议通过将井筒动态和气藏动态耦合的方式，优化非常规水平井完井设计。Alfonso Fragoso 等[54]近期又进一步实现了单井数模模型、井筒动态模型和地面设备的简单耦合，开展非常规储层的提高采收率研究。在今后的页岩气地质工程一体化建模和数值模拟研究中，需要进一步考虑井筒水动力学与气藏动态的相互作用以及对生产的影响，通过井筒模型和气藏模型的耦合，指导单井初期返排和优化生产制度管理、优选（中后期）人工举升方法、优化多井甚至多平台生产协同和调度。

在进一步发展和完善各个专业领域的建模及数值模拟技术的基础上，将一体化共享模型与数字化、智能化页岩气田相连，是地质工程一体化建模及数模的终极追求[55]。基于数字化、智能化的页岩气田可以实现：气藏—井筒—地面系统一体化建模及优化；气藏管理—气井动态—举升优化—流动保障—测量与监测等各个领域相互耦合；各种模型与作业管理、作业支持、资产管理、风险预警、决策支持的业务流程和作业进程相联系；页岩气田全生命周期管理及优化。充分运用基于大数据、多学科的一体化模型驱动数字化、智能化页岩气田的建设，将数据转化为信息、信息转化为知识、知识转化为决策和财富，真正发挥数字化、智能化的巨大推动作用，实现更充分的增效降本。

5 结束语

地质工程一体化是高效开发复杂油气藏的必由之路，贯穿项目全过程的地质工程一体化综合研究起到重要的技术支撑作用。在以三维共享综合模型为基础的地质工程一体化综合研究中，以单井或平台为目标的钻井、完井与压裂、返排测试及生产制度的工程设计及优化，仍然是一种局部优化。随着开发规模的不断扩大，加密钻井和立体布井等技术需求的出现，地质工程一体化需要考虑全局优化。因此，以整个气田、区块或项目资产为目标优化的开发工程，应该纳入地质工程一体化的核心内涵。

一体化建模及数模是地质工程一体化研究的核心内容之一。与各自的页岩气地质工程特点和数据基础相适应，北美和中国在非常规油气藏一体化建模及数模领域的进展各具特色、各有千秋。北美在非均质非常规油气藏多尺度（地震、测井、岩心）一体化表征、（多期次 / 多套）复杂天然裂缝系统储层水力缝网建模及模拟、基于云计算和大数据分析的压裂优化参数研究等方面，值得参考和借鉴。

复杂天然裂缝系统定量表征、多场和多尺度耦合井筒完整性、平面或立体加密钻井或立体布井、复杂渗流机理条件下的高效全气田数值模拟、井筒流体动力学和气藏动态耦合等五个挑战，是页岩气地质工程一体化建模及数模进一步发展和提高的重要研究领域。对它们的研究不断深入和发展，将进一步满足钻井、完井与压裂、生产和开发四大工程应用局部和全局优化的现实需求。

参考文献

[1] 吴奇，梁兴，鲜成钢，等．地质—工程一体化高效开发中国南方海相页岩气［J］．中国石油勘探，2015，20（4）：1–23.

[2] 谢军，张浩淼，佘朝毅，等．地质工程一体化在长宁国家级页岩气示范区中的实践［J］．中国石油勘探，2017，22（1）：21–28.

[3] 胡文瑞．地质工程一体化是实现复杂油气藏效益勘探开发的必由之路［J］．中国石油勘探，2017，22（1）：1–5.

[4] Suarez Rivera R，Handwerger D，Herrera A R，et al. Development of a heterogeneous earth model in unconventional reservoirs for early assessment of reservoir potential［C］. ARMA-2013-667，23–26 June 2013，San Francisco，CA，USA.

[5] Rodriguez Herrera A E，Suarez Rivera R，Handwerger D，et al. Field-scale geomechanical characterization of the Haynesville shale［C］. ARMA-2013-678，23–26 June 2013，San Francisco，CA，USA.

[6] Suarez Rivera R，Herring S，Handwerger D，et al. Integrated analysis of core geology，rock properties，well logs，and seismic data provides a well constrained geologic model of the bossier/Haynesville system［C］. SPE-167204-MS，5–7 November 2013，Calgary，Alberta，Canada.

[7] Suarez Rivera R，Dahl G V，Borgos H G，et al. Seismic-based heterogeneous earth model improves mapping reservoir quality and completion quality in tight shales［C］. SPE-164544-MS，10–12 April

2013, The Woodlands, Texas, USA.
[8] Marino S, Herring S, Stevens K, et al. Integration of quantitative rock classification with core-based geological studies : Improved regional-scale modeling and efficient exploration of tight shale plays [C] . SPE-167048-MS, 11–13 November 2013, Brisbane, Australia.
[9] Waters G, Dean B, Downie R, et al. Simultaneous hydraulic fracturing of adjacent horizontal wells in the Woodford shale [C] . SPE-119635-MS, 19–21 January 2009, The Woodlands, Texas, USA.
[10] Weng Xiaowei, Kresse O, Cohen C, et al. Modeling of hydraulic fracture network propagation in a naturally fractured formation [C] . SPE-140253-MS, SPE Hydraulic Fracturing Technology Conference and Exhibition, 24–26 January 2011, The Woodlands, Texas, USA.
[11] Olson J E, Wu Kan. Sequential versus multi-zone fracturing in horizontal wells : Insights from non-planar, multi-frac numerical model [C] . SPE-152602-MS, 6–8 February 2012, The Woodlands, Texas, USA.
[12] Sesetty V, Ghassemi A. Simulation of simultaneous and zipper fractures in shale formations [C] . ARMA-2015-558, 28 June-1 July 2015, San Francisco, CA, USA.
[13] Qiu Fangda, Porcu M M, Xu Jian, et al. Simulation study of zipper fracturing using an unconventional fracture model [C] . SPE-175980-MS, 20–22 October 2015, Calgary, Alberta, Canada.
[14] Pankaj P, Geetan S, MacDonald R, et al. Reservoir modeling for pad optimization in the context of hydraulic fracturing [C] . SPE-176865-MS, 9–11 November 2015, Brisbane, Australia.
[15] Liang Baosheng, Khan S, Puspita S D, et al. Improving unconventional reservoir factory model development by an integrated workflow with earth model, hydraulic fracturing, reservoir simulation and uncertainty analysis [C] . URTEC-2461423-MS, 1–3 August 2016, San Antonio, Texas, USA.
[16] Patel H, Cadwallader S, Wampler J. Zipper fracturing taking theory to reality in the eagle ford shale [C] . URTEC-2445923-MS, 1–3 August 2016, San Antonio, Texas, USA.
[17] Marongiu-Porcu M, Lee D, Shan D, et al. Advanced modeling of interwell-fracturing interference : An eagle ford shale-oil study [J] . SPE Journal, 2016, 21 (5) .
[18] Pankaj P, Gakhar K, Lindasay G. When to refrac ? Combination of reservoir geomechanics with fracture modeling and reservoir simulation holds the answer [C] . SPE-182161-MS, 25–27 October 2016, Perth, Australia.
[19] Suarez M, Pichon S. Combining hydraulic fracturing considerations and well spacing optimization for pad development in the Vaca Muerta shale [C] . URTEC-2436107-MS, 1–3 August 2016, San Antonio, Texas, USA.
[20] Gakhar K, Rodionov Y, Defeu C, et al. Engineering and effective completion and stimulated strategy for in-fill wells [C] . SPE-184835-MS, 24–26 January 2017, The Woodlands, Texas, USA.
[21] Edouard C C, Olga K, Xiaowei W. Stacked height model to improve fracture height growth prediction, and simulate interactions with multi-layer DFNs and ledges at weak zone interfaces [C], SPE-184876-MS, 24–26 January 2017, Thc Woodlands, Texas, USA.
[22] Bommer P, Bayne M, Mayerhofer M, et al. Re-designing from scratch and defending offset wells : Case study of a six-well Bakken zipper project, McKenzie County, ND [C] . SPE-184851-MS, 24–26

January 2017, The Woodlands, Texas, USA.

[23] Jacobs T. Oil and gas producers find frac hits in shale wells a major challenge [J]. Journal of Petroleum Technology, 2017, 69 (4).

[24] Algarhy A, Soliman M, Heinze L, et al. Increasing hydrocarbon recovery from shale reservoirs through ballooned hydraulic fracturing [C]. URTEC-2687030-MS, 24–26 July 2017, Austin, Texas, USA.

[25] Liang Baoshen, Shahzad K, Dewi P S. An integrated modeling workflow with hydraulic fracturing, reservoir simulation, and uncertainty analysis for unconventional-reservoir development [J]. SPE Reservoir Evaluation & Engineering, 2017, 21 (2).

[26] Sangnimnuan A, Li Jianwei, Wu Kan, et al. Application of efficiently coupled fluid flow and geomechanics model for refracturing in highly fractured reservoirs [C].SPE-189870-MS, 23–25 January 2018, The Woodlands, Texas, USA.

[27] Xu Tao, Lindsay G, Baihly J, et al. Proposed refracturing methodology in the Haynesville shale [C]. SPE-187236-MS, 9–11 October 2017, San Antonio, Texas, USA.

[28] Fan Li, Thompson J W, Robinson J R. Understanding gas production mechanism and effectiveness of well stimulation in the Haynesville shale through reservoir simulation [C]. SPE-136696-MS, 19–21 October 2010, Calgary, Alberta, Canada.

[29] Fan Li, Luo Fang, Lindsay G, et al. The bottom-line of horizontal well production decline in the Barnett shale [C]. SPE-141263-MS, 27–29 March 2011, Oklahoma City, Oklahoma, USA.

[30] Thompson J W, Fan Li, Grant D , et al. An overview of horizontal well completions in the Haynesville shale [J]. Journal of Canadian Petroleum Technology, 2011, 50 (6).

[31] Fan Li, Martin R, Thompson J, et al. An integrated approach for understanding oil and gas reserves potential in eagle ford shale formation [C]. SPE-148751-MS, 15–17 November 2011, Calgary, Alberta, Canada.

[32] Cohen C E, Abad C, Weng X, et al. Analysis on the impact of fracturing treatment design and reservoir properties on production from shale gas reservoirs [C]. IPTC 16400, the International Petroleum Technology Conference, 26–28 March 2013, Beijing, China.

[33] Azad A, Somanchi K, Brewer J R, et al. Accelerating completions concept select in unconventional plays using diagnostics and frac modeling [C]. SPE-184867-MS, 24–26 January 2017, The Woodlands, Texas, USA.

[34] Abhishek G, Gibbon E J, Roberson T M. Asset evaluation utilizing multi-variate statistics integrating data-mining, completion optimization, and geology focused on multi-bench shale plays [C]. SPE-185018-MS, 15–16 February 2017, Calgary, Alberta, Canada.

[35] Ge Y, Dwivedi P, Kwok C Ka, et al. The impact of increase in lateral length on production performance of horizontal shale wells [C]. SPE-185768-MS, 12–15 June 2017, Paris, France.

[36] Eburi S, Jones S, Houston T. Analysis and interpretation of Haynesville shale subsurface properties, completion variables, and production performance using ordination, a multivariate statistical analysis technique [C]. SPE-170834-MS, 27–29 October 2014, Amsterdam, The Netherlands.

[37] Xing Liang, Xiao Liu, Honglin Shu, et al. Characterization of complex multiscale natural fracture

systems of the Silurian Longmaxi gas shale in the Sichuan Basin，China [C] . SPE-176938-MS，SPE Asia Pacific Unconventional Resources Conference and Exhibition，9–11 November 2015，Brisbane，Australia.

[38] Qin Jun，Xian Cheng Gang，Liang Xing，et al. Characterizing and modeling multi-scale natural fractures in the ordovician-silurian Wufeng-Longmaxi shale formation in south Sichuan Basin [C] . URTeC : 2691208，24–26 July 2017，Austin，Texas，USA.

[39] Liang Xing，Xian Chenggang，Shu Honglin，et al. Three-dimensional full-field and pad geomechanics modeling assists effective shale gas field development，Sichuan Basin，China [C] . IPTC-18984-MS，International Technology Conference，14–16 November 2016，Bangkok，Thailand.

[40] 鲜成钢，张介辉，陈欣，等 . 地质力学在地质工程一体化中的应用 [J] . 中国石油勘探,2017,22 (1): 75–88.

[41] Xie Jun，Qiu Kaibin，Zhong Bing，et al. Construction of a 3D geomechanical model for development of a shale gas reservoir in the Sichuan Basin [C] . SPE-187828-MS，16–18 October 2017，Moscow，Russia.

[42] Wang Weixu，Xian Chenggang，Liang Xing，et al. Production controlling factors of the Longmaxi shale gas formation——A case study of Huangjingba shale gas field [C] . SPE-186874-MS，17–19 October 2017，Jakarta，Indonesia.

[43] 谢军 . 关键技术进步促进页岩气产业快速发展——以长宁—威远国家级页岩气示范区为例 [J] . 天然气工业，2017，37 (12) : 1–10.

[44] Pitakbunkate T，Balbuena P B，Moridis G J，et al. Effect of confinement on pressure/volume/temperature properties of hydrocarbons in shale reservoirs [J] . SPE Journal，2016，21 (2) .

[45] Qiu Fangda，Yuan Ge，Porcu M M，et al. Sinuosity of the hydraulic fractured horizontal well impact on production flow assurance an Eagle Ford case [C] . SPE-179168-MS，9–11 February 2016，The Woodlands，Texas，USA.

[46] Browning S，Jayakumar R. Effects of toe-up vs toe-down wellbore trajectories on production performance in the Cana Woodford [C] . URTEC-2461175-MS，1–3 August 2016，San Antonio，Texas，USA.

[47] Brito R，Pereyra E，Sarica C. Existence of severe slugging in toe-up horizontal gas wells [C]，SPE-181217-MS，25–27 October 2016，The Woodlands，Texas，USA.

[48] Lu Haidan，Olatunbosun A，Xu Lili. Understanding the impact of production slugging behavior on near-wellbore hydraulic fracture and formation integrity [C] . SPE-189488-MS，7–9 February 2018，Lafayette，Louisianan，USA.

[49] Brito R，Pereyra E，Sarica C. Effect of well trajectory on liquid removal in horizontal gas wells [C] . SPE-181423-MS，26–28 September 2016，Dubai，UAE.

[50] Peter O. Artificial lift selection strategy to maximize unconventional oil and gas assets value [C]，25–27 October 2016，The Woodlands，Texas，USA.

[51] Wilson A. Artificial-lift selection strategy to maximize value of unconventional oil and gas assets [J] . Journal of Petroleum Technology 2017，69 (7) .

[52] Yuan Ge，Diego A. Narvaez，Han Xue，et al. Well trajectory impact on production from ESP-lifted shale

wells：A case study [C] . SPE-185145-MS，24–28 April 2017，The Woodlands，Texas，USA.
[53] Jain R，Syal S，Long T，et al. An integrated approach to design completions for horizontal wells for unconventional reservoirs [J] . SPE Journal，2013，18（6）.
[54] Fragoso A，Trick M，Harding T，et al. Coupling of wellbore and surface facilities models with reservoir simulation to optimize recovery of liquids from shale reservoirs [C] . SPE-185079-MS，15–16 February 2017，Calgary，Alberta，Canada.
[55] 鲜成钢 . 长期低油价下油气技术创新目标与方向探讨 [J] . 石油科技论坛，2017，36（4）：49–56.

Mutiscale pore structure and its effect on gas transport in organic-rich shale

Wu Tianhao[1], Li Xiang[1], Zhao Junliang[1], Zhang Dongxiao[2]

1 Department of Energy and Resources Engineering, College of Engineering, Peking University, Beijing, China;

2 BIC-ESAT, ERE, and SKLTCS, College of Engineering, Peking University, Beijing, China

Abstract: A systematic investigation of multiscale pore structure in organic-rich shale by means of the combination of various imaging techniques is presented, including the state-of-the-art Helium-Ion-Microscope (HIM). The study achieves insight into the major features at each scale and suggests the affordable techniques for specific objectives from the aspects of resolution, dimension and cost. The pores, which appear to be isolated, are connected by smaller pores resolved by higher resolution imaging. This observation provides valuable information, from the microscopic perspective of pore structure, for understanding how gas accumulates and transports from where it is generated. A comprehensive workflow is proposed based on the characteristics acquired from the multiscale pore structure analysis to simulate the gas transport process. The simulations are completed with three levels: the microscopic mechanisms should be taken into consideration at level I; the spatial distribution features of organic matter, inorganic matter, and macropores constitute the major issue at level II; and the micro-fracture orientation and topological structure are dominant factors at level III. The results of apparent permeability from simulations agree well with the values acquired from experiments. By means of the workflow, the impact of various gas transport mechanisms at different scales can be investigated more individually and precisely than conventional experiments.

Key Points

- A combination of various imaging techniques is applied to investigate the multiscale pore structure in organic-rich shale
- The pores, which appear to be isolated, are connected by smaller pores resolved by higher resolution imaging
- A multi-level simulation workflow is proposed to simulate the gas transport process, which agrees well with the experimental result

Gas transport mechanism in the matrix of shale oil/gas reservoir is one of the most critical issues in the development of unconventional oil and gas resources, which has significant effects on shale gas production. It also constitutes the theoretical basis of shale gas resources assessment and long-term production prediction [1]. However, as a complex issue of multiscale and multi-physics processes, microscopic mechanisms are not yet well understood and remain as a major challenging topic.

A common characteristic of the organic-rich shale matrix is extremely low intrinsic permeability and apparent permeability [2]. Many research efforts have been directed at this topic with conventional experimental investigations of shale core plug with the pulse-decay method [3-5]. These studies have involved measuring the impact of confining pressure, pore pressure, and anisotropy [2, 6-8]. Several other studies propose rapid permeability estimation using crushed shale samples. This method only determines the apparent permeability of shale matrix, which eliminates the presence of micro-fractures and the impact of stress [3, 9-10]. However, due to limited precision, irreversible process and time-consuming experimental design, permeability can only be investigated phenomenologically with conventional experimental methods.

On the other hand, characterization of the complex pore structure in shale matrix is the foundation for investigating the fluid transport mechanism. The pore size distribution (PSD) in shale matrix ranges from sub-nanometer to micrometer, and many core samples even contain micro-fractures in the scale of hundreds microns [11]. Therefore, shale matrix contains micropores (<2nm), mesopores (2~50nm), and macropores (>50nm) at the same time [12-14]. It should be noted that the pore-size can also be classified as picopores (<1nm), nanopores (1nm to 1μm) and micropores (1~65.5μm) [15]. In this study, the former nomenclature is applied. In order to describe the complex pore structure, pore-scale imaging and modelling, i.e., "digital rock" technology [16], has been introduced to the investigation of shale matrix. By means of modern imaging methods, such as X-ray computed tomography (e.g., Micro-CT, Nano-CT) and Focused-Ion-Beam/Scanning-Electron-Microscope (FIB/SEM), many efforts have been made for the acquisition of three-dimensional (3D) pore structure under various resolutions, especially for organic matter (kerogen) [17-21]. At the same time, some two-dimensional (2D) imaging methods can also be applied in a supplementary manner based on their unique advantages, such as the Scanning Electron Microscope (SEM) with Back Scattered Electron (BSE) imaging mode for the rapid identification of organic matter (OM) and inorganic matter (IOM), and the Helium-Ion-Microscope (HIM) for ultra-high resolution imaging [22-23]. It should be noted that the HIM can also be integrated with FIB as FIB/HIM to obtain a high resolution 3D pore structure model. Based on the above methods, the 3D model for numerical simulation can be directly constructed from 3D imaging model with pore segmentation and extraction [24-25], or reconstructed with cross-correlation simulation (CCSIM) to obtain stochastic realizations [26]. On the other hand, it can also be reconstructed with 2D images by means of stochastic methods, such as the truncated Gaussian random function method, the simulated annealing method, the Markov Chain Monte Carlo (MCMC) method, the sequential indicator simulation method (SISIM), and the multiple-point statistics (MPS) method [27-31].

However, many pores appear to be isolated or locally connected because Micro-CT and Nano-CT can only resolve the pore throats larger than~0.7μm and~50nm, respectively. This cannot explain that the shale matrix is permeable [11, 17]. The pore structure in kerogen and its adjacent area from FIB/SEM reveals good connectivity and larger local porosity than the bulk/total porosity, but it results in a large deviation from the actual macroscopic permeability because of heterogeneity by ignoring the other parts with poor connectivity and lower local porosity, such as inorganic matter [18, 20, 26]. In fact, due to the limited resolution and dimension of a single imaging method (Fig. 1), a trade-off must be made between the resolution and the dimension, which is insufficient to obtain the "global picture" of shale matrix. Pore structure was only investigated within a limited range in previous works with a single imaging method (Fig. 1), which is inadequate to describe the heterogeneity or the pore connectivity under different scales. Therefore, precisely how gas accumulates and transports from where it is generated from the microscopic perspective of pore structure in the organic-rich shale matrix remains to be determined.

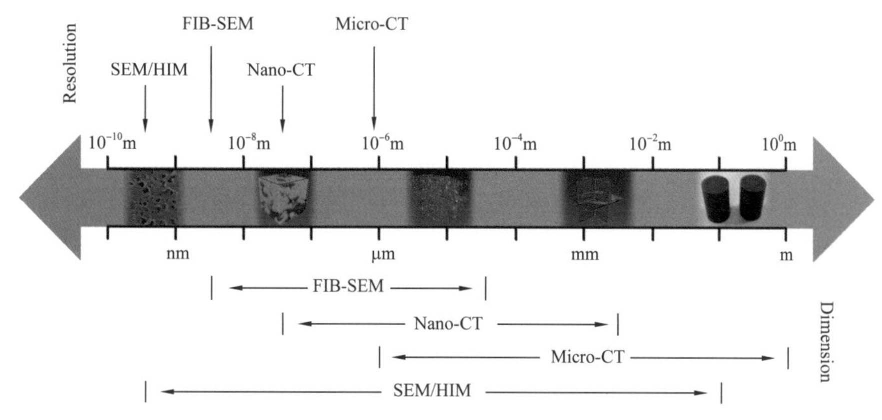

Fig. 1 Resolution and dimension of imaging facilities for digital rock

For conventional reservoirs, numerical simulations can be carried out based on the above 3D model with various methods, such as the Lattice-Boltzmann method (LBM), the pore-network method (PNM), and the equivalent permeability method (EPM) [16]. However, these methods have not been widely applied in numerical simulations of gas transport in the shale matrix. For example, the LBM is not applicable to large scale models with such complex pore structure in the shale matrix, due to computational capability [18]. In addition, it is difficult to extract the interconnected pore-network for PNM, due to the poor connectivity and the large range of pore size distribution [32]. The EPM is the most applicable method for shale matrix at present, and some conceptual macroscopic average models have been proposed [33-34]. On the other hand, numerous microscopic models have been proposed for gas transport in nanopores by taking into account the coupled impacts of adsorption and Knudsen diffusion, which have not yet been applied in the shale matrix [35, 36]. Therefore, it is essential to propose an applicable workflow for the simulation of gas transport in shale matrix by taking into account both the microscopic gas transport mechanisms and the multiscale pore structure properties.

In this work, we present a systematic study of multiscale pore structure properties in the organic-rich shale matrix, and propose a comprehensive workflow for gas transport simulation. First, we perform a "top-down" process of 3D and 2D imaging with various resolutions and dimensions, including Micro-CT, FIB/SEM, FIB/HIM, SEM, and HIM. To account for the characteristics at different scales, a "bottom-up" process is developed to compute multiscale apparent permeability. In the meantime, to validate the proposed workflow, the results from the simulations are compared with the experiments.

1 Multiscale pore structure analysis

Due to the wide pore size distribution ranging from sub-nanometer to micrometer, the combination of various imaging techniques is essential to obtain the "global picture" of the shale matrix. Therefore, according to the resolution and dimension presented in Fig. 1, characterization is performed with various 2D and 3D imaging techniques. The samples are taken from the same core for a series of experiments and simulations for each case, and the experimental design and the workflow for simulation are duplicated on the other cores. The cores are marine shales (S-001 to S-003) from the Sichuan Basin, China, as well as a terrestrial shale (S-004) from the Ordos Basin, China. The terrestrial shale [37-38] was also referred to as the continental shale in some previous studies. Additional information of the above samples can be found in the supporting information. In the following sections, we mainly demonstrate the multiscale pore structure analysis based on S-001.

1.1 Multiscale 3D imaging

Multiscale 3D imaging is performed on Micro-CT and FIB/SEM from low resolution to high resolution. The pore structure model from Micro-CT in Fig. 2a (resolution: ~13μm; dimension: 13.152mm × 13.152mm × 13.152mm) demonstrates the existence of micro-fractures and some isolated macropores. The porosity is 0.28% under this resolution. The micro-fractures may either exist naturally or be generated during coring. According to duplicated experiments on other cores, the geometry of the micro-fractures is found to be unique in different cores. Thus, the micro-fractures should be handled individually for each core sample. The pore structure model from Micro-CT in Fig. 2b (resolution: ~1μm; dimension: 968μm × 968μm × 968μm) reveals more details of the macropores than the low resolution model, except for the micro-fractures. In addition, no interconnected pores are detected. The porosity is 0.46% under this resolution, which is much lower than the conventional reservoirs. The pore structure model obtained from FIB/SEM in Fig. 2c (resolution: ~6.5nm; dimension: 3.25μm × 3.25μm × 3.25μm) presents the complex spatial distribution of OM and IOM. On the other hand, the preliminary results demonstrate that the characteristics of pore structures in OM and IOM are different from each other. Therefore, three important issues arise which require further investigation: the features of the macropores from Micro-CT with the resolution of ~1μm; the spatial distribution features of OM and IOM; and the pore structure characteristics in OM and IOM, respectively.

1.2 Macropores from Micro-CT

Comprehensive analysis of the representative elementary volume (REV) and statistical representative elementary volume (SREV) is carried out to determine whether this method is adequate to represent the entire core sample[39]. We choose porosity as the target quantity to identify the REV and SREV. For the REV, porosity is obtained as an average from a cube at the centre of the domain. For the SREV, the cube centre for each scale is taken from a random position for one sampling process, and we sampled 500 times in this case. The results in Fig. 3a show that, at small scales, porosity fluctuates obviously and has a large standard deviation. As the scale increases, however, the porosity approaches a constant, and the standard deviation decreases substantially. It can be concluded that the REV and SREV for macroscopic properties are around 500μm in this case, excluding the micro-fractures. On the other hand, the pore size distribution beyond 1μm can be calculated (Fig. 3b). The result shows that most of the macropores under this resolution range from 5μm to 20μm.

Fig. 2 Multi-scale 3D pore structure models. (a) Core model from Micro-CT with ~13μm resolution (gray: original gray-scale picture; gold: micro-fractures). (b) Core model from Micro-CT with ~1 μm resolution (gray: original gray-scale picture; orange: macropores). (c) Core model from FIB/SEM with ~6.5nm resolution (light gray: inorganic matter; dark gray: organic matter; black: pores)

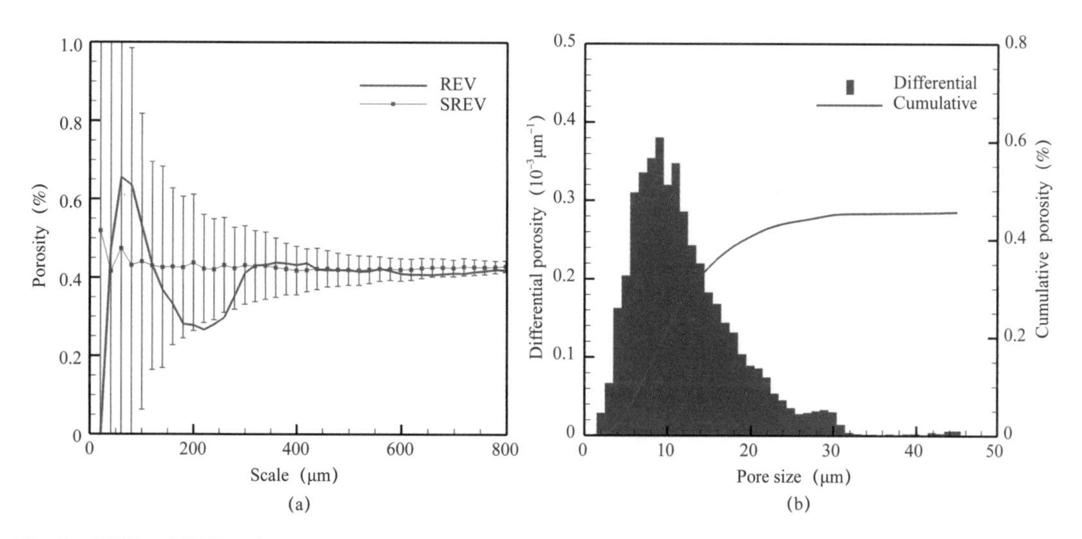

Fig. 3 REV and SREV of macropore porosity based on 3D pore structure model from Micro-CT with the resolution of ~1 μm. (a) REV and SREV analysis. The errors are obtained by adding or subtracting one standard deviation of the mean. (b) Pore size distribution

1.3 Spatial distribution of OM and IOM

In order to obtain the global feature of the spatial distribution of OM and IOM, enough images must be collected. However, FIB/SEM imaging is both time-consuming and expensive, and achieving a large number of pore structure models is neither practical nor affordable. Therefore, we applied 2D-SEM imaging with BSE mode to provide a large number of images concerning the spatial distribution of OM and IOM. To obtain the features along different directions, the sample surfaces for SEM are processed from the cross-section parallel and orthogonal to the bedding plane, respectively. Each surface is processed precisely by argon beam milling. For each surface, we obtain approximately 60 images at random positions within the area of approximately 10mm × 10mm. The dimension of the image is ~25μm × 25μm, the pixel size of which is ~30nm. Some examples are presented in Fig. 4a and Fig. 4b. With these images, analysis of REV and SREV is performed to determine whether it is adequate to represent the spatial distribution of OM and IOM. The segmentation of OM and IOM is based on grayscale threshold, the values of which are corrected with the grayscale value of OM between different images (see Text S1 in the supporting information). We choose the proportion of OM as the target quantity to identify the REV and SREV. For the REV, the OM proportion is obtained as an average from a square at the centre of each image for each scale [40]. For the SREV, the square for each scale is taken from a random image and a random position for one sampling process. We sampled 500 times in this case. The pore structure model obtained from FIB/SEM is also studied for comparison. In Fig. 4c, it can be seen that the average OM proportion has obvious fluctuation and large standard deviation at small scales, and approaches the global average values gradually with decreasing standard deviation as the scale increases. However, the OM proportion from FIB/SEM has large fluctuation all through the scales until approaching the whole domain of this pore structure model, which means that the pore structure model from FIB/SEM is smaller than the REV and SREV. Therefore, a single pore structure model from FIB/SEM cannot represent the entire shale. It can be concluded

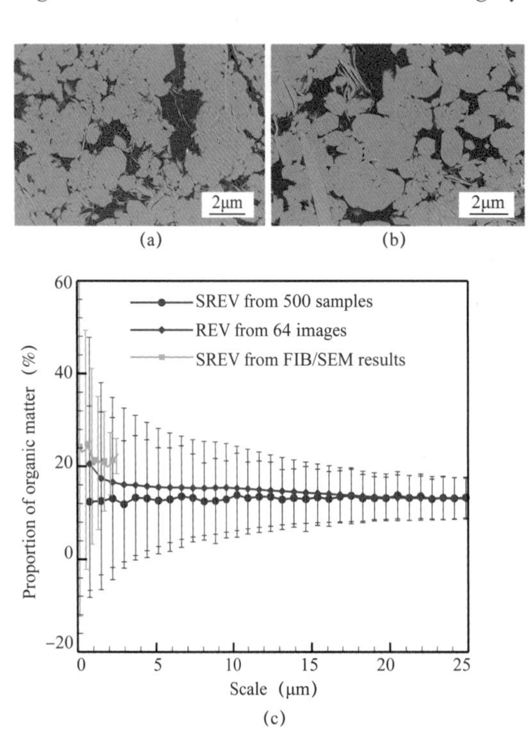

Fig. 4 SEM images and analysis of REV and SREV based on OM proportion from images. (a) From the surface parallel to the bedding plane. (b) From the surface orthogonal to the bedding plane. (Light gray: inorganic matter; dark gray: organic matter). (c) REV and SREV analysis based on OM proportion. The errors are obtained by adding or subtracting one standard deviation of the mean

that the REV and SREV for OM and IOM spatial distribution are above 10μm, at which the average of OM proportion can approach a constant. However, a large number of images must be taken to decrease the statistical error, and sample from various positions at a large area to improve the representativeness for the entire core.

On the other hand, anisotropy can be detected qualitatively from the images (Fig. 4a and Fig. 4b). Then, the indicator variogram is calculated to investigate the anisotropy quantitatively, which can be expressed as follows:

$$\gamma(\boldsymbol{h})=\frac{1}{2n(\boldsymbol{h})}\sum_{i=1}^{n(\boldsymbol{h})}\left\{I(x_i)-I(x_i+\boldsymbol{h})\right\}^2 \tag{1}$$

where $n(\boldsymbol{h})$ is the number of pairs of data locations separated by a vector $\boldsymbol{h}$; x denotes spatial location; and $I(x)$ is the indicator function, which is assigned to 1 for OM and 0 for IOM. We fitted the indicator variogram function with the exponential variogram model for three marine shale samples and a terrestrial shale sample, and obtain the ranges of OM aggregate along different directions (Tab. S1 in the supporting information). It can be seen that the samples have different levels of anisotropy, which may be related to the sedimentary environment. The anisotropy of spatial distribution of OM and IOM may be one of the reasons that caused the anisotropy of permeability.

1.4 Pore structure in OM and IOM

First, in order to extract the precise pore structure, we carried out FIB/SEM imaging on an area filled by kerogen aggregate (Fig. 5a), with the resolution of 6.5nm. It can be seen that the porosity increases up to 4.05% in this pore structure model. In addition, the pore size mainly ranges from 20nm to 200nm (Fig. 5d), and many interconnected pores can be detected. Then, FIB/HIM imaging was also performed to increase the precision of pore structure, which has the resolution of 1.46nm × 1.46nm × 2.43nm per voxel (Fig. 5b). More details about the pores below 10nm appear, but some information of pores larger than 50nm is lost (Fig. 5e), which may result from the heterogeneity. Some pores can be detected that are connected under this resolution, but not connected under a lower resolution (e.g., 6.5nm). 2D HIM imaging was also applied to further increase the resolution, which can approach 0.5nm (Fig. 5c). The pore size distribution is calculated based on 66 images from the 2D HIM and 99 images from the 2D SEM. The porosity is found to increase up to 21.18% (Fig. 5f). It is seen that many isolated macropores detected from the low resolution techniques may be connected by the smaller pores beyond the resolution. Although more interconnected pores can be found as the resolution increases, large pores may be lost due to the decreasing of dimension and the heterogeneity. In addition, the 3D and 2D images show that the pore geometry is close to the complex tube, which is important for selecting appropriate numerical models. The level of anisotropy of the pores in kerogen is also investigated with the indicator variogram based on the 2D HIM images. The indicator function is assigned to 1 for pores and 0 for kerogen skeleton. The indicator variogram is calculated along different directions and fitted with the exponential variogram model. The results indicate that there is no clear anisotropy of the pores and skeleton in kerogen (Tab. S2 in the supporting information) .

Regarding inorganic matter, micropores and mesopores mainly exist in clays [2, 12, 41]. However, due to the fact that the marine shale has relatively low clay content, which often distributes dispersively, it is difficult to acquire a pure clay model in marine shale. Here, we pick a clay model from a terrestrial shale that has clay content up to 50%. The pore structure model is acquired from FIB/SEM with the resolution of 6.5nm (Fig. 6a), and the clay mainly consists of illite. The results show that the pores may have obvious orientations, and the pore size is ~50nm to ~200nm (Fig. 6b) . From the 2D SEM and the 2D HIM images, it can be seen that the pore geometry in clays is close to slit (Fig. 6c and Fig. 6d), which is related to the molecular structure of clay [42, 43]. The principal direction is correlated to the features of IOM spatial distribution, which is affected by the sedimentary environment. Thus, the pores in IOM have obvious anisotropy.

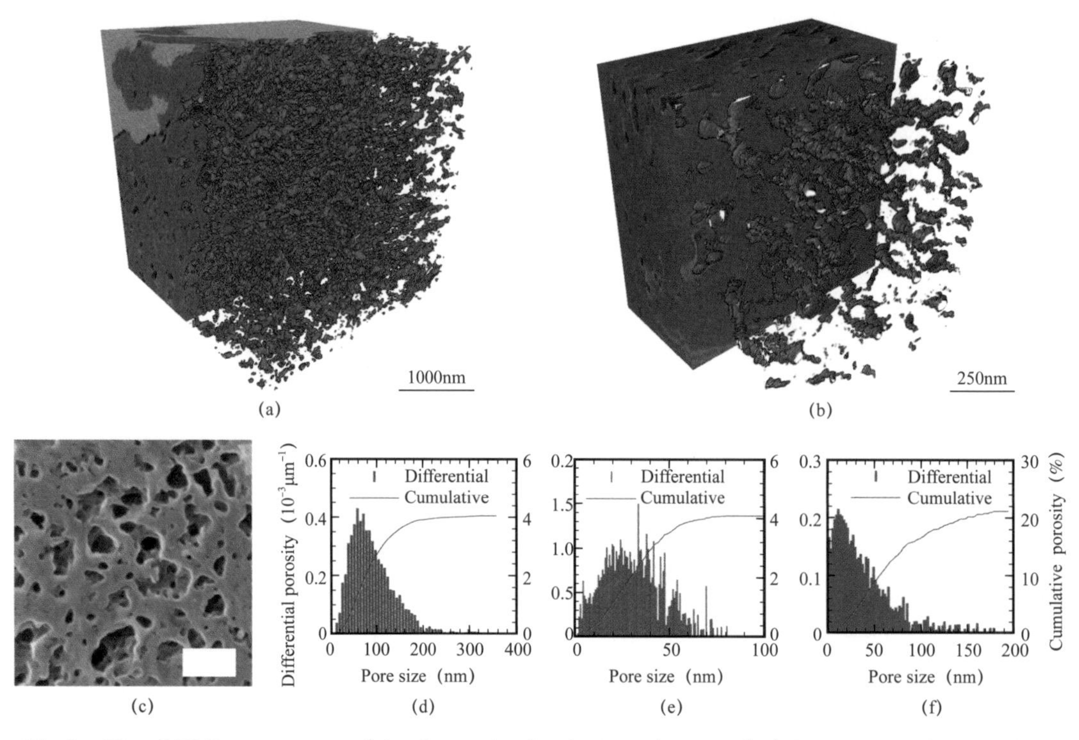

Fig. 5 3D and 2D Pore structure model and pore size distribution in kerogen. (a) Pore structure from FIB/SEM with ~6.5nm resolution. (b) Pore structure from FIB/HIM with ~1.5nm resolution. (c) 2D HIM image with the resolution of ~0.5nm. (d) Pore size distribution of the pore structure model in (a) . (e) Pore size distribution of the pore structure model in (b) . (f) Pore size distribution based on a series of 2D HIM images. (In (a) and (b), gray : original gray-scale picture ; red : pores)

2 Multi-scale gas transport simulation

Based on the features acquired from the multiscale pore structure analysis, a workflow can be constructed specifically to implement the "bottom-up" process. To account for the characteristics of different scales, the proposed workflow consists of three levels (Fig. 7). At level I, the permeability distributions for OM and IOM are determined, respectively, which can be calculated with an appropriate permeability model and permeability correction model by means of porosity, PSD, total organic carbon (TOC), and other basic properties (see Fig. S1, Tab. S3 and Tab. S4 in the supporting information). At level II, the stochastic ternary component models containing OM, IOM and macropores (>1μm) are constructed to represent the shale solid excluding the micro-fractures, based on the large numbers of SEM images and the Micro-CT pore structure model. Then, the equivalent apparent permeability can be calculated, which is also an important parameter for the next level. At level III, the shale core model is constructed according to the micro-fracture structure from the low resolution Micro-CT pore structure model, and the macroscopic shale matrix permeability can be determined with a numerical simulator. The crucial methods and models are described in the following sections.

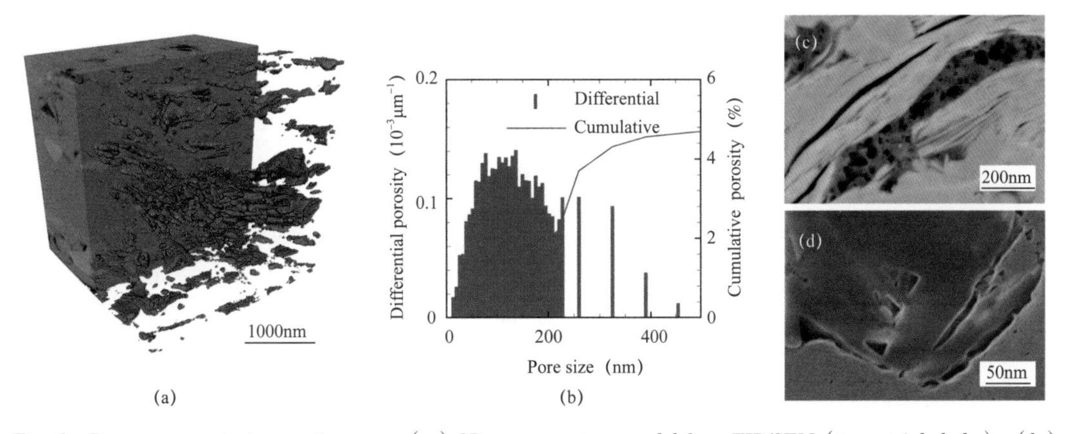

Fig. 6 Pore structure in inorganic matter. (a) 3D pore structure model from FIB/SEM (terrestrial shale). (b) Pore size distribution of the pore structure model in (a). (c) SEM image from marine shale. (d) HIM image from marine shale

2.1 Apparent permeability model

From the macroscopic perspective, the apparent permeability k can be expressed as the intrinsic permeability k_∞ multiplied by the correction factor f[44]. At level I, we applied the Kozeny-Carman model to determine the intrinsic permeability for OM and IOM, which can be expressed as[45]:

$$k_\infty = \frac{\phi^3}{(1-\phi)^2}\frac{d^2}{180} \tag{2}$$

where ϕ is the porosity; and d is the characteristic diameter, which can be represented by the pore size as in previous works[45]. Because of the different features in OM and IOM, the correction model should be chosen based on pore geometry. In addition, the permeability correction model should take the coupled effect of adsorption and Knudsen diffusion into account. Therefore, in the OM, the Dual Region (DR) model can be applied for the tube-like pore [36]:

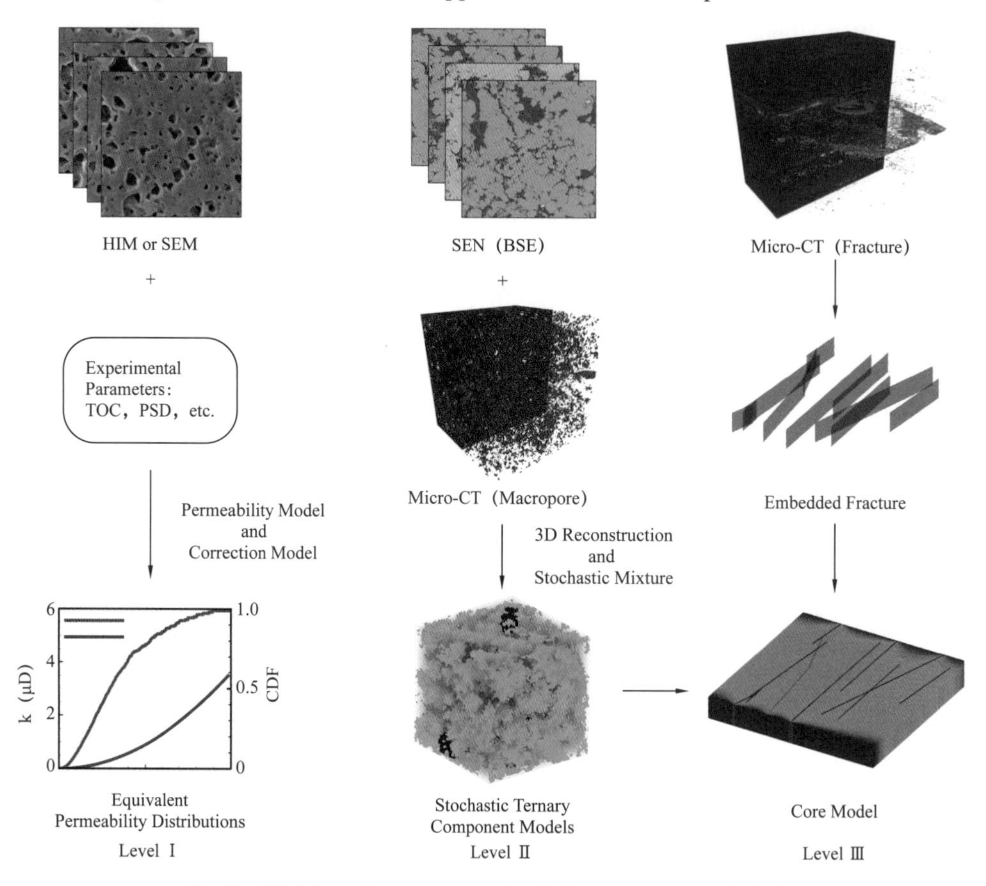

Fig. 7 Workflow of gas transport simulation in organic-rich shale

$$f=(1+aKn)\left\{-\left(r^{*4}+2w_1r^{*2}\right)-\frac{\alpha^2}{\beta}\left[\left(1-r^{*4}\right)+\frac{4}{3}\left(1-r^{*3}\right)w_2+2\left(1-r^{*2}\right)w_3\right]\right\} \tag{3}$$

where a is a rarefaction coefficient for viscosity correction; Kn is the Knudsen number; r^*, α, and β are the ratio parameters of pore radius, gas density, and viscosity, respectively, related to the adsorption effect; and w_1, w_2, and w_3 are parameters in terms of r^*, α, and β. While in the IOM, the DR model can be applied for the slit-like pore [36]:

$$f=(1+aKn)\left\{-\left(\frac{1}{2}h^{*3}+6w_1h^*\right)-\frac{\alpha^2}{\beta}\left[\frac{1}{2}\left(1-h^{*3}\right)+\frac{3}{2}\left(1-h^{*2}\right)w_2+6\left(1-h^*\right)w_3\right]\right\} \tag{4}$$

where h^* is a ratio parameter of slit pore width related to the adsorption effect. The expressions of the coefficients in the above equations can be found in the referenced paper [36]. On the other hand, the OM can be treated as an isotropic porous medium, while the IOM is anisotropic. In this study, the anisotropy coefficient for permeability is obtained from the ranges along different directions presented in see Tab. S1 in the supporting information. Other parameters can be found in Tab. S4 in the supporting information.

2.2 Stochastic ternary component model construction

In order to describe the spatial distribution of OM and IOM, stochastic methods are applied based on the large numbers of SEM images along different planes. In this study, we use the SISIM to generate binary component stochastic realizations using the parameters obtained from the variogram fitting (Tab. S1 in the supporting information). On the other hand, a sub-block with macropores (>1μm) is taken from the Micro-CT model (resolution: ~1μm) at a random position, which has the same size as the binary component model. Then, the binary component model is merged with the sub-block with macropores to generate the ternary component model (Fig. 8a). In this study, each ternary component model has 100 × 100 × 100 grids that represent a "shale solid" in the size of 50μm × 50μm × 50μm. Since the domain size of the shale solid is larger than the REV and SREV for the spatial distribution feature of OM and IOM, but smaller than the REV and SREV for the shale solid with macropores (>1μm), a large amount of stochastic realizations (100 realizations in this case) must be generated to acquire the statistical distribution of apparent permeability in the shale solid. The properties for every grid are determined according to its component type. For example, the apparent permeability in each grid is calculated with the method in the previous section for OM or IOM (Fig. S2 in the supporting information). It should be noted that porosity is treated as a constant in each component, while the characteristic pore size in each grid is a random variable following the PSD. In this case, the PSD in OM is acquired from the HIM and SEM images (Fig. 5c and Fig. 5f). The PSD in IOM is calculated by subtracting the PSD in OM according to the TOC from the PSD of the entire shale solid, which can be obtained from the N_2 and/or CO_2 adsorption experiment at low temperature and low pressure (Fig. S1 in the supporting information). The permeability of the macropores above 1 μm is estimated with the theoretical model as [46]:

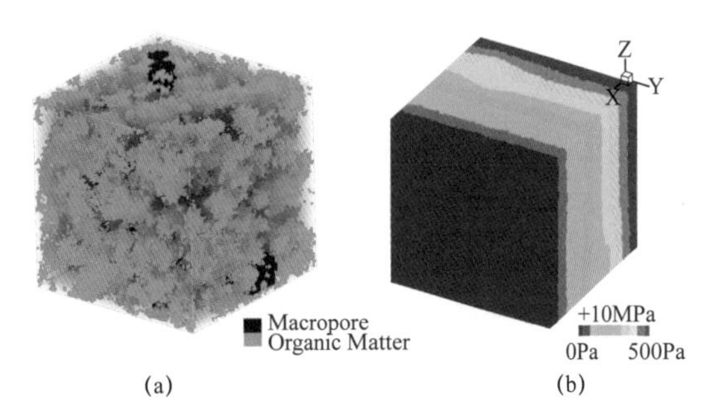

Fig. 8 Ternary component model and numerical simulation results for shale solid. (a) Ternary component model. The transparent parts are the inorganic matter. (b) The pressure distribution at steady-state of the numerical simulation based on the model in (a)

$$k = \frac{r^2}{8} \tag{5}$$

where r is the pore radius, which is determined according to its own actual geometry for every pore in the ternary component model. In general, the apparent permeability in macropores above 1μm is approximately several orders of magnitude higher than that in OM and IOM. The apparent permeability in OM is also higher than that in IOM because of the higher porosity and better connectivity.

2.3 Gas transport simulation and experimental verification in the shale solid

Based on the ternary component models of shale solid, numerical simulations are performed with the Unconventional Oil and Gas Simulator (UNCONG) [47]. The gas pressure gradient is set to 10Pa/μm along the target direction. The numerical tests are carried out along the three directions, and under various average pressures, respectively. In this study, the average pressure is set to 1MPa, 5MPa, 10MPa and 30MPa, respectively, which are common conditions in shale gas reservoirs. Thus, 12 numerical tests were carried out for each shale solid model. The apparent permeability can be calculated with Darcy's law when the flow reaches the steady-state. There is an example in Fig. 8b to demonstrate the pressure distribution at steady-state of the model in Fig. 8a. To verify this method, experimental tests were performed on micro-plugs with a diameter of approximately 2 mm to eliminate the micro-fractures. Information about the apparatus and the experimental procedure can be found in the Supporting Information (Fig. S3 and Text S2) . We applied the same methods of simulation and experiment on the four shale samples. The results are presented in Fig. 9. The values of apparent permeability from the simulations agree well with the results from experiments. The apparent permeability ranges from 0.2×10^{-6}mD to 20×10^{-6}mD, which is extremely low. This agrees with the values obtained from various experiments using different kinds of shale particles reported from other works [2, 48-49]. As the average pressure increases, the apparent permeability decreases, which results from the effects of adsorption and Knudsen diffusion, and also agrees with the macroscopic phenomenon [36]. In addition, the different extent of anisotropy can also be simulated.

2.4 Gas transport simulation in shale matrix with micro-fractures

At level III, the shale core model with micro-fractures is constructed to calculate the macroscopic apparent permeability of shale matrix. Limited by computational capability, we only demonstrate the viability of this method with a slice model. The model has $200 \times 200 \times 25$ grids that represent a shale core in the size of $1\text{cm} \times 1\text{cm} \times 0.25\text{cm}$, which is close to the size for shale core plug experiments. The apparent permeability for each grid is assigned as the result from level II (the case in Fig. 9a) . The micro-fractures are extracted and simplified from the low resolution Micro-CT pore structure model with ~13μm resolution (Fig. 2a and Fig. 10a), the width of which is set to 26.49μm. It should be noted that the model does not contain throughout fractures (with lengths larger than, or as large as, the domain size) in this case. The permeability for the micro-fracture is estimated with the theoretical model as [46]:

$$k = \frac{h^2}{12} \qquad (6)$$

where h is the width of the micro-fracture. In addition, since the width of micro-fractures may change due to the confining pressure, we also simulated the cases to test the sensitivity with the fracture width of 50μm, 10μm, and 1μm, respectively. The gas pressure gradient is set to 10Pa/μm along the x-direction and the y-direction to calculate their permeability, respectively. The average pressure is kept consistent with the cases at level II. The simulations are carried out with the embedded discrete fracture model (EDFM) in UNCONG[47]. The results are presented in Fig. 10. Because the micro–fractures have an obvious dominant orientation and topology structure, the apparent permeability is enhanced to a different extent along the x and y direction, respectively. The apparent permeability ranges from 6.77 to 7.13 times as large as the shale solid permeability along the x–direction, while it only increases up to 1.14–1.19 times of the shale solid permeability along the y–direction. The apparent permeability approaches the range of the values obtained from shale plug pulse–decay experiments[2, 8]. However, under the same average pressure and the same direction, the values of macroscopic permeability with different micro–fracture widths are almost identical. In the meantime, for the same micro–fracture geometry along the same direction, the macroscopic permeability exhibits the same trend as the shale solid permeability under different pressures. Therefore, it can be concluded that the orientation and topological structure of the micro–fractures are dominant factors for macroscopic apparent permeability, while the effect of fracture width may not be as important ; under a certain orientation and topological structure of the micro–fractures, the permeability variation is largely determined by the properties of the shale solid as a limiting factor. It should be noted that the scenario of containing the throughout fractures and the effect of fracture closure are not included in this study.

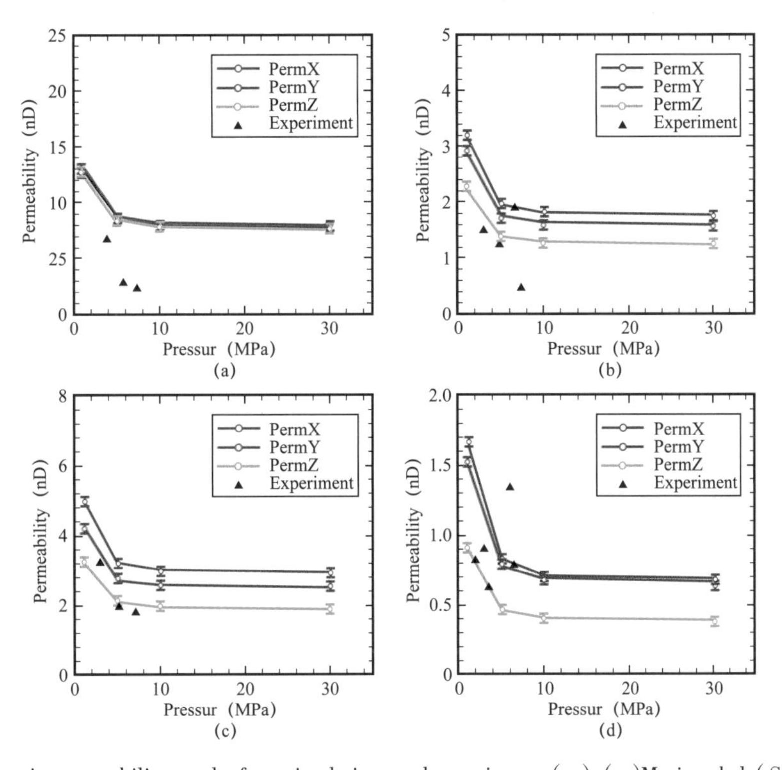

Fig. 9 Shale matrix permeability results from simulations and experiments.(a)–(c)Marine shale(S–001–S–003). (d) Terrestrial shale (S–004) . The data point is the median, and the error bars are lower quartile and upper quartile, respectively

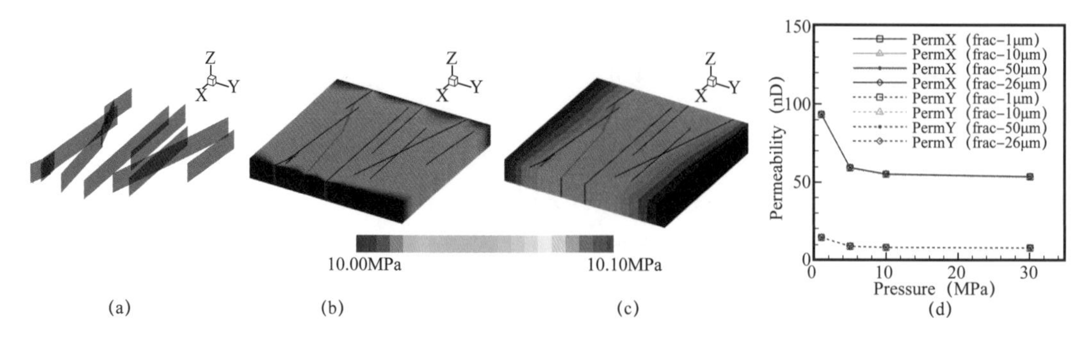

Fig. 10 Gas transport simulation in shale matrix with micro-fractures. (a) The topological structure of the micro-fractures. An example of pressure distribution at steady-state, which pressure gradient is along the x-direction (b) and along the y-direction (c). (d) The apparent permeability of shale matrix with micro-fractures

3 Discussion

The development of modern high precision imaging techniques has provided an opportunity to investigate the complex pore structure in such ultra-tight porous media as organic-rich shale. By means of the combination of various advanced imaging techniques, multiscale pore structure analysis provides insights into the key features of organic-rich shale at each scale and reveals the major path for gas transport, especially the micropores and mesopores interconnecting the macropores. It also suggests the affordable techniques for specific objectives based on REV and SREV analysis. To account for the characteristics of various scales, the proposed workflow for gas transport simulation consists of three levels: the microscopic mechanisms should be taken into consideration at level I; the spatial distribution feature of OM, IOM, and macropores is the major issue at level II; and the fracture orientation and topological structure are dominant factors at level III. By means of the comprehensive workflow, the impact of various mechanisms at different scales can be investigated more individually and precisely than with conventional experimental tests. In addition, this study offers alternative models based on the workflow at each scale, which can be updated according to further understanding of each mechanism in future work. In a broad sense, this study shows that the multi-scale imaging method and the multi-level simulation workflow can be powerful tools to study the global pore structure and the dominant factors of fluid transport under different scales for complex porous media.

Acknowledgements

This work is partially funded by the National Natural Science Foundation of China (Grant No. U1663208 and 51520105005) and the National Science and Technology Major Project of China (Grant No. 2016ZX05037-003 and 2017ZX05009005). The authors would like to thank Zheng Jiang, Su Jiang, and Xiaohan Li for constructive discussions on various parts of the manuscript. The data generated in the study can be obtained from the authors upon request.

References

[1] Yang T, Li X, Zhang D. Quantitative dynamic analysis of gas desorption contribution to production in shale gas reservoirs [J]. Journal of Unconventional Oil and Gas Resources, 2015, 9: 18–30.

[2] Heller R, Vermylen J, Zoback M. Experimental investigation of matrix permeability of gas shales [J]. AAPG Bulletin, 2014, 98 (5): 975–995.

[3] Cui X, Bustin A, Bustin R M. Measurements of gas permeability and diffusivity of tight reservoir rocks: different approaches and their applications [J]. Geofluids, 2009, 9 (3): 208–223.

[4] Dicker A, Smits R. A practical approach for determining permeability from laboratory pressure-pulse decay measurements, paper presented at International Meeting on Petroleum Engineering [C] // International Meeting on Petroleum Engineering, 1–4 November 1988, Tianjin, China. DOI: https://doi.org/10.2118/17578-MS.

[5] Jones S. A technique for faster pulse-decay permeability measurements in tight rocks [J]. SPE Formation Evaluation, 1997, 12 (01): 19–26.

[6] Kwon O, Kronenberg A K, Gangi A F, et al. Permeability of Wilcox shale and its effective pressure law[J]. Journal of Geophysical Research: Solid Earth, 2001, 106 (B9): 19339–19353.

[7] Ma Y, Pan Z, Zhong N, et al. Experimental study of anisotropic gas permeability and its relationship with fracture structure of Longmaxi Shales, Sichuan Basin, China [J]. Fuel, 2016, 180: 106–115.

[8] Pan Z, Ma Y, Connell L D, et al. Measuring anisotropic permeability using a cubic shale sample in a triaxial cell [J]. Journal of Natural Gas Science and Engineering, 2015, 26: 336–344.

[9] Luffel D, Guidry F. New core analysis methods for measuring reservoir rock properties of Devonian shale[J]. Journal of Petroleum Technology, 1992, 44 (11): 1184–1190.

[10] Luffel D, Hopkins C, Schettler Jr P. Matrix permeability measurement of gas productive shales [C] // SPE Annual Technical Conference and Exhibition, 3–6 October 1993, Houston, Texas. DOI: https://doi.org/10.2118/26633–MS.

[11] Josh M, Esteban L, Delle P C, et al. Laboratory characterisation of shale properties [J]. Journal of Petroleum Science and Engineering, 2012, 88–89: 107–124.

[12] Chalmers G R, Bustin R M, Power I M. Characterization of gas shale pore systems by porosimetry, pycnometry, surface area, and field emission scanning electron microscopy/transmission electron microscopy image analyses: Examples from the Barnett, Woodford, Haynesville, Marcellus, and Doig units [J]. AAPG Bulletin, 2012, 96 (6): 1099–1119.

[13] Clarkson C R, Solano N, Bustin R, et al. Pore structure characterization of North American shale gas reservoirs using USANS/SANS, gas adsorption, and mercury intrusion [J]. Fuel, 2013, 103: 606–616.

[14] Rouquerol J, Avnir D, Fairbridge C, et al. Recommendations for the characterization of porous solids[J]. Pure Appl. Chem., 1994, 66 (8): 1739–1758.

[15] Loucks R G, Reed R M, Ruppel S C, et al. Spectrum of pore types and networks in mudrocks and a descriptive classification for matrix-related mudrock pores [J]. AAPG bulletin, 2012, 96 (6): 1071–1098.

[16] Blunt M J, Bijeljic B, Dong H, et al. Pore-scale imaging and modelling [J]. Advances in Water

Resources, 2013, 51: 197–216.

[17] Bai B, Zhu R, Wu S, et al. Multi-scale method of Nano (Micro) -CT study on microscopic pore structure of tight sandstone of Yanchang Formation, Ordos Basin [J] . Petroleum Exploration and Development, 2013, 40 (3): 354–358.

[18] Chen C, Hu D, Westacott D, et al. Nanometer-scale characterization of microscopic pores in shale kerogen by image analysis and pore-scale modeling [J] . Geochem. Geophys. Geosyst., 2013, 14 (10): 4066–4075.

[19] Curtis M E, Sondergeld C H, Ambrose R J, et al. Microstructural investigation of gas shales in two and three dimensions using nanometer-scale resolution imaging [J] . AAPG Bulletin, 2012, 96 (4): 665–677.

[20] Kelly S, El-Sobky H, Torres-Verdín C, et al. Assessing the utility of FIB-SEM images for shale digital rock physics [J] . Advances in Water Resources, 2016, 95: 302–316.

[21] Zhou S, Yan G, Xue H, et al. 2D and 3D nanopore characterization of gas shale in Longmaxi formation based on FIB-SEM [J] . Marine and Petroleum Geology, 2016, 73: 174–180.

[22] King Jr H E, Eberle A P, Walters C C, et al. Pore architecture and connectivity in gas shale [J] . Energy & Fuels, 2015, 29 (3): 1375–1390.

[23] Peng S, Yang J, Xiao X, et al. An integrated method for upscaling pore-network characterization and permeability estimation : example from the Mississippian Barnett Shale [J] . Transport in Porous Media, 2015, 109 (2): 359–376.

[24] Dong H, Blunt M J. Pore-network extraction from micro-computerized-tomography images [J] . Phys. Rev. E, 2009, 80 (3): 036307.

[25] Silin D, Patzek T. Pore space morphology analysis using maximal inscribed spheres [J] . Physica A : Statistical mechanics and its applications, 2006, 371 (2): 336–360.

[26] Tahmasebi P, Javadpour F, Sahimi M. Multiscale and multiresolution modeling of shales and their flow and morphological properties [J]Scientific Reports, 2015, 5: 16373.

[27] Hazlett R D. Statistical characterization and stochastic modeling of pore networks in relation to fluid flow[J]. Mathematical Geology, 1997, 29 (6): 801–822.

[28] Keehm Y, Mukerji T, Nur A. Permeability prediction from thin sections : 3D reconstruction and Lattice-Boltzmann flow simulation [J] . Geophys. Res. Lett., 2004, 31 (4): 1668.

[29] Quiblier J A. A new three-dimensional modeling technique for studying porous media [J] . J. Colloid Interface Sci., 1984, 98 (1): 84–102.

[30] Tahmasebi P, Hezarkhani A, Sahimi M. Multiple-point geostatistical modeling based on the cross-correlation functions [J] . Computational Geosciences, 2012, 16 (3): 779–797.

[31] Wu K, Dijke M I J, Couples G D, et al. 3D stochastic modelling of heterogeneous porous media–applications to reservoir rocks [J] . Transport in Porous Media, 2006, 65 (3): 443–467.

[32] Mehmani A, Prodanović M, Javadpour F. Multiscale, multiphysics network modeling of shale matrix gas flows [J] . Transport in porous media, 2013, 99 (2): 377–390.

[33] Akkutlu I Y, Fathi E. Multiscale gas transport in shales with local Kerogen heterogeneities, SPE J., 17

(04), 1, 002–1, 011, doi : 10.2118/146422–PA.

[34] Naraghi M E, Javadpour F. A stochastic permeability model for the shale-gas systems [J] . International Journal of Coal Geology, 2015, 140: 111–124.

[35] Falk K, Coasne B, Pellenq R, et al. Subcontinuum mass transport of condensed hydrocarbons in nanoporous media [J] . Nat. Commun., 2015, 6: 6949.

[36] Wu T, Zhang D. Impact of adsorption on gas transport in nanopores [J] . Scientific Reports, 2016, 6: 23629.

[37] Dyni J R. Geology and resources of some world oil-shale deposits [J] . Oil Shale, 2003, 20 (3): 193–252.

[38] Mahlstedt N, Horsfield B. Metagenetic methane generation in gas shales I [J] . Screening protocols using immature samples, Marine and Petroleum Geology, 2012, 31 (1) : 27–42.

[39] Zhang D, Zhang R, Chen S, et al. Pore scale study of flow in porous media : Scale dependency, REV, and statistical REV [J] . Geophys. Res. Lett., 2000, 27 (8): 1195–1198.

[40] Saraji S, Piri M. The representative sample size in shale oil rocks and nano-scale characterization of transport properties [J] . International Journal of Coal Geology, 2015, 146: 42–54.

[41] Heller R, Zoback M. Adsorption of methane and carbon dioxide on gas shale and pure mineral samples[J]. Journal of Unconventional Oil and Gas Resources, 2014, 8: 14–24.

[42] Jin Z, Firoozabadi A. Methane and carbon dioxide adsorption in clay-like slit pores by Monte Carlo simulations [J] . Fluid Phase Equilib., 2013, 360: 456–465.

[43] Jin Z, Firoozabadi A. Effect of water on methane and carbon dioxide sorption in clay minerals by Monte Carlo simulations [J] . Fluid Phase Equilib., 2014, 382: 10–20.

[44] Civan F. Effective correlation of apparent gas permeability in tight porous media [J] . Transport in Porous Media, 2010, 82 (2): 375–384.

[45] Chen L, Kang Q, Pawar R, et al. Pore-scale prediction of transport properties in reconstructed nanostructures of organic matter in shales [J] . Fuel, 2015, 158: 650–658.

[46] Bear J. Dynamics of fluids in porous media [M] . New York : Courier Corporation, 1988.

[47] Li X, Zhang D, Li S. A multi-continuum multiple flow mechanism simulator for unconventional oil and gas recovery [J] . Journal of Natural Gas Science and Engineering, 2015, 26: 652–669.

[48] Guidry K, Luffel D, Curtis J. Development of Laboratory and Petrophysical Techniques for Evaluating Shale Reservoirs [R] . Houston : Gas Research Institute, GRI–95/0496, 1996.

[49] Qu H, Pan Z, Peng Y, et al. Controls on matrix permeability of shale samples from Longmaxi and Niutitang formations, China [J] . Journal of Natural Gas Science and Engineering, 2016, 33: 599–610.

页岩气开发概述

齐亚东[1]，刘志远[2]，瞿云华[3]

（1. 中国石油勘探开发研究院；
2.《科技导报》编辑部；
3. 中国石油集团川庆钻探工程有限公司井下作业公司）

摘　要：页岩气作为一种资源量巨大的非常规天然气，具有“自生自储，原地成藏”的特征，其储层致密，开采难度大，气井投产后“初产高，递减快，长期低产”的特征明显。对全球页岩气的储量、开发技术及意义进行综述，并介绍了目前中国页岩气的开发现状：中国页岩气资源丰富，地质资源量 $134.42\times10^{12}m^3$，技术可采资源量 $25.08\times10^{12}m^3$，“水平井＋体积压裂＋工厂化作业”是页岩气成功开发的基本模式，目前在长宁—威远、昭通和涪陵 3 个南方海相页岩气示范区实现了初步规模开发，已经具备了年产气 $97\times10^8m^3$ 的生产能力，合理高效地开发页岩气对于缓解中国能源供需矛盾、调整能源消费结构、增强国际天然气定价影响力以及促进区域经济发展具有重要意义。

关键词：页岩气；能源结构；资源开发

页岩气是一种非常规的资源，其开采发端于美国，此后美国在页岩气的勘探、开采以及商业化应用上走在了全球的前列。1821 年，美国纽约州弗里多尼亚天然气矿井中首次开采出页岩气。20 世纪 30 年代，水平钻探技术应用于页岩气开采中。1947 年，美国泛美石油股份公司在页岩气气井中第一次使用压裂法，这成为以后页岩气开采最常用的方法。但是直到 20 世纪 70 年代之前，美国的页岩气开采仍没达到工业生产的规模。20 世纪 80 年代开始，美国政府开始加大对页岩气开采的政策支持，先后投入 60 多亿美元进行非常规天然气的勘探开发，其中大部分资金用于页岩气的相关研究，极大地推进了页岩气开采技术的进步，大大降低了开发成本，刺激了页岩气工业的发展。进入 21 世纪以来，随着水平井大规模压裂技术的成功应用，美国页岩气工业获得快速发展。鉴于页岩气巨大的储藏量和广阔的开发利用前景，美国乃至全球正在掀起一场“页岩气革命”[1]。

2011 年底，国务院正式批准页岩气成为中国第 172 个独立矿种，中国油气资源家族增添了一位新成员[2]。自此，页岩气的探测、开发、应用频频见诸媒体。那么，什么是页岩气？页岩气开发有何意义？如何有效开发页岩气？本文对此做一简要介绍，并综述国内外页岩气开发应用的最新进展。

1　页岩气定义

认识页岩气之前需先对天然气有所了解。狭义上讲，天然气是指天然蕴藏于地层中的烃类和非烃类气体的混合物，有“常规”和“非常规”之分。

常规天然气是指能够用传统的油气地质理论解释成因，并能够用常规技术手段开发的天然气。传统的油气地质理论认为[3]：富含有机质的母岩（烃源岩）经过一系列的地质作用过程，有机质逐渐达到成熟并生成甲烷等烃类气体，气体在多种力的作用下从母岩向储集性能良好的岩层（储层）运移，当遇到渗透性极差的遮挡物（盖层）时，气体无法继续前行而被圈闭，最终形成天然气藏（图 1、图 2[4]）。通常，常规天然气在储层中以自由气的形式存在，而且开发过程中无需对储层进行任何形式的改造。

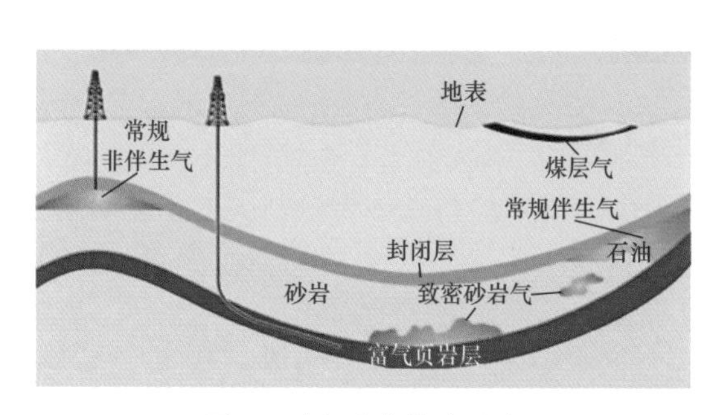

图 1　油气成藏模式示意

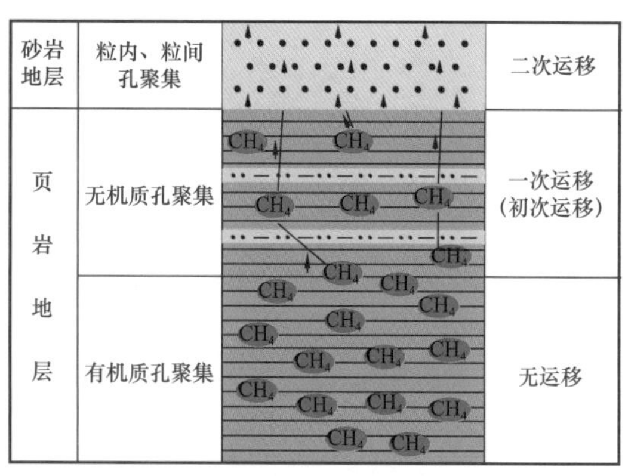

图 2　页岩气“原地滞留”机理

而非常规天然气则不同，它的形成难以用上述油气地质理论解释，并且不能用常规技术手段开采，页岩气便是一类典型的非常规天然气。

与常规天然气相比，页岩气“个性”鲜明。第一，页岩气“自生自储，原地成藏”（图 1、图 2），即母岩中的有机质在一定的温压条件下成熟并生烃，烃类气体基本无运移或极短距离运移而原地成藏，因其母岩类型主要为富有机质泥页岩及其夹层，因而称之为“页岩气”；第二，页岩气“赋存类型多样”，页岩气的主要成分是甲烷，特殊的储集条件决定了甲烷的 3 种赋存相态：天然裂缝和孔隙中的游离态（图 3）、干酪根和黏土颗粒表面的吸附态（图 3）以及干酪根和沥青中的溶解态，其中，吸附态甲烷含量在 20%～85%，一般为 50%[5]，溶解态仅有少量存在；第三，页岩气“储层超致密，不经人工改造无法贡献工业气流”，页岩储集空间以微—纳米级孔隙为主（中国寒武系—志留系页岩气储层孔隙直径主体为 80～200nm[6]）（图 4），通常孔隙度小于 10%，渗透率数量级为 10^{-8}～10^{-4}mD[7]，气体流动阻力巨大，必须进行大型人工改造在三维空间形成网状裂缝系统，分散在纳米孔隙中的天然气才能产出，实现工业开发；第四，页岩气单井初

图 3　页岩气主要赋存状态
（红色：游离态，黄色：吸附态）

期产量高，递减迅速，但低产生产周期长（一般为30～50年），以美国Haynesville典型页岩气井为例，单井初期日产普遍在$25\times10^4m^3$以上，第一年递减率达80%，生产2～3年后，日产降至$5\times10^4m^3$，生产10年后，日产基本保持在$1\times10^4m^3$左右，并持续低产生产。

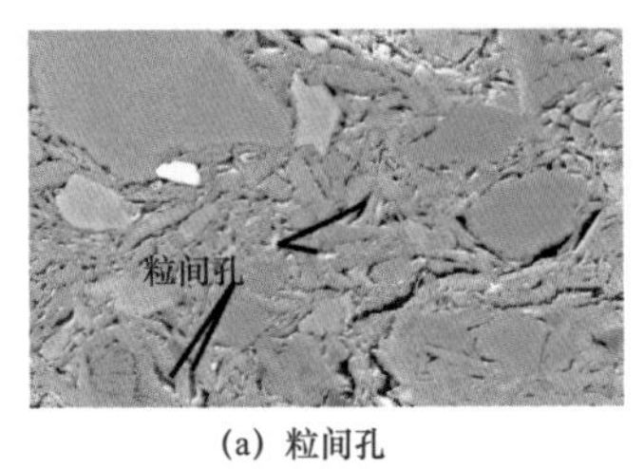

(a) 粒间孔

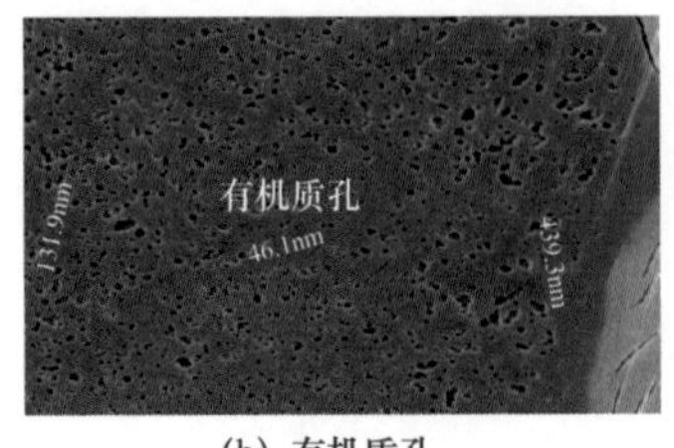

(b) 有机质孔

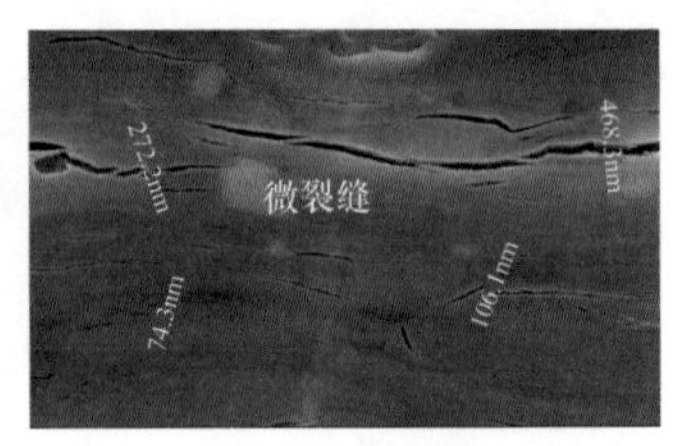

(c) 微裂缝

图4　页岩储层主要孔隙类型

2　页岩气资源量

页岩气“自生自储”的特性使得在母岩（烃源岩）中找寻天然气成为可能，这极大地拓展了油气勘探开发的领域与范围。从理论上讲，凡是有天然气生成的盆地必有母岩，亦应存在页岩气，特别是海相、湖相、海陆过渡相及煤系地层[8]。由此不难判断，世界页岩气资源量将非常巨大。根据美国能源信息署（EIA）2015年9月发布的数据[9]（表1），世界10个地理区域的46个国家95个页岩气盆地137个层位，页岩气地质资源量$1064.46\times10^{12}m^3$，技术可采资源量$214.54\times10^{12}m^3$，主要分布在北美、亚洲、欧洲、非洲、南美等地区。从表2的数据分析来看，页岩气技术可采储量排名前五位的国家占据了世界已知总量的半壁江山，其中，中国以$31.58\times10^{12}m^3$的技术可采资源量位列榜首，占目前世界已知总量的14.72%，其次为阿根廷和阿尔及利亚，技术可采资源量分别为$22.70\times10^{12}m^3$和$20.02\times10^{12}m^3$，美国以$17.63\times10^{12}m^3$位列第四，第五位为加拿大，技术可采资源量$16.22\times10^{12}m^3$。

表1　世界页岩气资源量调查统计（据EIA 2015年报告[9]）

地区	国家	地质资源量（$10^{12}m^3$）	技术可采资源量（$10^{12}m^3$）	更新时间
北美	加拿大	68.34	16.22	2013-05-17
	墨西哥	63.24	15.44	2013-05-17
	美国	131.50	17.63	2015-04-14
澳洲	澳大利亚	56.86	12.16	2013-05-17
南美	阿根廷	91.86	22.70	2013-05-17
	玻利维亚	4.36	1.03	2013-05-17
	巴西	36.23	6.94	2013-05-17
	智利	6.44	1.37	2013-05-17

续表

<table>
<tr><th>地区</th><th>国家</th><th>地质资源量（$10^{12}m^3$）</th><th>技术可采资源量（$10^{12}m^3$）</th><th>更新时间</th></tr>
<tr><td rowspan="4">南美</td><td>哥伦比亚</td><td>8.72</td><td>1.55</td><td>2013-05-17</td></tr>
<tr><td>巴拉圭</td><td>9.91</td><td>2.13</td><td>2013-05-17</td></tr>
<tr><td>乌拉圭</td><td>0.72</td><td>0.13</td><td>2013-05-17</td></tr>
<tr><td>委内瑞拉</td><td>23.08</td><td>4.74</td><td>2013-05-17</td></tr>
<tr><td rowspan="7">东欧</td><td>保加利亚</td><td>1.87</td><td>0.47</td><td>2013-05-17</td></tr>
<tr><td>立陶宛 / 加里宁格勒</td><td>0.69</td><td>0.07</td><td>2013-05-17</td></tr>
<tr><td>波兰</td><td>20.90</td><td>4.13</td><td>2013-05-17</td></tr>
<tr><td>罗马尼亚</td><td>6.60</td><td>1.44</td><td>2013-05-17</td></tr>
<tr><td>俄罗斯</td><td>54.38</td><td>8.06</td><td>2013-05-17</td></tr>
<tr><td>土耳其</td><td>4.63</td><td>0.67</td><td>2013-05-17</td></tr>
<tr><td>乌克兰</td><td>16.20</td><td>3.62</td><td>2013-05-17</td></tr>
<tr><td rowspan="8">西欧</td><td>丹麦</td><td>4.49</td><td>0.90</td><td>2013-05-17</td></tr>
<tr><td>法国</td><td>20.58</td><td>3.87</td><td>2013-05-17</td></tr>
<tr><td>德国</td><td>2.25</td><td>0.48</td><td>2013-05-17</td></tr>
<tr><td>荷兰</td><td>4.28</td><td>0.73</td><td>2013-05-17</td></tr>
<tr><td>挪威</td><td>0</td><td>0</td><td>2013-05-17</td></tr>
<tr><td>西班牙</td><td>1.18</td><td>0.24</td><td>2013-05-17</td></tr>
<tr><td>瑞典</td><td>1.38</td><td>0.28</td><td>2013-05-17</td></tr>
<tr><td>英国</td><td>3.78</td><td>0.73</td><td>2013-05-17</td></tr>
<tr><td rowspan="7">北非</td><td>阿尔及利亚</td><td>96.82</td><td>20.02</td><td>2013-05-17</td></tr>
<tr><td>埃及</td><td>15.15</td><td>2.83</td><td>2013-05-17</td></tr>
<tr><td>利比亚</td><td>26.68</td><td>3.44</td><td>2013-05-17</td></tr>
<tr><td>摩洛哥</td><td rowspan="3">2.70</td><td rowspan="3">0.58</td><td>2013-05-17</td></tr>
<tr><td>西撒哈拉</td><td>2013-05-17</td></tr>
<tr><td>毛里塔尼亚</td><td>2013-05-17</td></tr>
<tr><td>突尼斯</td><td>3.23</td><td>0.64</td><td>2013-05-17</td></tr>
<tr><td rowspan="2">亚撒哈拉地区</td><td>乍得</td><td>12.42</td><td>1.26</td><td>2014-12-29</td></tr>
<tr><td>南非</td><td>44.14</td><td>11.03</td><td>2013-05-17</td></tr>
<tr><td>亚洲</td><td>中国</td><td>134.40</td><td>31.58</td><td>2013-05-17</td></tr>
</table>

续表

地区	国家	地质资源量（$10^{12}m^3$）	技术可采资源量（$10^{12}m^3$）	更新时间
亚洲	印度	16.54	2.73	2013-05-17
	印度尼西亚	8.58	1.31	2013-05-17
	蒙古	1.56	0.12	2013-05-17
	巴基斯坦	16.60	2.98	2013-05-17
	泰国	0.62	0.15	2013-05-17
里海	哈萨克斯坦	7.17	0.78	2014-12-29
中东	约旦	0.99	0.19	2013-05-17
	阿曼	8.92	1.37	2014-12-29
	阿拉伯联合酋长国	23.45	5.81	2014-12-29
合计		1064.46	214.54	

表 2 世界主要国家页岩气技术可采资源量统计表（据 EIA 2015 年报告[9]）

编号	国家	技术可采资源量（$10^{12}m^3$）	所占比例（%）
1	中国	31.58	14.72
2	阿根廷	22.70	10.58
3	阿尔及利亚	20.02	9.33
4	美国	17.63	8.22
5	加拿大	16.22	7.56
6	墨西哥	15.44	7.20
7	澳大利亚	12.16	5.67
8	南非	11.03	5.14
9	俄罗斯	8.06	3.76
10	巴西	6.94	3.23
11	其他国家	52.76	24.59
合计		214.54	100.00

按照 EIA 发布的数据，世界油气资源大国俄罗斯、加拿大、巴西 3 国的页岩气技术可采资源量之和为 $31.22\times10^{12}m^3$，较中国的 $31.58\times10^{12}m^3$ 尚低 1%，由此可见，EIA 所发布的中国页岩气技术可采资源量多少有一些“忽悠”的成分，需冷静思考、小心求证。

多年来，国内外不同科研机构对中国页岩气资源量进行了多次评估，结果间差异较大（表 3），业内对国土资源部 2012 年的调查结果[10]认同度较高。通过对全国 4 个大区的

41个盆地和地区、87个评价单元、57个含气页岩层系的系统评价，得到全国页岩气技术可采资源量为 $25.08\times10^{12}m^3$（不含青藏区）。其中，上扬子及滇黔桂区 $9.94\times10^{12}m^3$，占全国总量的39.63%；华北及东北区 $6.70\times10^{12}m^3$，占全国总量的26.70%，中下扬子及东南区 $4.64\times10^{12}m^3$，占全国总量的18.49%，西北区 $3.81\times10^{12}m^3$，占全国总量的15.19%。

在 $25.08\times10^{12}m^3$ 的页岩气可采资源量中，现实可转入勘探开发的、可靠程度较高的资源量为 $15.95\times10^{12}m^3$。

表3　国内外机构对中国页岩气资源量估算

时间	预测机构及专家	预测范围	地质资源量（$10^{12}m^3$）		技术可采资源量（$10^{12}m^3$）	
			区间值	期望值	区间值	期望值
2009	中国石油勘探开发研究院[4, 11]	中国陆上	86～166	100	15～32	20
2010	中国石油勘探开发研究院[12]	中国陆上			15.1～33.7	24.5
2010	中国石油勘探开发研究院[4]	中国陆上	30～100	50	10～15	
2010	中国石油勘探开发研究院[13]	中国陆上			21.4～45	30.7
2010	中国地质大学（北京）[14]	中国陆上			15～30	26.5
2011	国土资源部油气战略中心[15]	中国陆上				31
2011	美国能源信息署[16]	四川、塔里木盆地		144.4		36.1
2011	中国石油勘探开发研究院[17]	中国陆上				15.2
2012	中国工程院[17]	中国陆上		83.3		11
2012	国土资源部油气战略中心[10]	中国陆上		134.42		25.08
2013	美国能源信息署（EIA）[18]	四川、塔里木盆地		134.3		31.58

3　页岩气开发的意义

作为一种低碳、清洁的新兴能源，页岩气在美国的成功开发已经引发全球能源领域的一场革命，可以预见，合理高效地开发利用中国丰富的页岩气资源意义重大。

3.1　有利于缓解能源供需矛盾，保证能源供应安全

进入21世纪以来，随着国内经济的强劲增长，能源消费需求不断攀升。根据国家统计局《2015中国统计年鉴》[19]中公布的数据（图5），中国的能源消费总量整体上呈现迅猛增长的态势，2002年以来，平均年增长率8.16%，至2014年底，中国能源消费总量已达 42.6×10^8t 标准煤（BP统计[20]为 29.70×10^8t 油当量），占全球能源消费总量的22.81%，已连续6年超过美国成为世界第一能源消费大国。当前，尽管中国经济增速放缓进入了新常态，相应地，能源需求也进入了中速增长阶段，但能源需求势头依然旺盛。预计2016—2020年间，能源消费总量年均增长3%左右，到2020年能源消费总量将达到 50×10^8t 标准煤，到2030年可能达到 57×10^8t 标准煤[21]。

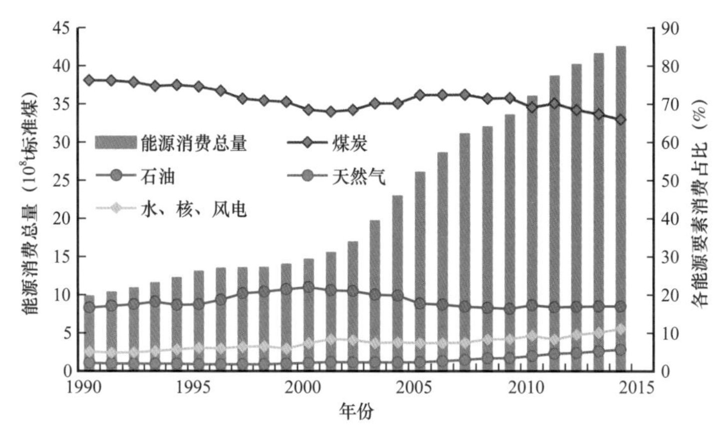

图 5　中国能源消费总量及结构的变化情况

能源需求的不断攀升导致能源对外依存度逐年升高，能源安全形势严峻（图 6、图 7）。中国 1993 年首次成为石油净进口国，2006 年又跨入天然气净进口国之列。石油对外依存度由 1993 年的 1.2% 飙升至 2015 年的 61.66%，天然气对外依存度由 2006 年的 0.8% 攀升至 2015 年的 30.06%，成为全球能源市场中最重要的进口国。

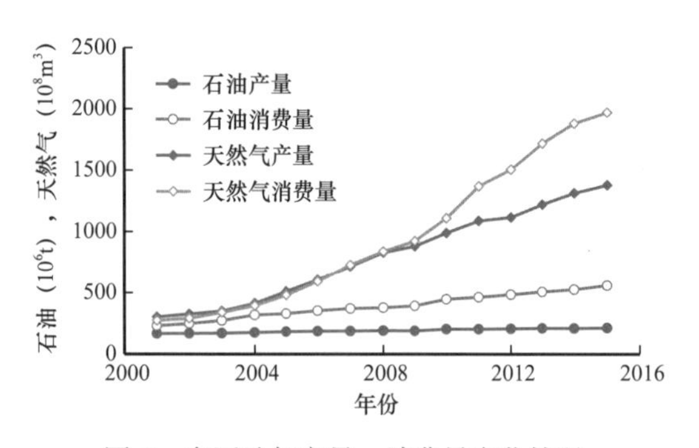

图 6　中国油气产量、消费量变化情况

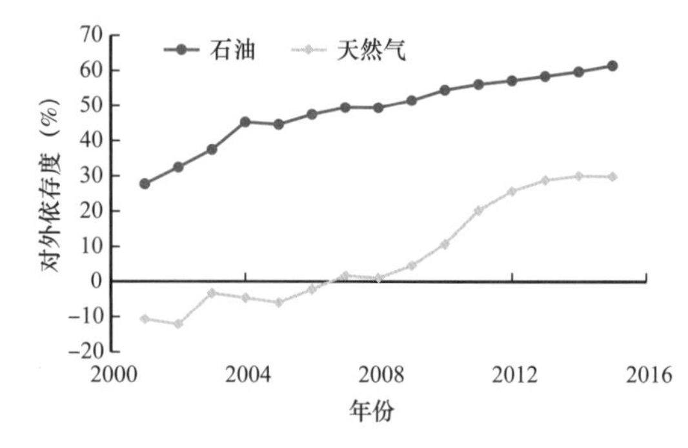

图 7　中国油气对外依存度变化情况

单从天然气供需看，2006 年以来，尽管中国的天然气产量年均增长在 10% 左右，至 2015 年已达 $1350 \times 10^8m^3$，成为全球天然气产量增速最快的国家之一，然而产量的攀升无法满足国内需求的强劲增长，国内的天然气消费量在以更为迅猛的态势增加，年均增长率在 13% 左右，天然气供需缺口从 2006 年的 $13 \times 10^8m^3$ 增至 2015 年的 $282 \times 10^8m^3$。当前，中国经济处于转型期，服务业发展、城市化推进以及大气污染治理将带动天然气消费持续增长。在当前政策环境下[21]，2020 年和 2030 年天然气消费量将达到 $3000 \times 10^8m^3$ 和 $4500 \times 10^8m^3$，而天然气产量分别达到 $2300 \times 10^8m^3$ 和 $3800 \times 10^8m^3$，天然气缺口 $700 \times 10^8m^3$；在加强环境监管和征收碳税的情况下，2020 年和 2030 年天然气消费量将达到 $3500 \times 10^8m^3$ 和 $5800 \times 10^8m^3$，而天然气产量分别达到 $2700 \times 10^8m^3$ 和 $4700 \times 10^8m^3$，天然气缺口 $800 \times 10^8m^3$。

由此可见，油气供需缺口将长期存在，特别是天然气，供需缺口的逐年增大已经给国内天然气增产带来巨大压力，亟须积极改善国内天然气供应结构，加快页岩气等非常规天然气的开发利用，使天然气供应格局多元化。

3.2 有助于调整能源消费结构，缓解节能减排压力

从能源消费结构上看（图 5、图 8），煤炭一直以来都占据国内能源消费的主体地位，近年来虽表现出占比下降的趋势，但依然保持在 65% 以上；其次是石油，比例维持在 15% 以上；天然气和水—核—风电等清洁能源的消费比例较低，至 2014 年分别达到 5.7% 和 11.2%。

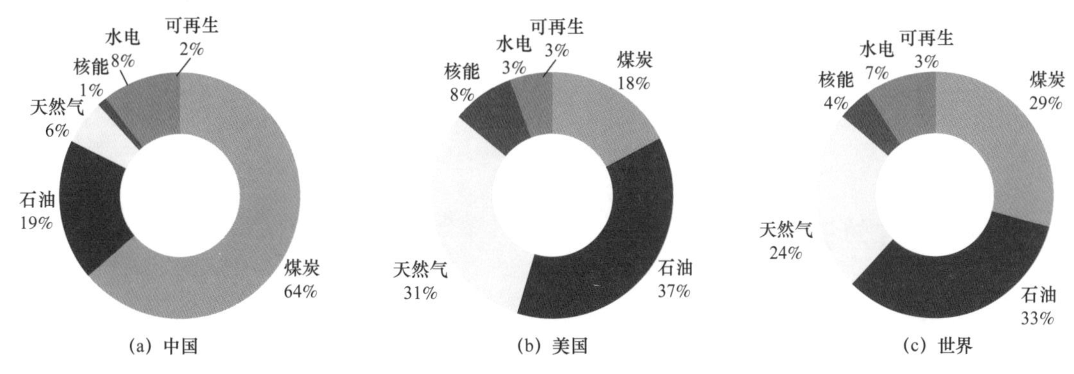

图 8 中国、美国、世界 2015 年能源结构对比

从世界能源消费构成上看，中国能源消费结构欠合理。2014 年，中国一次能源消费结构中，煤炭占 66.0%（为世界煤炭消费量的 50%），比美国高出 46%，比世界平均水平高出 35%；而油气消费占比仅为 24.5%，远低于美国的 68.6% 和 56.8% 的世界平均水平。

以煤为主的能源结构导致温室气体及其他污染物排放不断激增，使得环境保护、应对气候变化及节能减排面临巨大的国际压力和国内挑战。天然气清洁、高效和便于使用，无论是当前治理大气污染、提高能源系统效率的需要，还是应对未来可再生能源大发展带来的调峰需求，都需要大力发展天然气，这也是发达国家能源体系优化升级过程中的典型做法。

就目前国内天然气资源结构而言，加快页岩气开发是发展天然气产业最具潜力也是最现实的途径。定量分析表明[21]，在现有体制下，2020 年和 2030 年页岩气产量将达到 $400\times10^8m^3$ 和 $800\times10^8m^3$，分别占天然气总产量的 17.4% 和 21.1%；如果对体制进行改革，2020 年和 2030 年页岩气产量将达到 $600\times10^8m^3$ 和 $1500\times10^8m^3$，分别占天然气总产量的 22.2% 和 31.9%。

3.3 有助于扩大中国在全球天然气市场的份额，增强中国对天然气定价的影响力

20 世纪 70 年代的两次石油危机对世界造成的深远影响有两个[22]：美元与石油挂钩确立了美元的全球霸主地位，以沙特阿拉伯为首的 OPEC 取得了石油定价权，此后几十年“各取所需，相安无事”，但 2014 年 11 月 5 日，在美国达拉斯举行的第五届世界页岩油气峰会上，情况发生了变化，美国人不但要继续坚持石油与美元挂钩，还要从 OPEC 手中“夺回油气定价权”，美国之所以底气十足，根本原因在于美国的页岩油气革命确立了其世界油气生产大国的地位，改变了世界油气供应格局。

借鉴美国的经验，坐拥巨大页岩油气资源量的中国也不应该是国际油气价格的被动接

受者，中国应通过加大国内页岩油气勘探开发的方式更加积极地参与国际油气市场，逐步影响国际油气定价的主导权。

3.4 有利于拉动相关产业发展及基础设施建设，促进区域经济发展

美国页岩油气的成功开发带动了美国装备制造业、化工业，以及天然气发电等基础产业的发展，支撑了美国制造的回归和“再工业化”，促进了各州的经济增长。可以预见，中国页岩气的规模开发利用同样也将拉动装备制造、化工业及工程建设等领域的发展，增加劳动力需求，增加税收，促进区域经济发展；更重要的是，国内的页岩气资源主要分布在交通不便的山地区域或经济欠发达地区，页岩气开发可改善当地基础设施建设，拉动经济增长。

4 页岩气有效开发技术

页岩孔隙极其微小，储层渗透性极差，导致气体流动的阻力巨大，自然条件下的页岩气井几乎无法产生工业气流，因而必须采取工程措施对页岩储层进行改造才能有效开发页岩气，其基本思路是“降低气体渗流阻力，扩大气体泄流面积，同时还要降低改造成本”，就目前的国内外经验，“水平井 + 体积压裂 + 工厂化作业”的开发模式可有效开发页岩气。

4.1 体积压裂水平井

水平井，顾名思义，就是井眼轨迹在目的层中保持近水平状态延伸一定长度的一种定向井。水平井开采页岩气的优势在于：（1）水平井在页岩储层中钻进长度普遍在 1500m 甚至更长，可以控制地下较大的储层面积，与直井相比，节省了钻井数量和地面施工，同时有利于环境保护；（2）水平井施工成本是直井的 1.25～1.5 倍，但日产气量及最终产气量是直井的 3～5 倍，产气速率则可以提高 10 倍；（3）水平井与储层中发育的天然裂缝相交的概率较大，裂缝与井筒沟通，天然气渗流环境明显改善，泄流面积增大；（4）水平井在地层中延伸，而地面设施相对较少，因而可降低地面不利地形对施工的影响，这一点在中国南方复杂山地页岩气发育区表现得最为明显。

典型的页岩气水平井施工分为以下几个步骤（图 9）：（1）直井段导眼井施工和表层套管固井；（2）在进入目的层位之前的适当层位开始造斜井段钻进和生产套管固井；（3）水平井段钻进施工；（4）洗井、射孔或压裂完井。

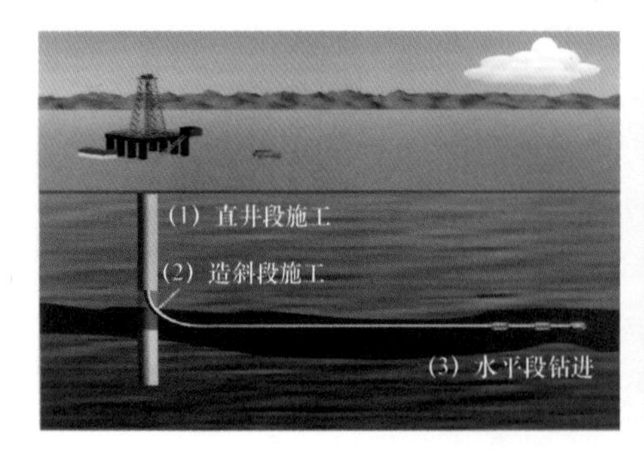

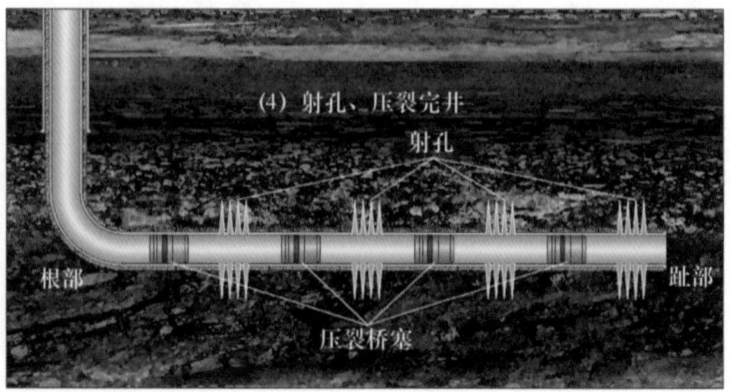

图 9 页岩气水平井施工基本步骤

为了引导钻头在复杂页岩储层中的目标区域内安全稳定地钻进，并形成规则的井眼以利于后续的压裂作业，旋转地质导向技术（图 10）至关重要。该技术是在钻具底部配备能测试页岩层电性信号（如自然伽马和电阻率等参数）的随钻测井仪器，依据判识标准快速识别目标地层，如果钻井轨迹偏离目的层，导向系统便会纠偏进行轨迹控制，提高钻遇目的层的精度。

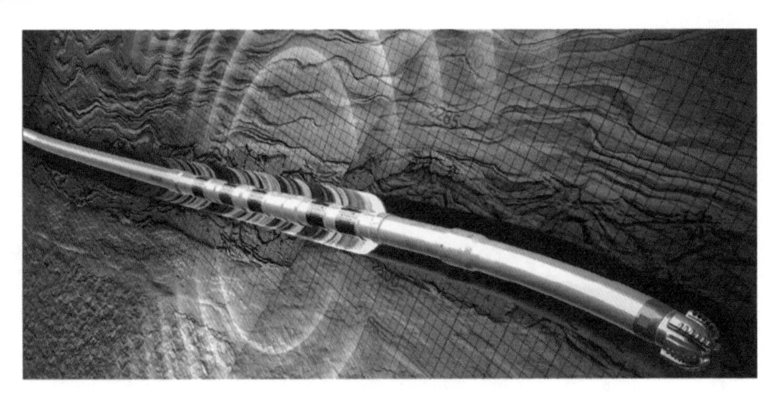

图 10　旋转地质导向示意

以中国石油黄金坝地区页岩气水平井地质导向钻井为例[23]，井下工具主要包括：旋转导向系统 + 随钻成像测井系统 LWD+ 随钻测量系统 MWD。旋转导向系统主要用于造斜、纠偏，精确控制井斜调整，复合式旋转导向系统造斜率最大可达 15°/30m；随钻成像测井系统 LWD 可将井筒周围的页岩储层伽马特征以图形的形式显示出来，有助于判断钻井轨迹与地层的上、下切关系；随钻测量系统 MWD 可在钻进过程中随钻测量工具面、井斜、方位及自然伽马等数据；三系统相互配合，完成旋转地质导向钻井，确保经验轨迹始终在目标层段钻进。

水平井钻成后，便是射孔与压裂（图 9），依照射孔方案通过射孔枪在井筒中射孔，之后向井筒中泵入远超过地层破裂压力的高压低黏液体，液体通过射孔孔眼压入地层，地层破裂产生裂缝。

页岩气有效开发普遍采用体积压裂，所谓体积压裂[24]是指通过压裂的方式对储层实施改造，在形成一条或多条主裂缝的同时，通过分段多簇射孔、高排量、大液量、低黏液体及转向材料及技术的应用，实现对天然裂缝、岩石层理的沟通，以及在主裂缝的侧向强制形成次生裂缝，并在次生裂缝上继续分支形成二级次生裂缝，其余类推（图 11）。让主裂缝与多级次生裂缝交织形成裂缝网络系统，将可以进行渗流的有效储集体“打碎”，使裂缝壁面与储层基质的接触面积最大，使得油气从任意方向的基质向裂缝的渗流距离最短，极大地提高储层整体渗透率，实现对储层在长、宽、高三维方向的全面改造。该技术可以大幅提高单井产量，最大程度提高储层动用率和采收率。

水平井体积压裂主要包含 3 个方面的内容：分段多簇射孔、快速可钻式桥塞工具以及大型滑溜水压裂。

通俗讲，分段多簇射孔就是将水平井段分成若干段，每段跨度控制在 100～150m，分 2～6 簇射孔；施工时，依照设计方案，用电缆将射孔管串和复合桥塞（图 12）输送至目的层[25]，完成桥塞坐封与多簇射孔联作。具体来说，主要的作业流程为：（1）采用油管输送的方式进行第 1 段射孔作业，然后进行压裂使地层与井筒连通；（2）采用泵送的方式将复合桥塞与射孔管串输送至第 2 个改造目的段，坐封桥塞，然后进行第 2 段的多簇射孔（图 13）；（3）进行第 2 段的压裂施工；（4）将复合桥塞与射孔管串输送至第 3 个改造目的段，坐封桥塞，然后进行第 3 段的多簇射孔和压裂施工，以此类推，直至完成方案设计的所有改造目的段；（5）压裂完后，一次性钻掉所有的复合桥塞，使井筒完全贯通，完成井筒与地层的有效沟通。

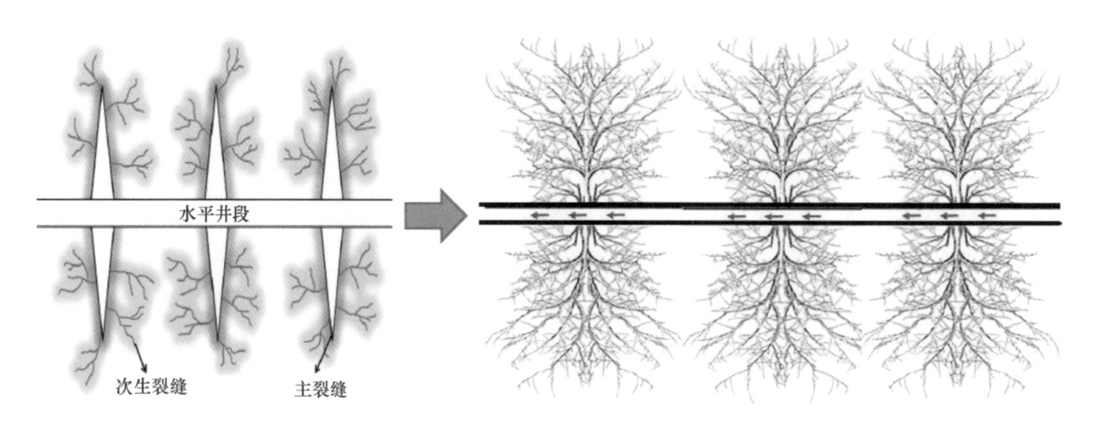

图 11　水平井体积压裂示意

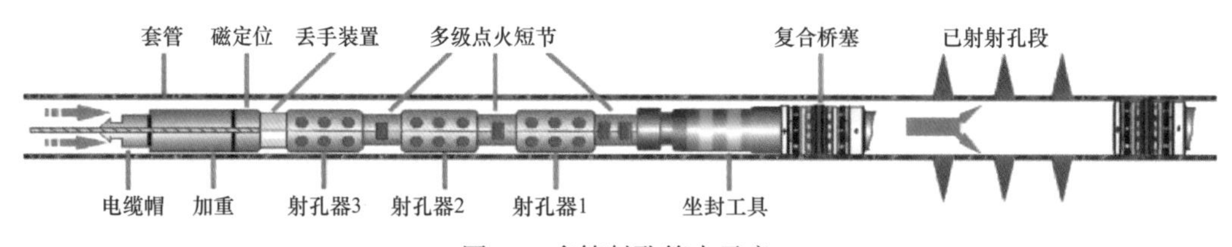

图 12　多簇射孔管串示意

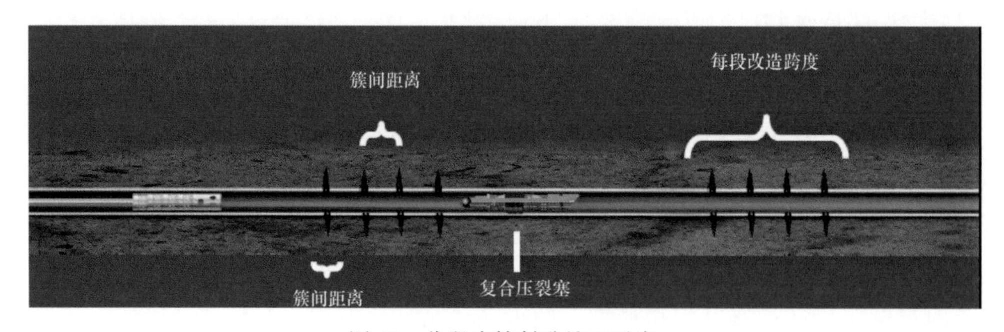

图 13　分段多簇射孔施工示意

桥塞的作用是在水平井分段压裂过程中暂时封堵已压裂井段，减小干扰，以达到目的井段压裂的最优化；前文述及，压裂完成后，所有桥塞均需钻掉，因而，桥塞工具的优劣同样影响到体积压裂的时率与效果。快速可钻式桥塞工具具有节省钻时，易钻，易排出（钻掉时间<35 分钟，常规铸铁桥塞>4 小时），适用于套管压裂，可满足多种尺寸套管的需要。

页岩储层压裂所用的液体多为滑溜水，主要是因为滑溜水摩阻低，作业强度小，对储层伤害率低；施工规模多为“千方砂，万方液”的大型压裂。大型滑溜水压裂的基本特点是：大液量、大排量、大砂量、小粒径、低砂比；具体说来，主要技术参数有：水平段长1000～1500m，分8～15段，每段4～6簇，每簇长度0.46～0.77m，簇间距20～30m，排量10m^3/min以上，平均砂比3%～5%，每段压裂液量1000～1500m^3，每段支撑剂量100～200t。

为了监测体积压裂施工中裂缝延伸的走向和形态，评价压裂效果并优化压裂方案，需要应用微地震技术对裂缝网络进行成像和监测。基本原理是[7]：水力压裂过程中，水力裂缝周围的天然裂缝、层理面等薄弱面的稳定性受到影响，发生剪切滑动，引起小量级的微小地震，其释放出的地震能量能够被监测井的地震波检波器探测到，通过数据处理，可以得到有关震源（即裂缝）的信息。压裂施工时，在压裂井的邻井下入一组检波器，对压裂产生的微地震事件进行接收和处理，确定震源在时间和空间上的分布，由此直观反映出地下裂缝网络的走向和形态（图14），每一种颜色的点群各代表一级压裂，点在空间上的展布（长、宽、高）反映了裂缝网络的走向和形态。

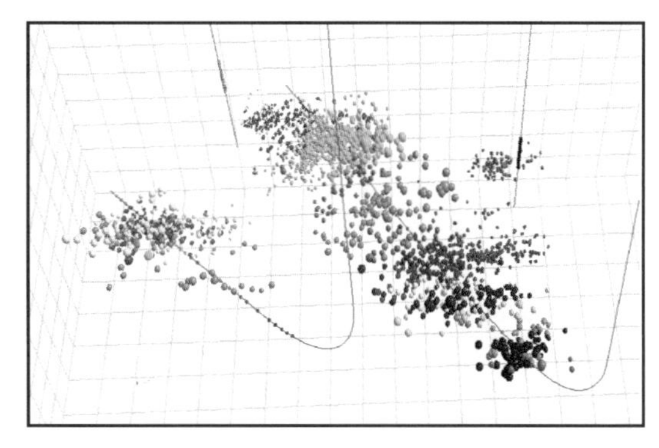

图14　微地震监测体积压裂水平井的缝网形态

4.2　工厂化作业

页岩气田为了保持稳产规模，需要大量的生产井，开发成本很高，为节约成本，减少占地，方便气井管理，多采用丛式水平井进行开发（图15），尤其中国南方山地页岩气田，受地形的限制，可部署井场的位置非常有限，因而丛式水平井的优势更为显著，由此发展出了工厂化作业模式。

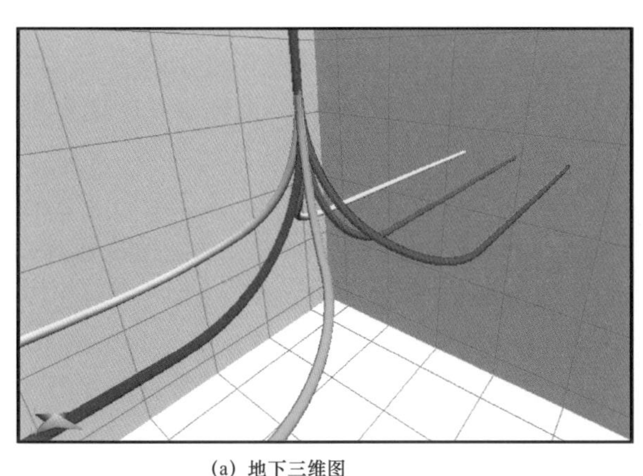

(a) 地下三维图

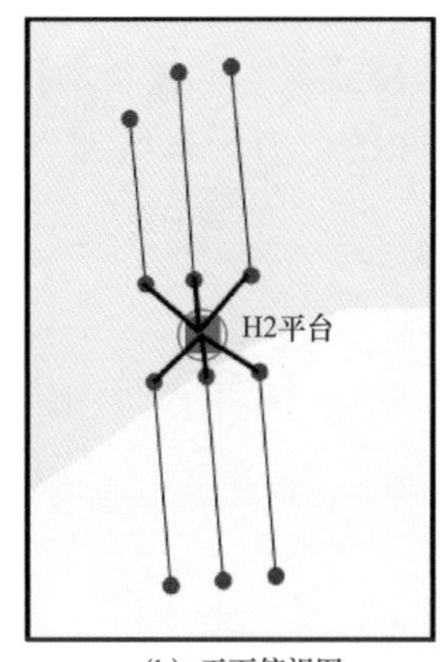

(b) 平面俯视图

图15　丛式水平井示意

工厂化作业[26]是指在开发阶段，在对油气分布认识清楚的情况下，为降低钻完井、井场建设成本，提高作业效率，以批量钻丛式井、批次完成完井及压裂、批次生产为手段的钻完井模式，主要包括工厂化钻井和工厂化压裂。图16[27]为加拿大ENCANA公司工厂化作业现场。

图 16　加拿大 ENCANA 公司工厂化作业现场

工厂化钻井[28]是指在同一地区集中布置大批相似井、使用大量标准化的装备和服务，以生产或装配流水线作业的方式进行钻井和完井的一种高效低成本的作业模式，该模式即利用快速移动式钻机对丛式井场的多口井进行批量钻完井，一种是批量钻完井后钻机搬走，采用工厂化压裂模式进行压裂、投产；另一种模式是以流水线的方式，实现边钻井、边压裂、边生产，钻完一口压裂一口，以 Bakken 致密油田 6 口井井场为例，这种流水线作业模式可节省 62.5% 的时间，作业效率进一步提升。

工厂化压裂[28]即所有的压裂装备都布置在井场中央区，不需要移动设备、人员和材料，就可以对多口井进行压裂。页岩气工厂化压裂模式最主要的方法就是“拉链式水力压裂”，这种模式减少了设备动迁次数，提高了压裂效率，降低了施工的成本，同时可以利用裂缝之间的应力干扰增加改造体积和裂缝网络的复杂性，大幅提高初期产量和最终采收率。

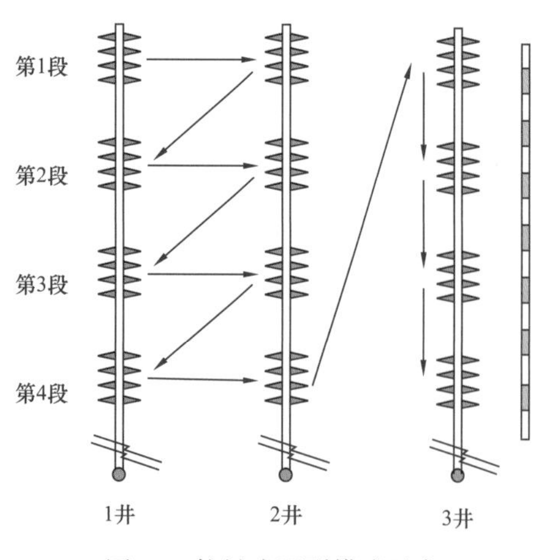

图 17　拉链式压裂模式示意

“拉链式水力压裂”[29]即同一井场中，一口井进行压裂时，另一口井进行电缆桥塞射孔联作，为压裂做准备，两项作业在两口井间逐段交替进行并无缝衔接。以长宁区块某平台为例（图 17[28]），拉链式压裂的具体作业流程是：1 井第 1 段进行压裂时，2 井第 1 段进行电缆桥塞射孔联作；2 井第 1 段进行压裂时，1 井第 2 段进行电缆桥塞射孔联作；以此类推，直至 1 井、2 井完成压裂，在此期间，3 井实施微地震监测；最后，单独对监测井（3 井）进行压裂，3 井压裂时，1 井、2 井开始钻磨桥塞，完毕后，钻磨 3 井桥塞；钻磨完 1 口井的桥塞后即开始放喷排液，最终实现 3 口井放喷排液试采。该平台 24 段“拉链式”压裂平均每天压裂 3.16 段，最多 1 天压裂 4 段，段与段之间的准备时间在 2～3 小时，完成设备保养、燃料添加等工作，施工效率比传统压裂方式提高了 78%，极大提高了作业时效。

5 中国页岩气勘探开发现状

中国页岩气的勘探开发大致经历了 3 个阶段[30]：（1）2004 年以前为引入阶段，主要是介绍和引入国外的页岩气基础理论、勘探开发经验和技术；（2）2005—2008 年为基础研究阶段，主要是针对中国页岩气地质特征进行基础研究；（3）2009 年之后为勘探开发阶段，相继开展了全国页岩气资源潜力评估、有利区带优选和勘探区块招标，并在四川盆地、鄂尔多斯盆地取得重大突破。

从页岩气资源所属沉积环境而言，中国页岩气资源主要分布于三大沉积相[31]，分别为海相、海陆过渡相和陆相。海相页岩主要分布在中国南方、华北、塔里木和羌塘等地；海陆过渡相页岩主要分布在渤海湾、鄂尔多斯、扬子地区、塔里木和准噶尔—吐哈等地；陆相页岩主要分布在松辽盆地、渤海湾盆地、鄂尔多斯盆地和四川盆地等地。

2010 年，中国第一口页岩气勘探评价井——威 201 井在四川盆地奥陶系五峰组—下志留统龙马溪组海相页岩中获得工业气流，自此，以南方下古生界五峰组—龙马溪组、筇竹寺组海相页岩为重点，开展页岩气勘探评价与开发先导试验，陆续在四川盆地、渝东鄂西、滇黔北、湘西地区的五峰组—龙马溪组发现页岩气，并在四川盆地的长宁—威远、富顺—永川、涪陵等地区获得工业页岩气产量，截至 2015 年底[32]，中国陆上累计设置页岩气探矿权区块 54 个，面积为 $17 \times 10^4 km^2$，相继建立了四川盆地长宁—威远、滇黔北昭通、重庆涪陵 3 个海相页岩气示范区和富顺—永川合作开发区以及鄂尔多斯盆地延长陆相页岩气示范区（图 18），其中，长宁—威远、昭通和涪陵已经取得了实质性开发进展，实现了初步规模开发，而延长陆相页岩气示范区尚未取得产量突破。

涪陵页岩气田是中国第一个投入大规模商业开发的页岩气田，其核心建产区为焦石坝区块。2012 年 11 月 28 日，中国石化在焦石坝地区钻探的焦页 1HF 井开始放喷试采，测试稳定产量在 $11 \times 10^4 m^3/d$，最高产量 $20.3 \times 10^4 m^3/d$，标志着中国页岩气勘探的重大突破；2013 年 1 月 9 日，该井投入试采，稳定在 $6 \times 10^4 m^3/d$，正式拉开中国页岩气商业开发的序幕；基于焦页 1HF 井的钻井、压裂、试采成果，2013 年 1—12 月，涪陵页岩气田开展了页岩气开发评价试验工作，并以此为基础于 2013 年底编制了涪陵页岩气田焦石坝区块一期开发方案[31]，拟用 2 年时间（2014—2015 年）建设 63 个平台，总井数 253 口，建成 $50 \times 10^8 m^3/a$ 产能的页岩气田；截至 2016 年 9 月底涪陵页岩气田完钻页岩气井 280 余口，具备 $56.4 \times 10^8 m^3$ 的生产能力，本年已经生产页岩气 $35 \times 10^8 m^3$。此外，涪陵页岩气田预计 2017 年和 2020 年将分别形成 $100 \times 10^8 m^3$ 和 $130 \times 10^8 m^3$ 产能。

长宁—威远页岩气示范区位于四川盆地西南部，其中，长宁区块横跨四川省宜宾市、长宁县、珙县、兴文县和筠连县，威远区块位于内江市威远县、资中县和自贡市的容县。2010 年 4 月 18 日，中国第一口页岩气勘探评价井——威 201 井完钻，井深 2840m，初期产气量 $2000m^3/d$，拉开了中国页岩气勘探开发的序幕；2011 年 2 月 13 日，中国第一口页岩气水平井——威 201-H1 井完钻，水平段长 1079m，压裂后日产气（1.15～1.34）×

10^4m^3，标志着中国页岩气分段压裂水平井开发技术取得实质性突破；2012 年 7 月，长宁区块第一口页岩气水平井宁 201-H1 井获得高产工业气流，日产气量 $15\times10^4m^3$，显示出了长宁—威远页岩气示范区良好的勘探开发前景。自此，中国石油加快推进该示范区的产能建设工作，截至 2016 年 9 月，长宁—威远区块共完钻页岩气井 140 余口，日产气 $636\times10^4m^3$，具备 $34\times10^8m^3/a$ 的产气能力。

图 18　中国页岩气勘探开发形势示意

昭通页岩气示范区横跨四川、云南、贵州等省，目前，该示范区的核心建产区主要为黄金坝区块。2011 年部署了昭通示范区第一口页岩气水平井 YSH1-1，水平井压裂试气效果良好，测试产量 $3.56\times10^4m^3/d$，为龙马溪组页岩气勘探开发奠定了基础；2013 年，产能评价井 YS108H1-1 获得 $20.86\times10^4m^3/d$ 的页岩气高产工业气流，确定了黄金坝龙马溪组页岩气甜点区，并于 2014—2015 年开展黄金坝区块页岩气 $5\times10^8m^3/a$ 产能建设。截至 2016 年 9 月，昭通页岩气示范区共完钻页岩气井 30 余口，日产气 $118\times10^4m^3$，具备 $6.5\times10^8m^3/a$ 的产气能力。

6　结论

在中国能源供需矛盾日益严峻、环境污染问题备受关注的大背景下，页岩气作为一种资源量丰富的清洁能源，具有巨大的勘探开发潜力，特别是南方海相页岩气的成功开发更坚定了中国快速推进页岩气产业发展的信心。据国家能源局2016年9月发布的《页岩气发展规划（2016—2020年）》，中国力争2020年实现页岩气产量$300\times10^8m^3$，2030年实现页岩气产量（800～1000）$\times10^8m^3$。为此，需积极总结当前阶段页岩气开发的经验和教训，努力攻关关键技术瓶颈，不断完善页岩气开发的理论和技术体系，支撑中国页岩气的规模开发。从产能规模扩建所需的资源保障角度而言，第一，应积极开展地质评价选区工作，及早储备建产接替区；第二，开展关键技术攻关，尽早实现3500m以深的页岩气开发。从经济效益开发角度而言，第一，应借鉴苏里格气田成功开发的经验，大力引进竞争机制，努力推进科技创新，不断降低钻井、压裂等施工成本；第二，应充分挖掘现有页岩气井的生产动态资料，开展开发技术政策优化工作，努力提高单井累计产量，最大限度地提高经济效益。

在正确的政策引导以及业界的努力攻关下，页岩气产业的蓬勃发展必将带给未来中国一个新的惊喜。

参考文献

［1］美国页岩气开发的历史［EB/OL］.［2016-05-31］. http：//www.csgcn.com.cn/news/show-5717.html.

［2］李响.我国第172个独立矿种诞生记——页岩气登上历史舞台的前前后后［J］.国土资源,2014（2）：27–29.

［3］陈昭年.石油与天然气地质学［M］.北京：地质出版社，2005.

［4］邹才能，董大忠，王社教，等.中国页岩气形成机理、地质特征及资源潜力［J］.石油勘探与开发，2010，37（6）：641–653.

［5］肖钢，唐颖.页岩气及其勘探开发［M］.北京：高等教育出版社，2012.

［6］邹才能，朱如凯，白斌，等.中国油气储层中纳米孔首次发现及其科学价值［J］.岩石学报，2011，27（6）：1857–1864.

［7］张东晓，杨婷云.页岩气开发综述［J］.石油学报，2013，34（4）：792–800.

［8］胡文瑞，鲍敬伟.探索中国式的页岩气发展之路［J］.天然气工业，2013，33（1）：1–7.

［9］U.S. Energy Information Administration . World Shale Resource Assessments，2015［EB/OL］.［2016–08–31］. https：//www.eia.gov/analysis/studies/worldshalegas/

［10］张大伟，李玉喜，张金川，等.全国页岩气资源潜力调查评价［M］.北京：地质出版社，2012.

［11］董大忠，邹才能，李建忠，等.页岩气资源潜力与勘探开发前景［J］.地质通报，2011，30（2/3）：324–336.

［12］邹才能，陶士振，侯连华，等.非常规油气地质［M］.北京：地质出版社，2011.

［13］刘洪林，王红岩，刘人和，等.中国页岩气资源及其勘探潜力分析［J］.地质学报，2010，84（9）：1374–1378.

［14］张金川，徐波，聂海宽，等.中国页岩气资源勘探潜力［J］.天然气工业，2008，28（6）：136–140.

［15］于猛.我国页岩气可采资源量约31万亿方［N］.中国国土资源报，2011–07–26.

［16］U.S. Energy Information Administration. World shale gas resources：an initial assessment of 14 regions outside the United States［EB/OL］.［2011–04–05］. https：//www.eia.gov/analysis/studies/

worldshalegas/archive/2011/pdf/fullreport_2011.pdf

[17] 董大忠，邹才能，杨桦，等 . 中国页岩气勘探开发进展与发展前景 [J] . 石油学报,2012,33 (增 1): 107–114.

[18] EIA. Technically Recoverable Shale Oil and Shale Gas Resources-An Assessment of 137 Shale Formations in 41 Countries Outside the United States [J/OL] . [2013–06–13] https : //www.eia.gov/analysis/studies/worldshalegas/archive/2013/pdf/fullreport_2013.pdf

[19] 中华人民共和国国家统计局 . 2015 中国统计年鉴 [M] . 北京：中国统计出版社，2015.

[20] BP Global. BP Statistical Review of World Energy 2016 [EB/OL] . [2016–08–31] . http : //www.bp.com/content/dam/bp/pdf/energy-economics/statistical-review-2016/bp-statistical-review-of-world-energy-2016-full-report.pdf

[21] 国务院发展研究中心，壳牌国际有限公司 . 中国天然气发展战略研究 [M]. 北京：中国发展出版社，2015.

[22] 胡文瑞 . 页岩油气峰会散发的信号 [J] . 中国石油化工，2015 (3) : 37.

[23] 陈志鹏，梁兴，王高成，等 . 旋转地质导向技术在水平井中的应用及体会——以昭通页岩气示范区为例 [J] . 天然气工业，2015，35 (12) : 64–70.

[24] 吴奇，胥云，刘玉章，等 . 美国页岩气体积改造技术现状及对我国的启示 [J] . 石油钻采工艺，2011，33 (2) : 1–7.

[25] 王海东，陈锋，欧跃强，等 . 页岩气水平井分簇射孔配套技术分析及应用 [J] . 长江大学学报 (自然版)，2016，13 (8) : 40–45.

[26] 李鹴，Hii King-Kai，Todd Franks，等 . 四川盆地金秋区块非常规天然气工厂化作业设想 [J] . 天然气工业，2013，33 (6) : 54–59.

[27] 许冬进，廖锐全，石善志，等 . 致密油水平井体积压裂工厂化作业模式研究 [J] . 特种油气藏，2014，21 (3) : 1–6.

[28] 郑新权 . 推进工厂化作业，应对低油价挑战 [J] . 北京石油管理干部学院学报，2016 (2) : 17–19.

[29] 钱斌，张俊成，朱炬辉，等 . 四川盆地长宁地区页岩气水平井组“拉链式”压裂实践 [J] . 天然气工业，2015，35 (1) : 81–84.

[30] 陆争光 . 中国页岩气产业发展现状及对策建议 [J] . 国际石油经济，2016，24 (4) : 48–90.

[31] 栾锡武 . 中国页岩气开发的实质性突破 [J] . 中国地质调查，2016，3 (1) : 7–13.

[32] 董大忠，邹才能，戴金星，等 . 中国页岩气发展战略对策建议 [J] . 天然气地球科学,2016,27 (3): 397–406.

二、开发地质篇

长宁地区优质页岩储层非均质性及主控因素

乔　辉，贾爱林，贾成业，位云生，袁　贺

（中国石油勘探开发研究院）

摘　要：通过大量实验测试数据的统计分析，明确了川南地区长宁区块优质页岩各小层间存在较强的非均质性，并分析了优质页岩储层发育的主控因素。研究结果表明，五峰组—龙马溪组下部储层的脆性矿物含量高，黏土矿物含量相对较低，平均为24.5%；储层的储集空间具有多种类型，主要以有机质孔隙为主，储层物性在纵向上具较强的非均质性，孔隙度主要在0.73%～10.25%，平均约为4.19%；五峰组和龙一$_1$亚段5个小层的总含气量在1.27～4.19m^3/t，平均达到2.74m^3/t，龙一$_1$亚段1小层含气量最高，其次为2小层和3小层，4小层含气量最低。沉积条件是优质页岩发育的物质基础，石英含量高和有机质发育有利于优质页岩储层的形成；压实作用与胶结作用是页岩储层发生致密化的主要机制，而溶蚀作用和有机质热成熟在一定程度上可改善页岩储层质量。

关键词：页岩气；储层非均质性；成岩作用；矿物成分；有机质含量；川南地区

近年来，北美地区在页岩成藏理论、“甜点区”选择、水平井钻井、压裂改造等方面取得了巨大的理论与技术进步，极大地促进了页岩气开发效率[1-5]。与之相比，中国的页岩气勘探开发起步较晚，但发展迅速，目前中国南方海相页岩已处于初步规模开发的阶段[6-7]。随着中国页岩气勘探开发的深入，开发中存在的一些问题日益突出，例如气藏微幅构造和非均质性导致区块内部，甚至同一个平台，气井产能和动态参数差异较大。目前，威远—长宁示范区水平井平均测试产量较高，但存在测试产量参差不齐、单井产量差异大等问题。除了工程上的影响因素外，地质上对储层认识不清是制约页岩气开发的重要因素。因此，研究页岩储层非均质特征及主控因素是后续页岩气赋存机理及产能差异研究的关键。前人对页岩储层的储层特征及主控因素开展了相关研究工作[8-14]。蒲泊伶等[8]通过大量实验对川南地区龙马溪组页岩有利储层的有机碳、矿物成分、储集空间类型及含气性进行了研究，并认为该页岩储层发育的控制因素为沉积环境、矿物组成和有机质发育特征。郭英海等[9]认为在微观尺度下，非均质性是页岩储层的重要特性，提出了“页岩储层微观非均质控气”理论。王秀平等[10]以川南及邻区龙马溪组为研究对象进行成岩作用研究，认为成岩作用对页岩储层具有重要影响。

鉴于研究区优质页岩储层及其主控因素的认识还不够全面清楚，本文利用大量的实验测试数据，对纵向上细分的不同小层优质页岩储层的非均质性特征进行研究，系统总结优质页岩的控制因素对油田开发显得尤为重要，可为后续研究及开发工作提供有力的技术支撑。

1 地质概况

长宁区块主体构造为长宁背斜，位于四川盆地南部的川南坳中隆低陡褶皱带（图1a）。纵向上，龙马溪组页岩沉积处于加里东构造运动时期，该期构造运动活跃，发育多期沉积旋回。页岩地层包括五峰组及龙马溪组沉积早期的龙一$_1$亚段及龙一$_2$亚段。龙一$_1$亚段存在较强的非均质性，自下至上又可依次划分为龙一$_1^1$、龙一$_1^2$、龙一$_1^3$及龙一$_1^4$4个小层（图1b）。

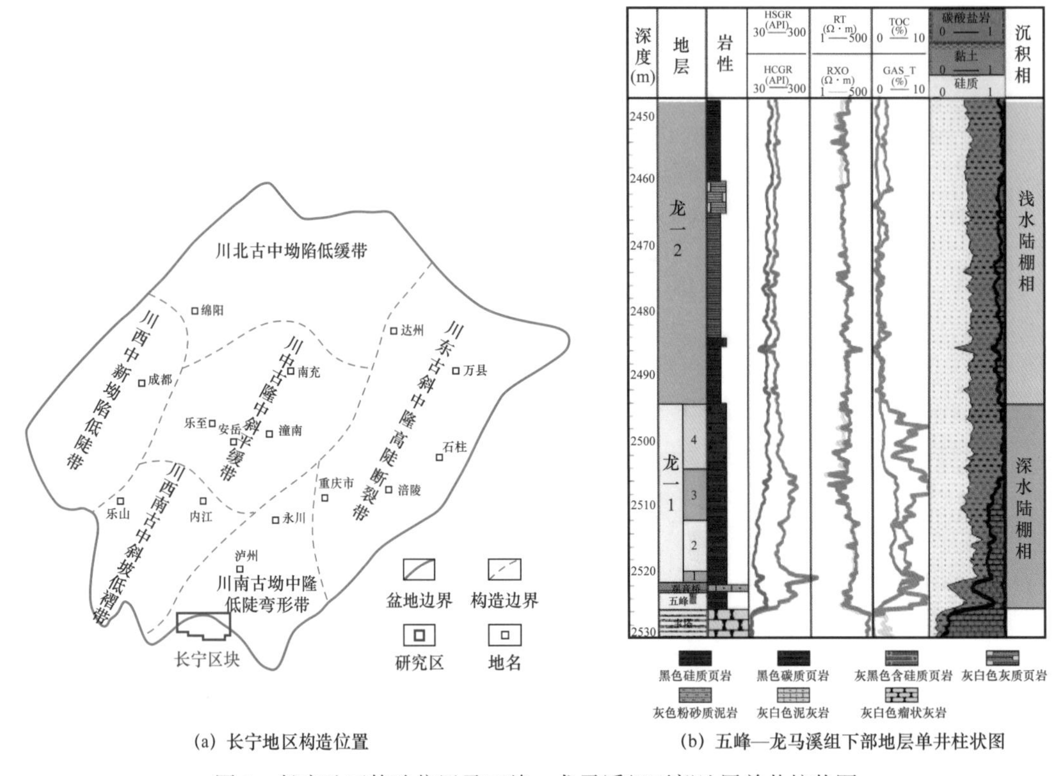

(a) 长宁地区构造位置　　(b) 五峰—龙马溪组下部地层单井柱状图

图1　长宁地区构造位置及五峰—龙马溪组下部地层单井柱状图

五峰组及龙马溪组沉积早期的龙一$_1$亚段沉积主体为半局限浅海相深水陆棚沉积环境，发育一套黑色碳质页岩、硅质页岩和黑色页岩沉积组合，有机质丰富，笔石发育，为优质页岩储层[8, 12]，厚度约36～48m。开发区块钻探资料显示，五峰—龙马溪组下部的龙一$_1$亚段的这套黑色碳质、硅质页岩分布稳定、有机质含量高、现场测试含气量高，主体埋深小于3000m，是目前中国南方海相页岩气开发的主力层系。龙一$_2$亚段沉积环境为浅水陆棚，主要岩性为深灰色泥页岩、灰色泥岩和粉砂质泥岩，笔石含量明显降低，有机质相对不发育地层厚度在105～200m。

2 优质页岩储层非均质性特征

2.1 岩石矿物学特征

通过对研究区2口单井N1井和N3井296块岩样X射线衍射分析表明，页岩矿物类型主要包括石英、长石、黏土矿物、方解石、白云石和黄铁矿等。五峰组—龙一$_1$亚段石英、长石及碳酸盐矿物单井平均值在70%以上，以硅质矿物及方解石为主，显示了良好的可压裂性特征；黏土矿物含量相对较低，平均为24.5%，伊利石、绿泥石和伊/蒙混层为研究区的主要黏土矿物。纵向上，各小层的矿物含量变化也较明显，说明纵向上储层的非均质性较强。其中，龙一$_1^1$、龙一$_1^2$小层硅质平均含量高达61.9%、61.6%，其他小层硅质含量为43.9%～59.0%（图2）。

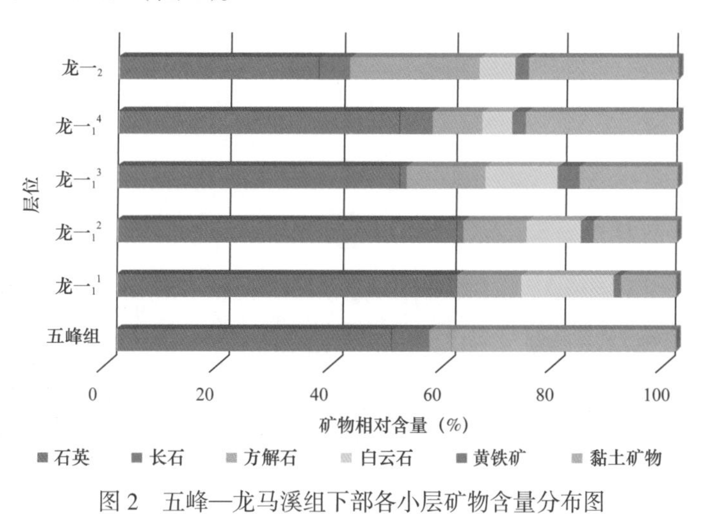

图2　五峰—龙马溪组下部各小层矿物含量分布图

2.2　有机地球化学特征

根据N1、N3、N9井82块样品的分析测试数据，长宁地区五峰—龙马溪组下部地层有机质含量TOC为0.2%～7.5%，平均为3.36%。纵向上，各小层的TOC含量差异较大，非均质性较强。自下至上五峰组、龙一$_1$亚段的1、2、3、4小层及龙一$_2$亚段的有机碳含量分布区间为：3.1%～4.7%、3.7%～7.5%、3.2%～4.1%、2.3%～5.3%、1.9%～2.8%、0.2%～1.6%。五峰组—龙一$_1$亚段TOC含量平均为3.4%，最高可达7.5%。而龙一$_2$亚段TOC值明显变小，TOC介于0.2%～1.6%，平均为0.8%。纵向上，龙一$_1^1$小层TOC值最高，其次为龙一$_1^3$小层，然后为五峰组和龙一$_1^2$小层，龙一$_1^4$小层最低（图3）。

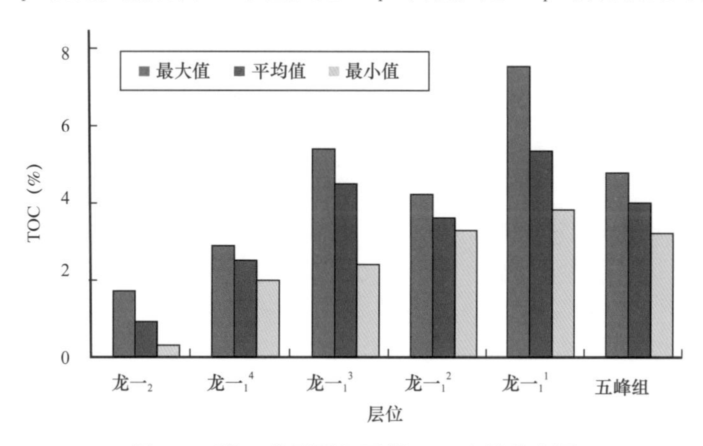

图3　五峰—龙马溪组下部TOC含量分布图

对 N1 井岩心样品进行干酪根镜检分析，平均腐泥组含量大于 80%，为典型的 I 型干酪根，局部 II_1 型。宁 201 井五峰组—龙一$_1$亚段干酪根碳同位素一般为 -27.92‰～-30.78‰，可以判断干酪根类型为 I 型，与干酪根结果一致。由于 I 型干酪根缺乏镜质组，常利用沥青反射率来换算镜质组反射率的方法测定，研究区五峰组—龙马溪组底部有机质成熟度平均为 2.60%。有机质反射率均达到高—过成熟阶段，以产干气为主。

2.3 储集空间及物性特征

利用岩心、露头、场发射扫描电镜及显微薄片镜下观察，发现五峰—龙马溪组底部的页岩储层发育多种类型孔隙，包括基质无机孔、有机孔和裂缝 3 大类。其中，无机孔主要有粒间孔、粒内孔、溶蚀孔和晶间孔等（图 4a–c），有机孔（图 4d）主要与有机质的含量及其热演化程度有关，裂缝主要有构造缝、页理缝、溶蚀缝及成岩收缩缝等多种类型（图 4e、f）。

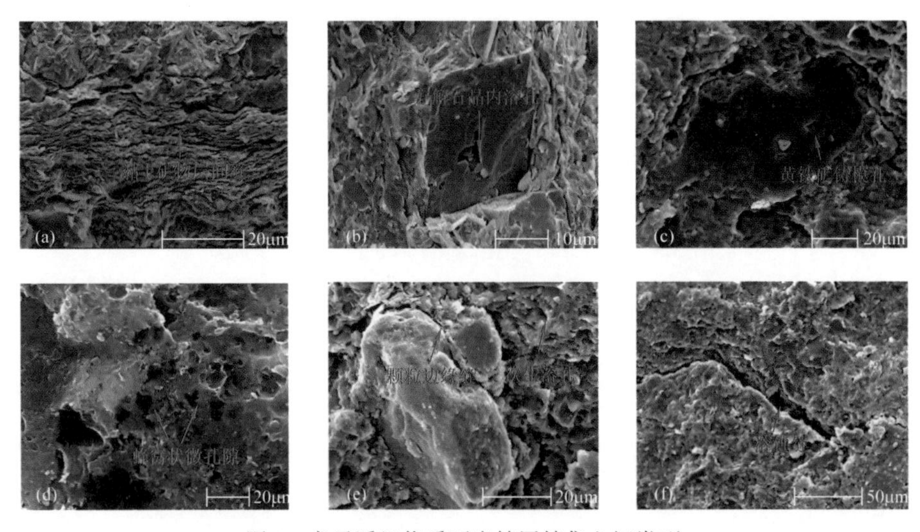

图 4　龙马溪组优质页岩储层储集空间类型

（a）N3 井，2385.3m，黏土矿物层间缝；（b）N1 井，2503.5m，方解石晶内溶孔；（c）Y2 井，2410.5m，黄铁矿铸模孔；（d）N9 井，3143.5m，有机物中蜂窝状微孔隙；（e）Y2 井，2433.66m，钾长石周围粒缘缝和次生溶孔；（f）Y2 井，2433.66m，溶蚀缝

N1 井和 N3 井 229 块岩心样品物性数据的统计分析结果显示，长宁地区五峰组到龙马溪组下部取心段的孔隙度介于 0.73%～10.25% 之间，平均为 4.19%。纵向上，该段储层的孔渗也存在较强的非均质性。从上至下，龙一$_2$亚段、龙一$_1^4$、龙一$_1^3$、龙一$_1^2$、龙一$_1^1$及五峰组的孔隙度分布区间分别为：0.89%～6.24%、1.96%～10.26%、3.83%～9.48%、3.01%～8.54%、3.43%～10.82%、0.73%～7.68%。五峰组—龙一$_1$亚段孔隙度平均为 5.44%，而龙一$_2$亚段孔隙度平均为 3.60%。龙一$_1$亚段龙一$_1^3$小层孔隙度值最高，其次为龙一$_1^2$、龙一$_1^4$小层，然后为龙一$_1^1$小层，五峰组孔隙度最低（图 5）。

2.4 含气性特征

在页岩储层中，天然气的赋存有 3 种状态：一是以吸附气的形式吸附在黏土颗粒与有机质表面；二是以游离气的形式游离于页岩储层基质孔隙和裂缝中；此外还有少量的气体以溶解气的形式存在于干酪根等物质中[15, 16]。页岩储层的总含气量主要借鉴煤层气的解吸法，分别测量解吸气量、残余气量和损失气量，三者总和相加即得到页岩总含气量。利用等温吸附实验确定页岩吸附气量[17]。由于页岩储层的特殊性，岩电实验很难进行，并且研究区黄铁矿相对发育，电阻率受黄铁矿导电性能的影响严重，采用阿尔奇公式计算游离气含量存在一定的问题，因此游离气含量一般采用解析法测得的总含气量减去等温吸附法计算的吸附气含量获得[17]。

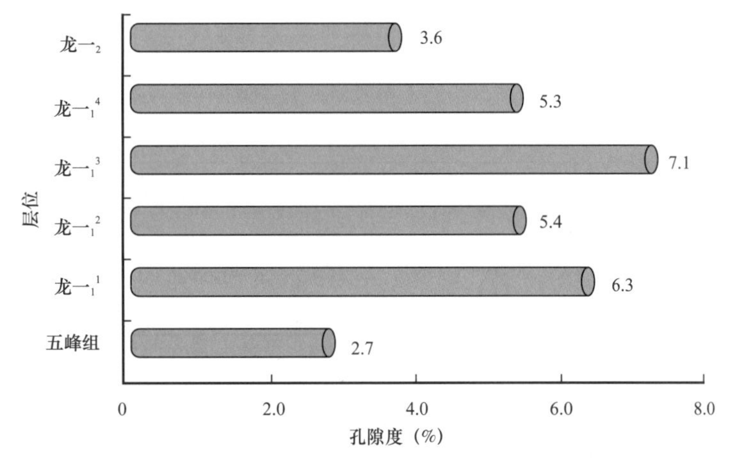

图 5　五峰—龙马溪组下部孔隙度分布图

现场测试和实验室测定结果均表明，五峰—龙马溪组下部储层均具有较好的含气性，且纵向上各小层的含气性差异较大（表 1）。

表 1　川南现场解析法总含气量测试数据表

地层	亚段	小层	样品数（块）	含气量（m^3/t）		
				最大值	最小值	平均值
龙马溪组一段	龙一$_2$	龙一$_2$	45	2.47	0.55	1.53
	龙一$_1$	龙一$_1^4$	12	3.68	1.27	2.36
		龙一$_1^3$	3	4.19	2.94	3.04
		龙一$_1^2$	10	4.06	2.09	3.03
		龙一$_1^1$	2	4.06	3.00	3.53
五峰组			2	3.21	2.65	2.93
五峰组—龙一$_1$亚段				4.19	0.55	2.74

长宁地区五峰—龙马溪组下部含气量为 0.55%～4.19%，其中龙一$_2$亚段总含气量在 0.55%～2.47%，平均为 1.53%，五峰组和龙一$_1$亚段 4 个小层含气量较高，总含气量为 1.27%～4.19%，平均为 2.74%。五峰组和龙一$_1$亚段各小层平均含气量均大于 2m^3/t，龙一$_1$亚段 1 小层含气量最高，其次为龙一$_1^2$和龙一$_1^3$小层，龙一$_1^4$小层含气量最低。

3 储层发育的主控因素

3.1 沉积条件

3.1.1 沉积环境

优质页岩储层的形成需要特定的沉积条件。王玉满等[18]通过分析地球化学资料总结了长宁地区页岩地层的沉积模式，即优质页岩均形成于持续缓慢沉降的深水陆棚中心区，海平面位置较高，沉积环境为弱—半封闭环境以确保古生产力保持较高水平。长宁地区地球化学资料显示，区块 P_2O_5/TiO_2 值在五峰—龙马溪组龙一$_1$亚段较高，介于0.20～0.85（平均为0.37），表明该地层具有较高的生产率。且该期构造稳定，沉积速度缓慢，为2.33～9.29m/Ma。志留纪早期该区域经历了两期海进—海退的沉积旋回变化，笔石等生物经历了繁盛—衰退变化过程。

岩心和测井资料分析可见，龙马溪组底部的龙一$_1$亚段主要发育两套富笔石的碳质页岩夹一套硅质页岩，为该区有利的页岩发育层段。该地层主体为深水陆棚沉积环境[8, 12]，水体稳定，黄铁矿较发育，为较强的还原环境，对有机质及页岩的页理发育都十分有利。经过统计分析发现，页岩层段中黄铁矿的含量与孔隙度、孔隙度与总含气量均呈正相关关系（图6a、b），且储层孔隙越发育，储层的总含气量越高，说明较强的还原环境有利于优质页岩储层的发育，为优质页岩储层的形成提供了良好的沉积条件。

沉积环境对页岩地层有机质含量及矿物成分均有较大的影响，进而影响储层的物性及含气性特征。前人研究认为有机碳含量是评价页岩储层的一个重要参数，其与页岩气总含气量、吸附气含量及储层的孔隙度之间都存在良好的正相关关系[12, 14, 17, 18]。有机质在达到成熟阶段以后，有机质孔隙随着干酪根的热分解增大[19]。有机质孔主要是生烃物质排烃后残留的孔隙和原油热裂解形成的沥青质内的微孔隙，它与其比表面为吸附态天然气的赋存提供了吸附剂，也为游离气的赋存提供了孔隙空间[20]。有学者通过对不同镜质组反射率的页岩岩样进行扫描电镜观测和孔隙度测试，发现页岩有机孔在生油窗内较少，进入生气窗后有机孔隙快速增加，且在 R_o 为3.6%时有机孔达到峰值，随后有机孔随着 R_o 的增加而减小[21]。

研究区目的层段有机质类型主要为I型，具良好的生烃潜力，镜质组反射率 R_o 介于2.3%～3.3%，为有机孔隙生成的最佳时期，同时氩离子抛光扫描电镜可观察到该区优质页岩储层段发育大量有机质孔（图4d）。通过N3井五峰组—龙马溪组底部TOC与孔隙度及其与总含气量之间的拟合发现，TOC与孔隙度及与总含气量之间均存在较好的正相关关系（图6c、d）。表明有机碳含量高的优质页岩储层，有机质纳米孔发育，含气量大，为优质页岩储集层段。

3.1.2 矿物成分

页岩储层中的矿物成分与含量对储层的孔隙度和渗透率及含气性影响较大。通过岩心测试样品的石英含量与实测孔隙度资料分析发现，孔隙度与石英含量呈正相关关系（图6e），且TOC含量随着石英含量的增大而增大（图6f），说明石英为生物成因石英矿物，来源于较为丰富的硅质生物，其间接增加了有机质的含量[8, 22]。研究区龙一$_1^1$、龙一$_1^2$

小层硅质平均含量高达61.9%、61.6%，其他小层硅质含量为43.9%～59.0%，丰富的有机质来源的石英和有机质伴生，发育丰富的微孔隙和较大的比表面积，增加了页岩中的可供页岩气吸附以及游离气赋存的空间，同时石英等脆性矿物越发育，岩石脆性越好，越易形成天然微裂缝，可有效改善储层物性且有利于后期页岩气的压裂改造[23]。统计发现，黏土矿物含量与实测孔隙度不具备相关性，说明五峰—龙马溪组下部页岩储层的孔隙度主要由有机质贡献，黏土矿物含量虽然可以影响页岩的吸附性及孔隙度，但由于研究区优质页岩储集层段的黏土矿物含量低，其对储层孔隙度及含气量影响不明显。

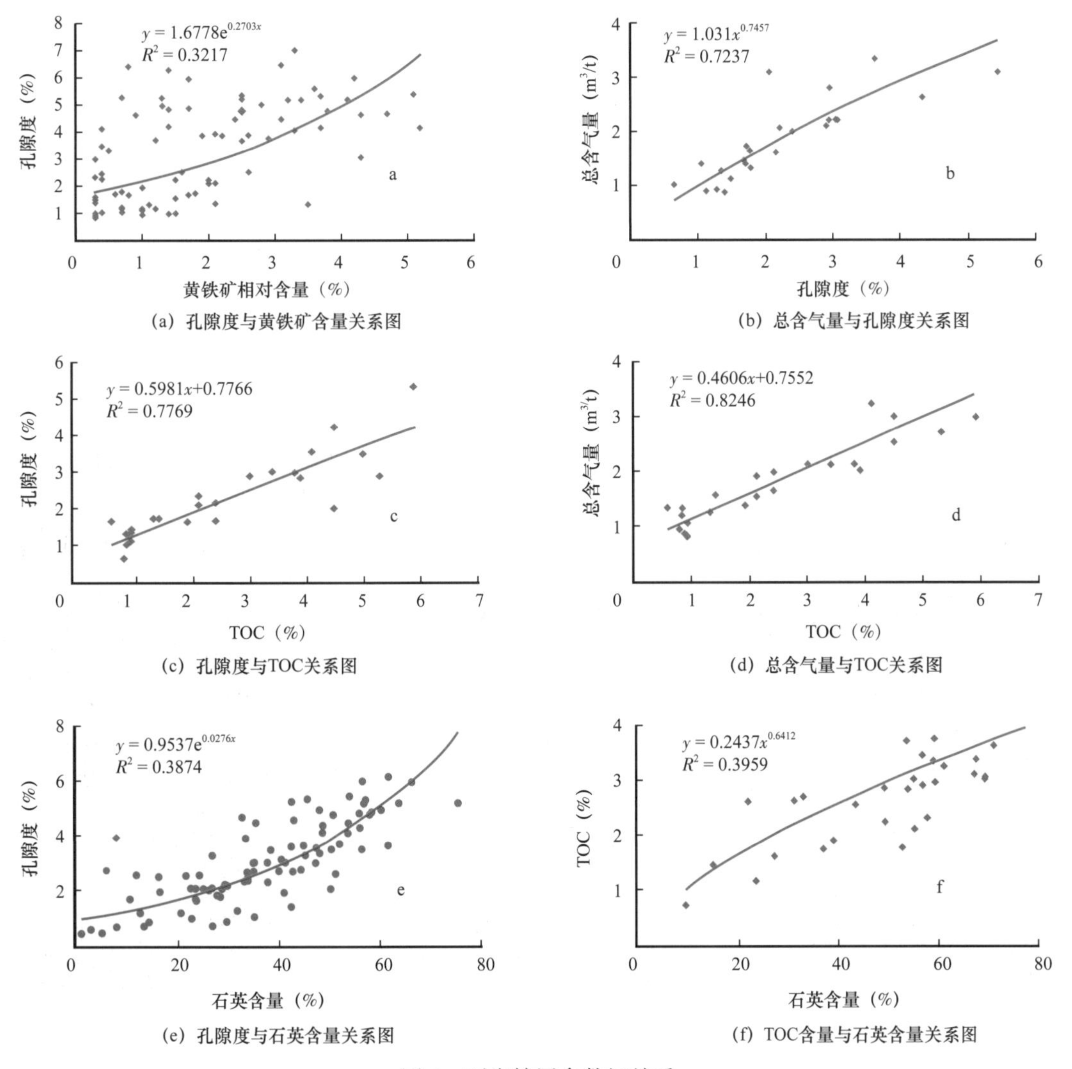

图6　页岩储层参数间关系

美国页岩气开发的经验表明，其产气层段储层的脆性矿物含量较高，页岩气产量随脆性矿物含量的增加而增大[4，8]。由于页岩储层黏土矿物成分与含量影响页岩的吸附性及储层的孔隙度，而脆性矿物也可影响储层的孔隙度及岩石的脆性，且脆性矿物含气量有利于形成天然裂缝及人工压裂后形成诱导缝，因此，不同的页岩气区块，黏土矿与脆性矿物的比例多少最有利于优质页岩的形成及压裂后天然气的产出，应与实际地质条件相结合确定。

3.2 成岩作用

成岩作用是影响储层发育的重要因素之一，它不仅控制储层孔隙的发育和保存，同时对岩石的力学性质也有一定的影响[10, 24-31]。有机质热成熟作用排出的天然气是页岩气的主要来源，该过程中生成的大量有机孔增大了储层的孔隙度，提高了储层的吸附能力[25]。在岩石薄片、扫描电镜观察及岩心描述等研究的基础上，认为五峰—龙马溪组页岩储层经历了多种类型的成岩作用，包括压实、胶结、交代、黏土矿物的转化、溶蚀及有机质热成熟作用等。其中，压实作用与胶结作用降低了页岩储层的孔隙度，有机质热成熟作用和溶蚀作用增大了储层孔隙度，而黏土矿物的转化作用和交代作用对储层孔隙度的影响相对较小。

3.2.1 压实作用与胶结作用

压实作用是五峰—龙马溪组页岩储层孔隙度低的最主要原因。泥页岩最开始沉积时主要为片状结构，其原始孔隙度可高达 70%～80%[26]。在压实过程中，黏土骨架垮塌，孔隙度迅速降低，随着埋藏深度的增加，孔隙度最终降低到仅百分之几[27]。五峰—龙马溪组下部页岩地层常见的压实作用识别标志有黏土矿物与片状矿物如云母等矿物的顺层定向排列及颗粒的压裂破碎等（图 7a）。

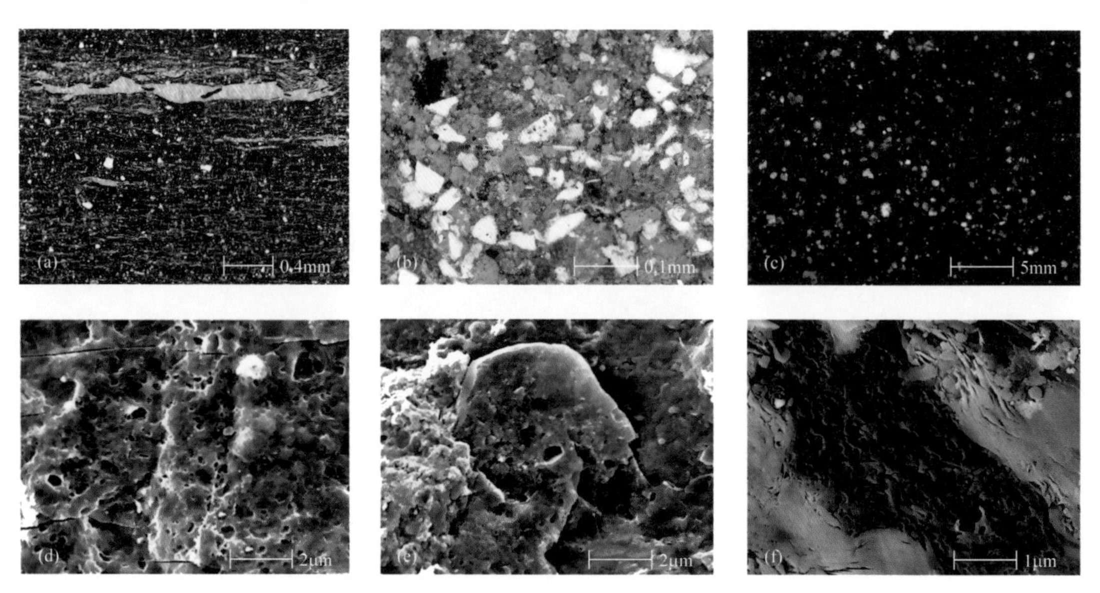

图 7 龙马溪组页岩主要成岩作用类型

（a）N1 井，2522m，×50，压实作用导致矿物顺层定向排列；（b）N3 井，2241.0m，×200，方解石基底式胶结；（c）N3 井，龙马溪组，2508.55m，×10，自形—半自形晶白云石胶结物；（d）Y2 井，2429.3m，×2500，钾长石成岩收缩缝和大量次生溶孔；（e）Y2 井，2455.11m，×2500，长石上的铸模孔及次生溶孔、粒缘缝；（f）N3 井，2378.02m，×5000，蜂窝状有机质孔

胶结作用是降低泥页岩孔隙度的另一重要影响因素，常见的胶结物有硅质胶结物、碳质胶结物和硫化物胶结物等。前人[27,28]研究认为硅质胶结物主要有 4 方面的来源。其中，硅质胶结物主要为石英，石英主要以自生石英或次生加大的形式存在，也可以填隙物的形

式出现。碳酸盐胶结物主要为方解石和白云石。方解石胶结物形成时间较早，主要充填于孔隙、裂缝或交代长石等矿物颗粒（图 7b）。白云石胶结物呈自形—半自形晶分散状产出，交代早期矿物或以裂缝充填物的形式出现（图 7c）。大量全岩 X 射线衍射实验结果发现，孔隙度与碳酸盐矿物含量呈微弱的反比关系，证明碳酸盐胶结物对页岩储层的孔隙度具有一定的消极影响。胶结作用一方面减小储层的孔隙度，造成储层进一步致密化，另一方面硅质胶结物和碳酸盐胶结物均可以提高页岩储层的脆性，从而有利于后期的压裂改造[27]。

3.2.2 溶蚀作用与有机质热成熟作用

溶蚀作用与有机质生烃排烃过程中产生的有机酸有关[10, 24]。有机质生烃排烃过程中，干酪根热裂解产生大量的羧酸和 CO_2 等酸性物质，导致地层流体的 pH 值降低，溶解储层中易溶解的矿物成分，形成次生溶蚀孔隙（图 7d、e）。五峰—龙马溪组下部页岩储层中含一定比例的长石和碳酸盐矿物等易溶组分，易被酸性物质溶解，形成溶蚀孔隙，在一定程度上增大了储层的孔隙度。但孔令明等[27]认为龙马溪组页岩储层的溶蚀作用总体并不十分发育，认为页岩地层是一个相对封闭的体系，其渗透率极低，地层中流体的流动不畅导致 H^+ 不能及时更新且流体中沉淀物质不能排出，阻碍了溶蚀作用的进一步发生。

有机质热成熟作用对页岩储层具有重要的意义。有机质在达到生油气门限条件后会大量的生烃，是页岩气的物质来源，同时有机质生排烃过程中形成有机孔（图 7f），不但增大了孔隙度，且提高了储层的吸附能力[10, 27]。因此在有机质富集的优质页岩储集层段，热成熟作用对页岩储层物性具有较大影响。但有机质孔隙的形成需要一定的条件，通过实验发现只有当镜质组反射率大于 0.6% 的时候，即有机质开始大量生油气阶段后，才会形成较多的有机孔。当有机质成熟度较低时，通常不存在或仅存在少量的有机孔[29]。Milliken K L 等[30]研究发现，有机孔的形成与保存同时受有机质热成熟作用与压实作用影响，当岩石中有机质含量较大且热演化程度较高，有机孔特别发育且相互连通时，在上覆地层的压实作用下，部分孔隙垮塌，总孔隙度反而降低。

4 结论

（1）川南地区优质页岩储层具有较强的非均质性。五峰—龙马溪组下部储层的脆性矿物含量高，黏土含量相对较低，平均为 24.5%。TOC 值普遍大于 2.0%，平均为 3.4%。龙一$_1$亚段 1 小层 TOC 值最高，其次为 3 小层，然后为五峰组和 2 小层、4 小层最低。五峰组—龙马溪组下部储层的孔隙类型多样，包括无机孔、有机孔和裂缝，孔隙度介于 0.73%～10.25% 之间，平均为 4.19%。纵向上，龙一$_1$亚段 3 小层孔隙度值最高，其次为 2、4 小层，然后为 1 小层，五峰组孔隙度最低。储层含气量介于 0.55%～4.19% 之间，五峰组和龙一$_1$亚段四个小层含气量较高，总含气量为 1.27%～4.19%，平均为 2.74%，各小层平均含气量均大于 $2m^3/t$，龙一$_1$亚段 1 小层含气量最高，其次为 2 和 3 小层，4 小层含气量最低。

（2）沉积条件是控制优质页岩储层发育的物质基础。通过统计分析，明确 TOC 与孔隙度及总含气量之间均存在较好的正相关关系，说明有机碳含量高的优质页岩储层中有机质纳米孔发育，含气量大。孔隙度与石英含量呈正相关且 TOC 含量随着石英含量的增大而增大，说明石英为生物成因石英矿物，来源于较为丰富的硅质生物残体。但黏土矿物含量与孔隙度不具备相关性，说明五峰—龙马溪组下部页岩储层的孔隙度主要由有机质贡献，黏土矿物含量低，其对储层孔隙度及含气量影响不明显。

（3）成岩作用是影响页岩储层发育的另一重要因素。压实作用和胶结作用降低了页岩储层的孔隙度和渗透率，但碳酸盐胶结物可提高页岩储层的脆性，有利于后期的压裂改造。而溶蚀作用和有机质热成熟作用在一定程度上改善了页岩储层的物性。

参考文献

[1] Scott L Montgomery，Daniel M Jarvie，Kent A Bowker，et al. Mississippian Barnett Shale，Fort Worth basin，north-central Texas：Gas-shale play with multitrillion cubic foot potential [J]. AAPG Bulletin，2006，89（2）：155–175.

[2] Roussel N P，Sharma M M. Optimizing fracture spacing and sequencing in horizontal-well fracturing [J]. SPE Production & Operations，2011，26（2）：173–184.

[3] Yu W，Wu K，Zuo L，et al. Physical Models for Inter-Well Interference in Shale Reservoirs：Relative Impacts of Fracture Hits and Matrix Permeability [C] //Unconventional Resources Technology Conference（URTEC），2016.

[4] Ross D J K，Bustin R M. The importance of shale composition and pore structure upon gas storage potential of shale gas reservoirs [J]. Marine and Petroleum Geology，2009，26（6）：916–927.

[5] Clarkson C R，Solano N，Bustin R M，et al. Pore structure characterization of North American shale gas reservoirs using USANS/SANS，gas adsorption，and mercury intrusion [J]. Fuel，2013，103：606–616.

[6] 邹才能，董大忠，王玉满，等. 中国页岩气特征，挑战及前景（一）[J]. 石油勘探与开发，2015，42（6）：689–701.

[7] 邹才能，董大忠，王玉满，等. 中国页岩气特征，挑战及前景（二）[J]. 石油勘探与开发，2016，43（2）：166–178.

[8] 蒲泊伶，董大忠，耳闯，等. 川南地区龙马溪组页岩有利储层发育特征及其影响因素 [J]. 天然气工业，2013，33（12）：41–47.

[9] 郭英海，赵迪斐. 微观尺度海相页岩储层微观非均质性研究 [J]. 中国矿业大学学报，2015，44（2）：300–307.

[10] 王秀平，牟传龙，王启宇，等. 川南及邻区龙马溪组黑色岩系成岩作用 [J]. 石油学报，2015，36（9）：1035–1047.

[11] 郭旭升，李宇平，刘若冰，等. 四川盆地焦石坝地区龙马溪组页岩微观孔隙结构特征及其控制因素 [J]. 天然气工业，2014，34（6）：9–16.

[12] 金之钧，胡宗全，高波，等. 川东南地区五峰组 - 龙马溪组页岩气富集与高产控制因素 [J]. 地学前缘，2016，23（1）：1–10.

[13] 李玉喜，何建华，尹帅，等. 页岩油气储层纵向多重非均质性及其对开发的影响 [J]. 地学前缘，2016，23（2）：118–125.

[14] 张晓明，石万忠，徐清海，等.四川盆地焦石坝地区页岩气储层特征及控制因素[J].石油学报，2015，36（8）：926–939.

[15] 关小旭，伊向艺，杨火海.中美页岩气储层条件对比[J].西南石油大学学报（自然科学版），2014，36（5）：33–39.

[16] Zhou Q，Xiao X M，Tian H，et al. Modeling free gas content of the lower Paleozoic shales in the Weiyuan area of the Sichuan basin，China [J]. Marine and Petroleum Geology，2014，56（3）：87–96.

[17] 万金彬，何羽飞，刘淼，等.页岩含气量测定及计算方法研究[J].测井技术，2015，39（6）：756–761.

[18] 王玉满，李新景，董大忠，等.上扬子地区五峰组—龙马溪组优质页岩沉积主控因素[J].天然气工业，2017，37（4）：9–20.

[19] Ross D J K，Bustin R M. Shale gas potential of the Lower ，Jurassic Gor-dondale Member，northeastern British Columbia，Canada [J]. Bulletinof Canadian Petroleum Geology，2007，55（1）：51–75.

[20] 张士万，孟志勇，郭战峰，等.涪陵地区龙马溪组页岩储层特征及其发育主控因素[J].天然气工业，2014，34（12）：16–24.

[21] Curtis M E，Cardott B J. Sondergeld C H，et al. Development of organic porosity in the Woodford Shale with increasing thermal maturity [J]. lnternational Journal of Coal Geology，2012，103（23）：26–31.

[22] 毕赫，姜振学，李鹏，等.渝东南地区龙马溪组页岩吸附特征及其影响因素[J].天然气地球科学，2014，25（2）：302–310.

[23] 贾爱林，位云生，金亦秋.中国海相页岩气开发评价关键技术进展[J].石油勘探与开发，2016，43（6）：949–955.

[24] 于雯泉，陈勇，杨立干，等.酸性环境致密砂岩储层石英的溶蚀作用[J].石油学报,2014,35（2）：286–293.

[25] 秦建中，付小东，申宝剑，等.四川盆地上二叠统海相优质页岩超显微有机岩石学特征研究[J].石油实验地质，2010，32（2）：164–170.

[26] 王勇，梁铭，达世攀.利用泥岩压实曲线特征预测靖边气田高产区带[J].天然气工业,2005,25(4)：47–49.

[27] 孔令明，万茂霞，严玉霞，等.四川盆地志留系龙马溪组页岩储层成岩作用[J].天然气地球科学，2015，26（8）：1547–1555.

[28] Peltonen C，Marcussen A，Bjorlykke K，et al. Clay mineral diagenesis and quartz cementation in mudstones：The effects of smectite to illite reaction on rock properties [J]. Marine and petroleum Geology，2009，26（6）：887–898.

[29] Loucks R G，Reed R M，Ruppel S C，et al. Spectrum of pore types and networks in mudrocks and a descriptive classification for matrix-related mudrock pores [J]. AAPG Bulletin，2012，96（6）：1071–1098.

[30] Milliken K L，Rudnicki M，Awwiller D N，et al. Organic matter-hosted pore system，Marcellus Formation（Devonian），Pennsylvania [J]. AAPG Bulletin，2013，97（2）：177–200.

Shale reservoir characteristics and exploration potential in the target : A case study in the Longmaxi Formation from the southern Sichuan Basin of China

Gai Shaohua[1], Liu Huiqing[1], He Shunli[1], Mo Shaoyuan[1], Chen Sheng[1, 2], Liu Ruohan[3], Huang Xing[1], Tian Jie[1], Lv Xiaocong[1], Wu Dongxu[4], He Jianglin[3], Gu Jiangrui[2]

1 China University of Petroleum ; 2 Research Institute of Petroleum Exploration and Development (Langfang Branch) ; 3 China University of Geosciences ; 4 PetroChina Hangzhou Institute of Geolog ; 5 Chengdu Institute of Geology and Mineral Resources, Ministry of Land and Resources

Abstract: The Lower Silurian Longmaxi shale Formation in southern China is experiencing a significant increase in gas exploration and development. The purpose of this paper focuses on the research of the reservoir characteristics and sweet spots inside the same local target. Based on the cores, field emission scanning electron microscopy (FE-SEM) observation, X-ray diffraction and organic geochemistry data, we determined that the organic-rich black shale in the Longmaxi Formation is deposited in the shelf. Organic geochemical characteristics (high TOC content, appropriate organic matter type, maturity and gas content) reveal that the Longmaxi shale has a good gas potential, especially at the bottom of the formation. There are four dominant reservoir space types, namely, organic matter pores, inter-particle pores, intra-particle pores and fractures. The medium pores are clearly dominant in the Longmaxi Formation. The Longmaxi Formation is dominated by quartz, calcite and clay, with a lesser amount of feldspar, dolomite and pyrite. The highly brittle mineral content indicates that the Longmaxi Formation is suitable for fracturing, especially at the bottom of the formation. By comprehensively analyzing gas potential generation, conservation and exploitation conditions, we divide the Changning gas demonstration zone into two subzones.

Keywords: Source rock ; Gas in shale ; Exploration potential ; Longmaxi Formation ; Sichuan Basin ; China

Shale-gas systems are characterized by self-generation, self-reservoir, self-seal [1]. A significant amount of research has been conducted on the shale. Recent studies have focused on the sediment environment, lithology and pore structure are well understood [2-7]. With the development of hydraulic fracturing, gas has been economical to produce from shale reservoirs in the USA .

In China, gas in shale exploration began in the Sichuan Basin [8], and the Lower Silurian Longmaxi Formation has been identified as a key exploration target [9]. Zhang et al. [10] suggested that the marine shale in the southern China was the gas exploration prospect by comparing the shale geological conditions of the United States and China. Pu et al. [11] pointed out that the southwest in Sichuan basin of the southern China were the preliminary forecast favorable areas for shale exploration, according to the analysis of strata thickness, organic carbon content, organic matter thermal evolution degree and hydrocarbon generation intensity indexes. Zou et al. [12], Huang et al. [13] and Li et al. [14] all suggested that Changning was the exploration targets in the favorable areas in the southwest of Sichuan Basin, according to the analysis of organic matter abundance, net thickness, buried depth, maturity, porosity, brittle mineral content. All of the above-mentioned studies are concerned with the evaluations of prospecting areas, favorable areas or targets in China.

However, little attention has been devoted to the evaluation of sweet spot inside the target. Only to find the sweet spots in the target, it can effectively realize the gas commercial exploitation.

The evaluation of sweet spot inside the target has been affected by shale mineral composition. Inside the Changning target, Chen et al. [15-16] and Chen et al. [17] suggested that the shale mineral composition is dominated by quartz and clay by the X-ray diffraction and studied pore structures by the mercury intrusion porosimetry (MIP) method from thirty-nine samples. However, thirty-nine samples may not fully reflect the mineral components and the MIP application cannot fully represent the nano pores and cannot provide direct information of the actual pore geometry and their connectivity [18].

To evaluate the sweet spot inside the Lower Silurian Longmaxi Formation in the Changning target, we present an integrated geological research method including depositional environment, organic geochemistry, reservoir properties and mineralogy, based on more samples and proper technological methods such as Nitrogen adsorption and Field Emission Scanning Electron Microscopy (FE-SEM). We attempt to provide a method or a theoretical support for the future gas exploration inside the target in China.

1 Geological setting

The Sichuan Basin is a large sedimentary basin in the middle of China, with an area of $18 \times 10^4 km^2$. It is a petroliferous basin and is rich in oil and natural gas. Currently, it is the largest gas production area in China. Today, the Sichuan Basin can be divided into six structural units : the Chuanbei low-flat fold belt, the Chuanxi low-steep fold belt, the Chuanzhong flat fold belt, the Chuanxinan low-steep fold belt, the Chuandong high-steep fold belt, and the Chuannan low-steep fold belt [19] (Fig. 1). Additionally, it is a rhombic basin, affected by multicycle tectonic movements. Its evolution process contains three stages, i.e., the Sinian-Early Paleozoic rift basin stage, the Late Paleozoic-Middle Triassic depression basin stage, and the Late Triassic-Quaternary foreland basin stage [20, 21]. Because of the tectonic movements at the late Ordovician, the Longmaxi Formation is formed in the low energy, anoxic deep-water marine sedimentary environment. During the formation period, fine clastic rocks, primarily black shale, were deposited[22]. The Longmaxi Formation is widely distributed in the Sichuan Basin and ranges from 200 to 600m in the southern Sichuan Basin.

In this paper, we focus on the Longmaxi Formation in the Early Silurian in the Changning area, with a stable thickness range of 200 to 400m. The thickness is relatively small in the center, due to denudation at the Quaternary[15] (Fig. 2).

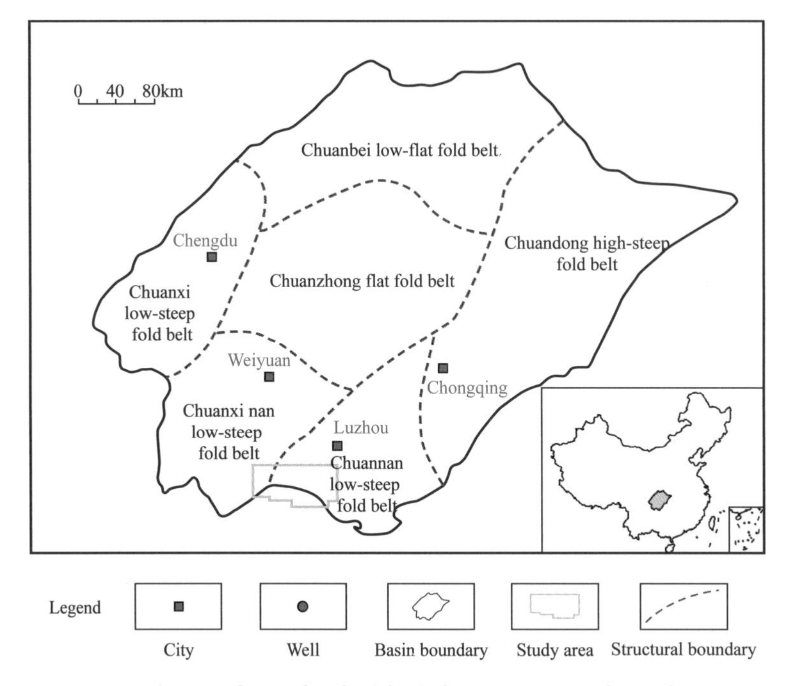

Fig.1 Structural map sketch of the Sichuan Basin, southern China

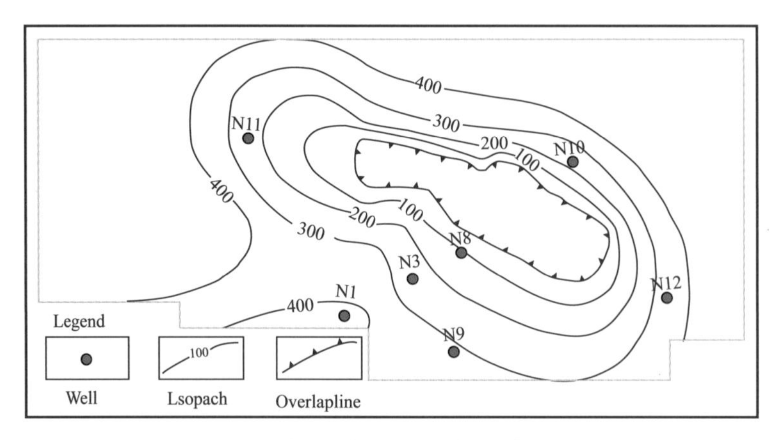

Fig.2 The study area and the well sites

2 Data and Methods

To evaluate the sweet spot inside the target, integrated geological research were carried out as follows:

The first step is depositional environment characterization which is to identify the face suited for the organic matter preservation through the core observation and wireline log. The second step is organic geochemical characteristics analysis which is to evaluate the hydrocarbon generation potential through TOC content, Organic matter type, Thermal maturity and Gas content. The third step is reservoir properties analysis which is to determine the horizon suited for fracturing and investigate the reservoir storage space. Finally, by combining gas generation potential, conservation and exploitation conditions, we evaluate two sweet spots inside the target and calculate the geological reserves. The workflow is shown in Fig.3.

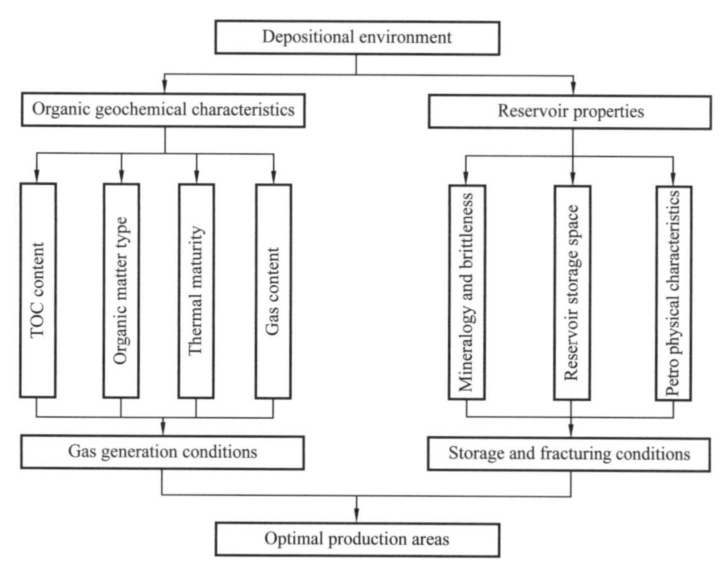

Fig.3 The schematic of integrated geological research method

The definite data and means are as follows:

Depositional environment is characterized by 68m core observation from Wells N1, N3 and N9; TOC content was analyzed using a Leco SC-632 instrument, with 105 samples data from Wells N1, N3, N8, N9 and N10.

Organic matter type is determined by Petrographic examinations and carbon isotope of kerogen, with 5 samples data from Well N1. Petrographic examinations were carried out the prepared shale samples using a Leica CTR 6000-M microscope with reflected and transmitted white and ultraviolet (UV) light sources and oil immersion objectives, and comparing the percentage of incident light reflected from the vitrinite particles in the samples to a known standard of 0.589%. These analysis methods are available in Hakimi et al.[23]. The stable carbon isotope values were measured on an Optima isotope ratio mass spectrometer (MS) with a Hewlett Packard 6890 II gas chromatograph (GC). The gas components were separated on a GC, converted into CO_2 in a combustion interface and then injected into the MS. These analysis processes are available in Cao et al.[24].

Thermal maturity is determined by vitrinite reflectance (R_o), with 9 samples data from Wells N1 and N3. The vitrinite reflectance(R_o)values were calculated using the formula: R_o=0.618B R_o + 0.4[25],

Where B R_0 is the bitumen reflectance. The analysis methods are available in Sun et al.[26].

For the mineral component, the shale powders were examined on an X-ray diffraction (XRD) instrument (Bruker D8 Advance X-ray diffractometer), with 205 samples data from Well N1 and N3. The data were measured in the 2θ range of 5°~80° at a rate of 4 (°) /min with a Cu radiation. The corresponding methods are available in Wang et al.[27].

To observe the storage space, we utilized a FE-SEM (Quanta 200F) equipped with an energy-dispersive spectrometer (EDS). The process was performed at a temperature of 24℃ and a humidity level of 35%. The corresponding data is from 6 samples of Well N1 and N3.

26 samples from Well N1 and N9 are used to measure petrophysical characteristics. Besides, low-pressure nitrogen adsorption is also carried out to analyzed petrophysical characteristics. The low-pressure N2 isotherm analyses were conducted on an Quadrasorb™ SI Surface Area Analyser and Pore Size Analyser. The prepared samples (60~80 mesh size) were outgassed at 378K for 24h under high vacuum. The data were measured at 77K under the relative pressures varying from 0.01 to 0.995. The specific surface area was interpreted with the multipoint Brunauer-Emmett-Teller (BET) method, and the pore size distribution (PSD) was calculated with the density functional theory (DFT) method. These analysis methods are available in Liang et al.[28].

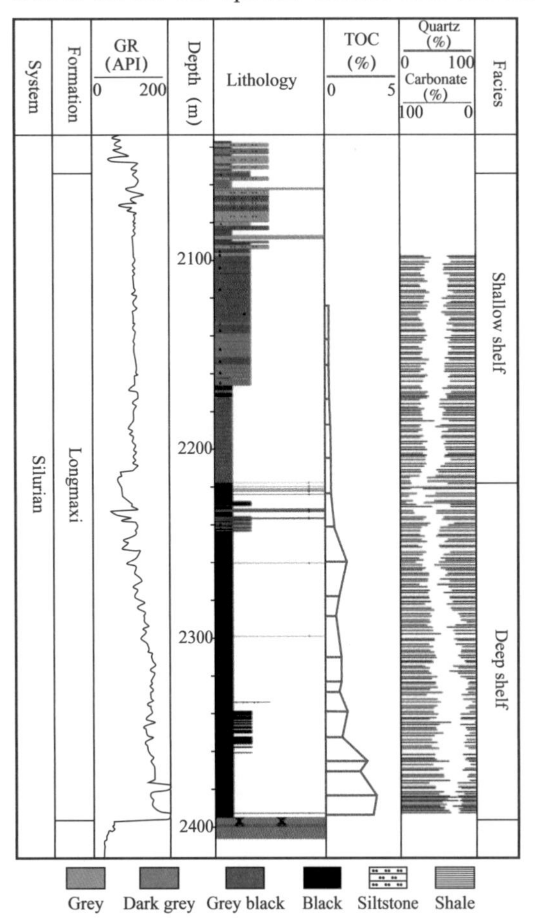

Fig.4 The synthesis column shows vertical characteristics of lithology, minerals, and TOC in the Longmaxi Formation in the Sichuan Basin

3 Results and Discussion

3.1 Depositional environment

In this study, one face (shelf) and two subfaces (as shown in Fig. 4, a shallow shelf at the top and a deep shelf at the bottom) have been identified based on their lithology, wireline log patterns and sedimentary structures.

Fig.4 indicates that shallow shelf is characterized by gray silty sandstone, dark gray-black gray sandy shale, gray-black shale and black sandy shale, which is intercalated with dark gray argillaceous limestone. Fig.5a illustrates the sedimentary structure of shallow shelf is primarily parallel bedding. Fig.5b shows the bioturbation. Parallel bedding and bioturbation imply intense water dynamic which indicates relatively poor preservation of organic matter.

Fig.4 indicates that deep shelf is characterized by black shale, intercalated with dark gray argillaceous siltstone. Pyrite is common (Fig 5c), and the deep shelf contains a large number of graptolite (Fig 5d). It indicates that deep shelf is deposited in quiet water with no benthic disturbance. Thus, deep shelf is suitable for preserving organic matter and is the most important developed section of hydrocarbon source rock.

(a) Parallel bedding, Well N1, 2483.74m

(b) Bioturbated texture, Well N3, 2101.36m

(c) Pyrite, Well N3, 2301.68m

(d) Graptolites, Well N3, 2387.32m

Fig. 5 Sedimentary structures

3.2 Organic geochemical characteristics

3.2.1 TOC content

As shown in Tab. 1 and Fig. 6, TOC content in this study ranged from 0.2% to 6.69%, with an average value of 1.72%. TOC content increases with the burial depth. Notably, 28 samples distributed at the bottom of deep shelf indicate TOC greater than 2.0% and average TOC of 3.17%. The corresponding shale thickness with TOC greater than 2% is approximately 30m. It implies the favorable potential of gas generation.

3.2.2 Organic matter type

With maceral and carbon isotope of kerogen, the dominant type of organic matter can be classified into four types: sapropelic (Ⅰ), humic-sapropelic (Ⅱ1), sapropelic-humic (Ⅱ2) and humic (Ⅲ), according to Tab. 2[29].

Tab. 1 TOC content, maturity and mineral content of the Longmaxi shale

Samples	Depth (m)	TOC (%)	R_o (%)	C (%)	Q (%)	PF (%)	PL (%)	Cal (%)	Dol (%)	Py (%)	TC (%)
N1-1	2492.0	1.26	1.85	39.0	30.0	5.0	10.0	10.0	4.0	2.0	14.0
N1-2	2505.2	4.66	2.01	36.0	37.0	3.0	6.0	9.0	4.0	5.0	13.0
N1-3	2516.6	3.10	1.86	13.0	48.0	1.0	3.0	20.0	12.0	3.0	32.0
N3-1	2124.5	0.23	/	22.2	37.6	0	0	40.2	0	0	40.2
N3-2	2142.5	0.20	/	33.5	34.7	0	0	27.7	3.1	1.0	30.8
N3-3	2156.4	0.24	/	31.8	28.1	0	0	34.5	4.1	1.5	38.6
N3-4	2174.0	0.27	/	28.5	26.0	0	0	43.3	1.8	0.4	45.1
N3-5	2187.4	0.35	/	28.0	28.0	0	0	39.2	4.8	0	44.0
N3-6	2204.2	0.35	/	26.6	34.1	0	0	39.0	0.3	0	39.3
N3-7	2223.8	0.39	/	26.8	21.8	0	0	45.5	5.9	0	51.4
N3-8	2241.6	0.59	/	26.7	28.2	0	0	45.1	0	0	45.1
N3-9	2259.9	1.50	/	32.0	20.0	0	0	42.2	4.6	1.2	46.8
N3-10	2278.4	0.93	/	22.9	38.2	0	6.1	32.8	0	0	32.8
N3-11	2288.8	0.69	/	31.8	42.2	0	7.0	15.3	3.7	0	19.0
N3-12	2310.4	1.09	/	32.7	48.9	0	0	13.7	4.7	0	18.4
N3-13	2323.5	1.11	1.6	39.9	46.9	0	6.3	6.9	0	0	6.9
N3-14	2329.0	0.94	/	34.7	56.5	0	0	8.8	0	0	8.8
N3-15	2339.3	1.50	1.84	29.3	44.4	0	19.7	6.6	0	0	6.6
N3-16	2352.0	1.12	1.96	28.7	45.2	0	13.0	8.8	4.3	0	13.1
N3-17	2365.4	2.77	2.48	22.8	64.8	0	0	0.1	7.6	4.7	7.7
N3-18	2370.2	2.33	2.7	24.0	51.4	0	4.6	9.5	6.9	3.6	16.4
N3-19	2383.8	3.44	3.07	9.3	65.6	0	0	13.1	9.5	2.5	22.6

Note: In the table, C = clay, Q = quartz, PF = potash feldspar, PL = plagioclase, Cal = calcite, Dol = dolomite, Py = pyriye and TC = total carbonate. Due to space limitation, only these data are listed.

Tab. 2 Organic matter type classified using maceral and carbon isotope of kerogen

Organic matter type	Carbon isotope of kerogen ($\delta^{13}C$) (‰)	Maceral of kerogen (TI value)
Sapropelic type (Ⅰ)	<-29	>80
Humic-sapropelic type (Ⅱ1)	-29～-27	40～80
Sapropelic-humic type (Ⅱ2)	-27～-25	0～40
Humic type (Ⅲ)	>-25	<0

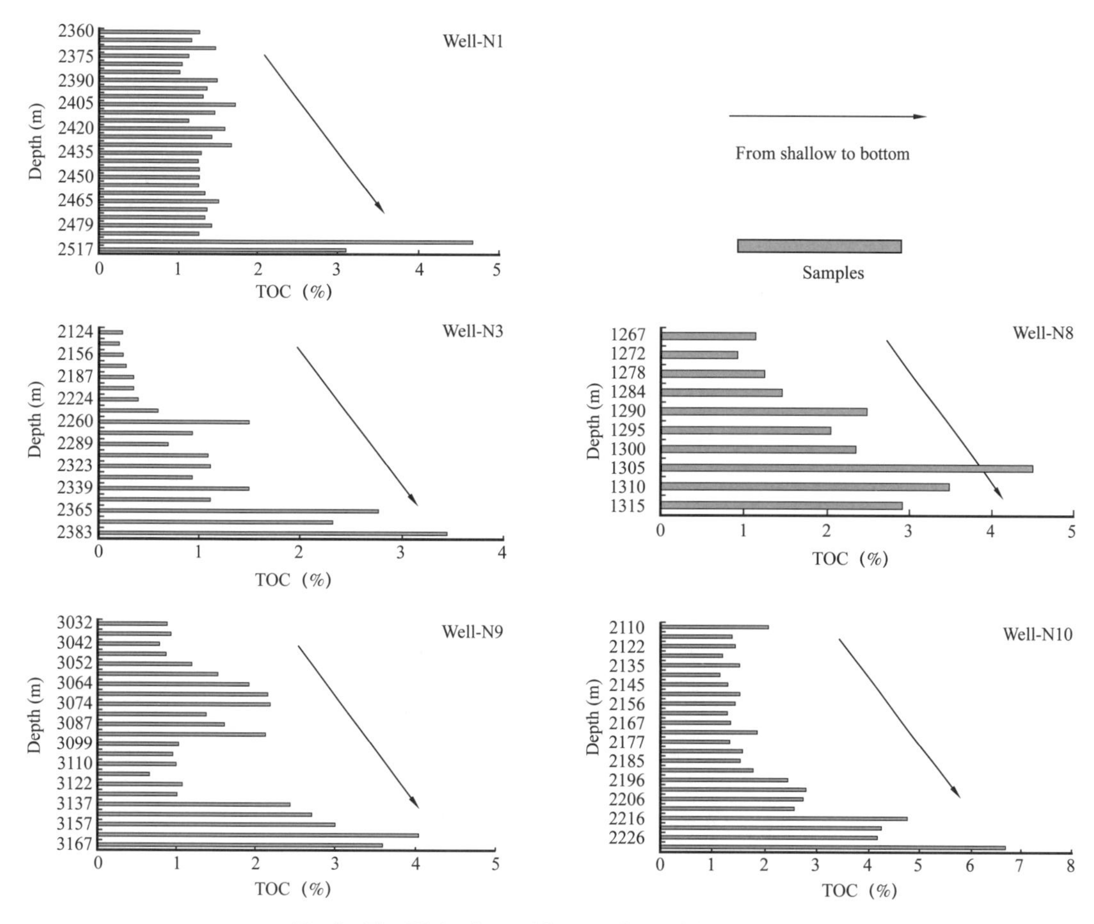

Fig.6 The TOC values of the samples with burial depth

As shown in Tab. 3, maceral for deep shelf and the kerogen $\delta^{13}C$ content ranges from −29.3‰ to −30.9‰, with an average value of −30.4‰. This determines that the organic matter type is I, which indicates a strong hydrocarbon generation capability.

Tab. 3 Content of maceral and carbon isotope of kerogen in the Changning area

Samples	Organic matter type	Carbon isotope of kerogen ($\delta^{13}C$) (‰)	Maceral of kerogen (TI value)	Sapropelinite	Asphalt	Vitrinite	Inertinite
N-a	I	-30.9	95	90	10	/	/
N-b	I	-30.7	94	88	12	/	/
N-c	I	-29.8	92.5	85	15	/	/
N-d	I	-29.3	93.5	87	13	/	/
N-e	I	-29.5	94	88	12	/	/

3.2.3 Thermal maturity

As shown in Tab. 1, equivalent vitrinite reflectance (R_o) in this study ranges from 1.6% to

3.07%, with an average value of 2.15%. This suggests that thermal maturity is high–over mature, and the source rock has reached the dry gas window. Namely, the target section has the greatest potential for gas generation indicated by higher thermal maturity.

3.2.4 Gas content

The measured values show that the gas content ranges from 0.5cm^3/g to 4.2cm^3/g, with an average value of 1.8cm^3/g in the Longmaxi Formation. As shown in Fig. 7, there is a positive correlation between gas content and TOC content. Gas is primarily adsorbed on kerogen surface within organic matter. High TOC content corresponds to abundant organic matter, implying a significant adsorption capacity and a large scale of gas [30]. The gas content value increases with TOC content and burial depth. There is more than 2cm^3/g of gas in black shale at the bottom of deep shelf, ranging from 2.0cm^3/g to 4.2cm^3/g (an average of 2.9cm^3/g), which is an ideal horizon for exploration and development.

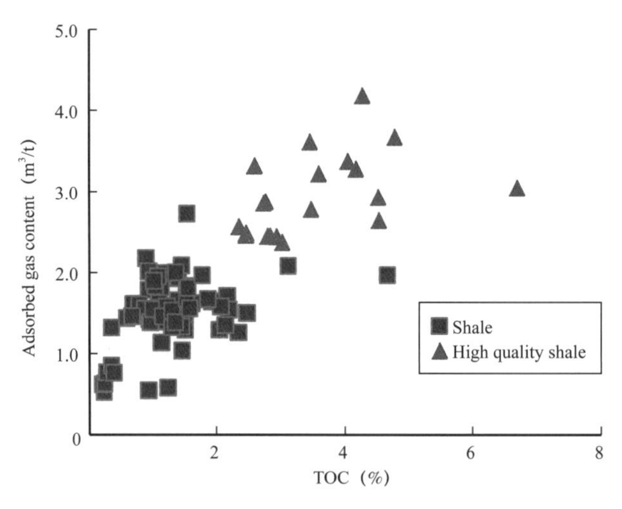

Fig. 7 Relationship between adsorbed gas content and TOC

All geochemical characteristics suggest that deep shelf has a good gas generation potential, especially the bottom with 30m thickness. TOC is greater than 2% and gas content is more than 2cm^3/g. We name it "high quality shale".

3.3 Reservoir properties

3.3.1 Mineralogy and brittleness

As shown in Tab. 1, the Longmaxi Formation is dominated by quartz, calcite and clay, with subordinate feldspar, dolomite and pyrite. Mineralogically, Fig. 8a illustrates that mineral composition of high quality shale is distributed in the region close to the point of quartz, feldspar and pyrite which are the brittle minerals. Fig. 8b illustrates that clay minerals of high quality shale is closer to illite with an average content over 80%. The high content of illite implies middle–late diagenetic stage of shale.

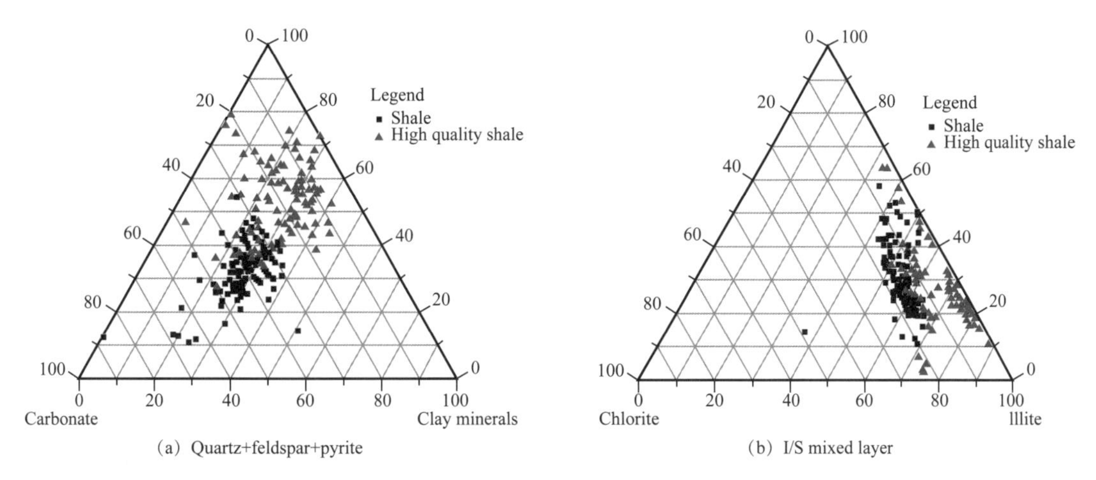

Fig. 8 Ternary plot of the entire rock mineral composition and clay mineral composition of shale in the Longmaxi Formation in the study area

The brittle mineral content is an essential assessment for artificial fracturing. A higher brittle mineral content indicates a better artificial fracturing effect. Fig.9 displays that the brittle mineral content tends to increase with the burial depth. And the content is greater than 40% at the bottom of deep shelf, namely in the high quality shale. In addition, the test data from Wells N1, N3, N9 and N10 reveal that the average Young's modulus value and Poisson's ratio are 47.7GPa (range: 20.5—75.6GPa) and 0.24 (range: 0.11~0.35), respectively. The higher Young's modulus and lower Poisson's ratio are indicative of greater brittleness. Thus, the high quality shale is suitable for the treatment of fracturing.

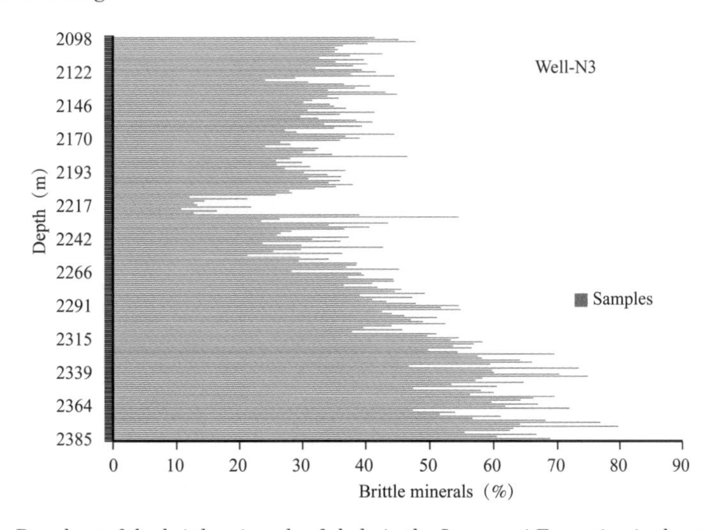

Fig. 9 Bar chart of the brittle minerals of shale in the Longmaxi Formation in the study area

3.3.2 Reservoir storage space

The observed cores and FE-SEM reveal that there are four dominant types of reservoir space in the Longmaxi Formation, namely, organic matter pores, inter-particle pores, intra-particle pores and fractures.

3.3.2.1 Organic matter pores

Organic matter pores are found within organic matter (OM). Fig. 10a and Fig. 10b illustrate that Organic matter pores have irregular, bubble-like, elliptical cross sections and, generally, range from 10 to 950nm in length. Fig. 10a and Fig. 10b also display dense organic matter pores occurring in organic particles. It is reported that organic matter pores would achieve approximately 50% volume of organic particles [31]. It means that contribution of organic pores to porosity would increase with the content of organic matter, which is also indicated by the positive correlation between TOC and porosity as shown in Fig.11. Beside, according to above geochemical characteristics, Type I kerogen and high thermal maturity signify a high conversion of organic matter, leading to more organic matter pores. Obviously, more organic matter pores would form more superficial area for gas to adsorb on. Thus, organic matter pores are dominant storage space for the high quality shale of Longmaxi formation.

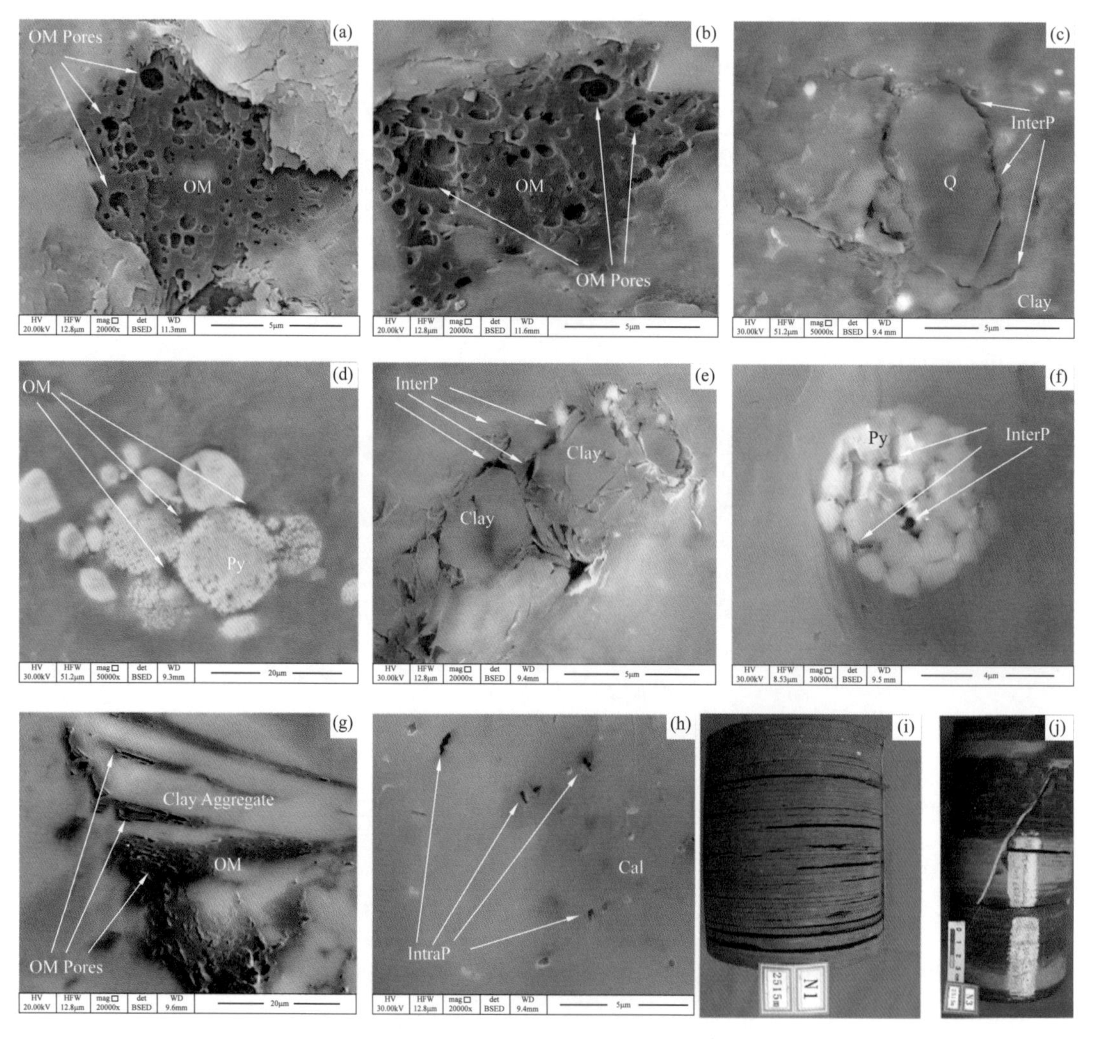

Fig. 10 SEM images of reservoir spaces in the Longmaxi Formation (OM-organic matter, Q-quartz, Py-pyrite, Cal-calcite)

3.3.2.2 Inter-particle pores

Inter-particle pores occur between grains, such as quartz (Fig. 10c), pyrite framboids (Fig. 10d), clay flocculates (Fig. 10e), and may occur at grain edges during compaction. Displayed in Fig.10c, pore structure is long and narrow indicating a lineal contact between grains. This contact makes the pore structure unstable and would benefit the potential of fracturing. As shown in Fig.10d, this inter-particles pore structure indicates point contact between grains. Point contact signifies the stable pore structure which can be occupied by organic matter. Fig.10e illustrates inter-particle pores between clay. Due to strong plastic deformation of clay, these pores would easily close during compaction, which would aggravate the connectivity.

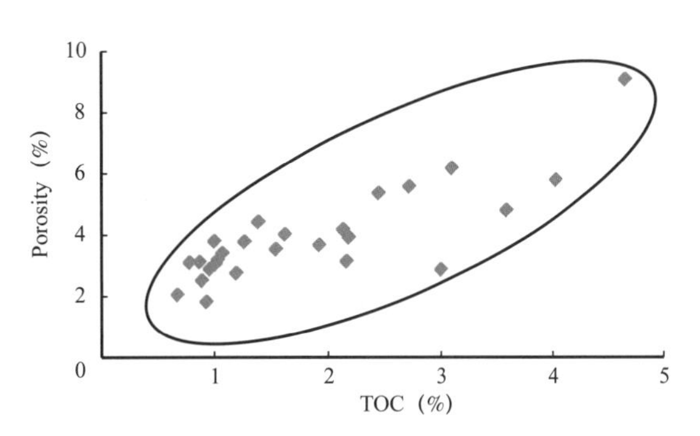

Fig. 11 Relationships of pore volume and TOC

3.3.2.3 Intra-particle pores

Intra-particle pores primarily occur in pyrite framboids, clay minerals and calcite. In the study area, pyrite is relatively abundant in the Longmaxi Formation, and inter-crystalline pores within pyrite framboids are the most common intra-particle pores, with pore diameters ranging from 20nm to 500nm (Fig. 10f) . Due to the brittleness of pyrite, these pores would contribute a lot to connectivity after fracturing. Intraplatelet pores within clay aggregates are very small due to severe compaction. Notably, organic matter can be also found in Intraplatelet pores as shown in Fig.10g. In addition, pores in calcite can be identified (Fig. 10h) . These pores are often meso- or macro-pores with diameters of approximately 1μm, which may benefit the potential of fracturing.

3.3.2.4 Fractures

In this study, fractures are dominated by interlaminated fractures and structural fractures. Interlaminated fractures form as a result of weak interfaces within shale (Fig. 10i) . Structural fractures occur due to tectonic stresses [32]. Some fractures are partially or completely filled with calcite (Fig. 10j) .

Fractures play a key role in shale-gas systems. They are important storage and migration pathways. Nevertheless, a large number of open natural fractures may lead to gas dispersion and prevent gas conservation [33].

All of the above-mentioned pores are large enough for storage and migration of hydrocarbon molecules. Organic matter pores provide important reservoir spaces, and fractures have a significant effect on hydrocarbon production. They all contribute to porosity network of the shale reservoir.

3.3.3 Petrophysical characteristics

As shown in Fig. 12, the Longmaxi shale porosity ranges from 1.8% to 9.1%(average : 3.9%). Permeability is very low, ranging from 0.0024×10^{-3}mD to 1.25×10^{-3}mD, with an average value of 0.0338×10^{-3}mD. It is noted that permeability is remarkable when the corresponding porosity over 4.5%. As shown in Tab. 4, BET specific surface ranges from 6.315m^2/g to 10.147m^2/g, with an average value of 7.801m^2/g. The total pore volume of the Longmaxi shale ranges from 6.26×10^{-3}mL/g to 11.75×10^{-3}mL/g, with an average value of 8.31×10^{-3}mL/g. The radius of the medium pore ranges from 1.954nm to 2.219nm, with an average of 2.069nm. According to the IUPAC classification [34], the pores have been classified into 3 types : micro-pores with radius less than 2nm, medium-pores with radius from 2nm to 50nm, macro-pores with radius over 50nm. For Longmaxi Formation, percentages of micro-pores, medium-pores and macro-pores are 3.19%, 91.24% and 5.57%, respectively. Medium pores are clearly dominant in the Longmaxi Formation. Moreover, Fig.13 illustrates that pores with radius of 3 to 5nm dominate the medium-pores. Due to the tiny dominant pores, fracturing is necessary to obtain more gas.

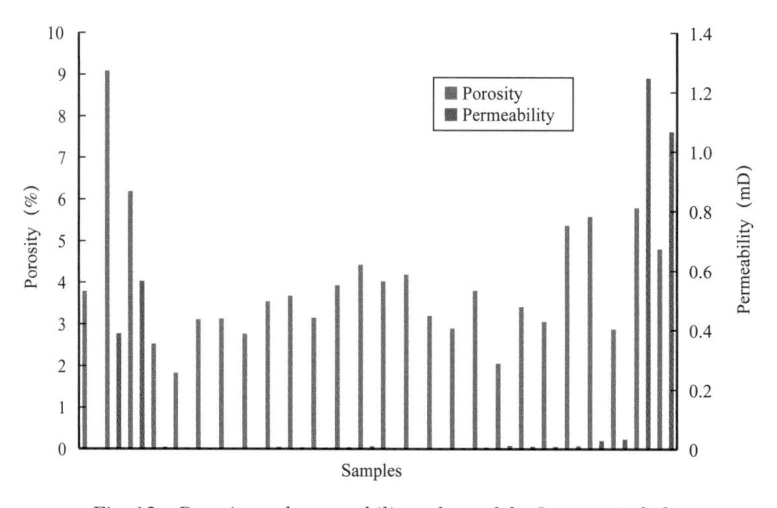

Fig. 12 Porosity and permeability values of the Longmaxi shale

Tab. 4 Characteristics of pore parameters of the Longmaxi shale

Samples	BET specific surface (m^2/g)	BJH total pore volume (mL/g)	Medium radius of pores (nm)	Micro-pore (<2nm) volume (%)	Medium-pore (2~50nm) volume (%)	Macro-pore (>50nm) volume (%)
N-1	6.315	0.0063	1.95	2.17	94.58	3.26
N-2	6.687	0.0072	2.05	5.17	89.80	5.03
N-3	8.214	0.0092	2.21	2.94	92.93	4.13
N-4	7.893	0.0078	1.96	2.06	91.40	6.54
N-5	7.549	0.0077	2.02	3.65	89.52	6.83
N-6	10.147	0.0118	2.22	3.15	89.21	7.64
Average	7.801	0.0083	2.07	3.19	91.24	5.57

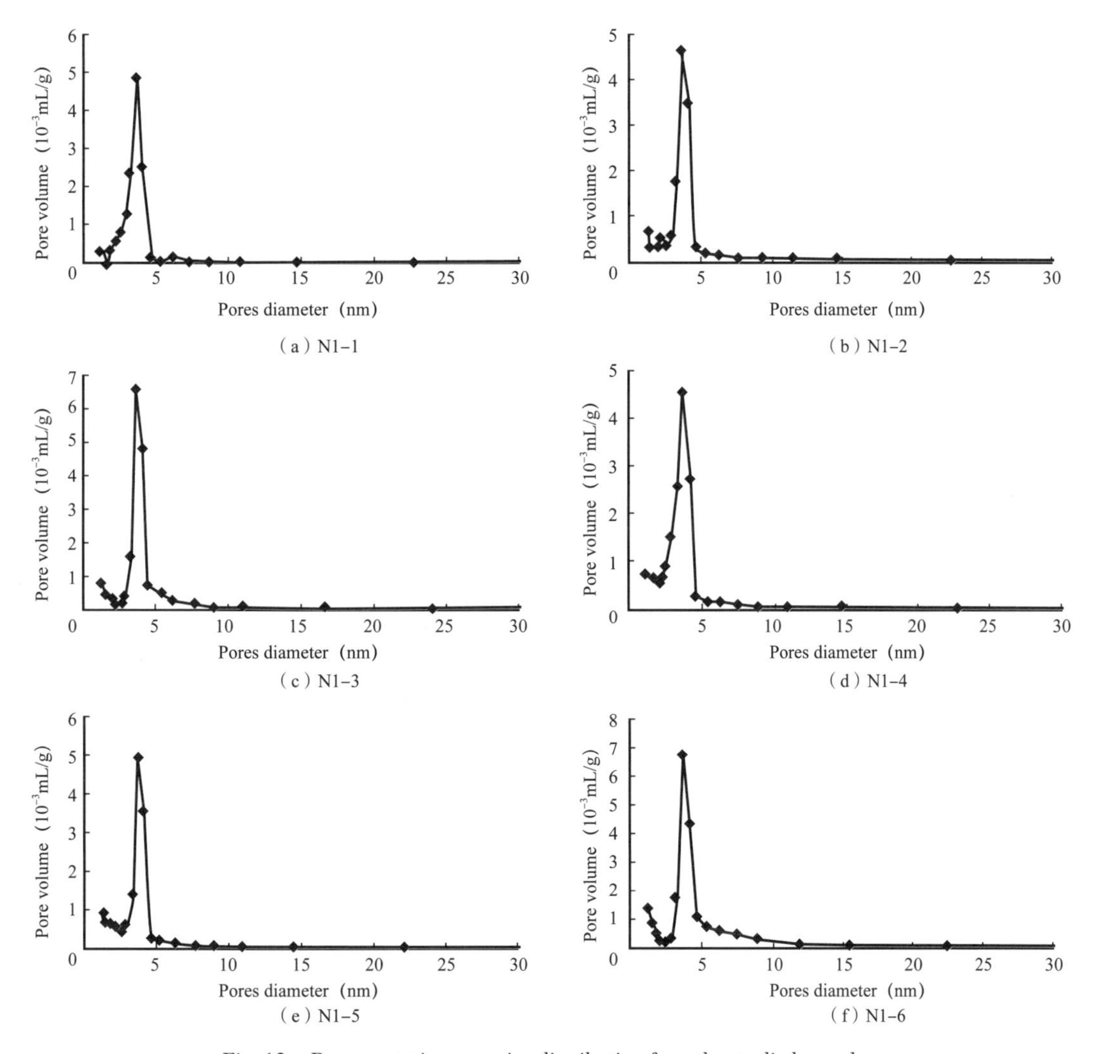

Fig. 13 Representative pore size distribution from the studied samples

3.4 Sweet spots

To achieve commercial development of gas, the formation must meet some requirements. Gas content increases with TOC content. For commercial development, TOC value should be over 2% according to the successfully developed gas in shale in North America. High quality shale of longmaxi formation with average TOC of 3.17% is completely satisfied.

Thermal maturity is used to estimate whether organic matter reached the dry gas window. Gas would surge if the organic matter reached the dry gas window. According to successful gas in shale in North America, R_o, the index of maturity, is over 1.1%. High quality shale of longmaxi formation with average R_o of 2.15% is completely satisfied. TOC and maturity is used to estimate the potential of gas generation.

Mineral component and brittleness have an impact on fracturing treatment. Despite of the great potential of gas generation, low potential productivity on account of poor fracturing is predictable, if brittle mineral content is less.

Porosity is an important parameter to reflect the storage capacity. Though the potential of gas generation in high quality shale is great, the potential productivity is likely low in part of this section with low storage capacity.

According to gas production test, we made the cross-plot to analyze above parameters and obtain the criteria of evaluation on high quality shale for a fast breakthrough exploring and developing gas in Changning (Tab. 5).

Tab. 5 Classification and evaluation criteria of sweet spots in Changning

Category	I	II
TOC (%)	≥3	3～2
Gas content (cc/g)	≥3	3～2
R_o (%)	≥2.0	<2.0
Total Porosity (%)	≥6.0	6.0～4.0
Brittle minerals content (%)	≥60	60～50

According to the evaluation criteria, as shown in Tab. 5, the Changning gas demonstration zone is divided into two subzones, as shown in Fig. 14.

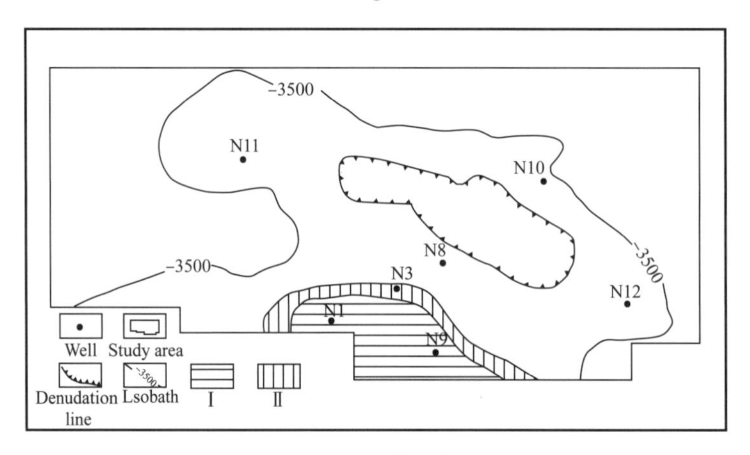

Fig. 14 Sweet spots in the Changning target

4 Conclusions

Longmaxi Formation is identified as one face (shelf) and two subfaces. The deep shelf deposited in quiet water is suitable for preservation of organic matter and is the most important section of hydrocarbon source rock. Geochemical characteristics demonstrate that 30m thickness at the bottom of deep shelf has a strong potential of gas generation, with average TOC content of 3.17%, average R_o of 2.15% and Type I Kerogen. This section is named as high quality shale. Mineral component analysis of high quality shale demonstrates that brittle mineral is dominant.

The brittleness of deep shelf is over 40%. It indicates favorable conditions for fracturing treatment. The storage space of high quality sale is dominated by organic matter pores. The size of pore mainly belongs to medium type. It reveals a good storage capacity. Gas generation potential, storage capacity and fracturing conditions demonstrate favorable exploration and development potential of high quality shale in Longmaxi Formation. For a fast breakthrough exploring and developing gas, criteria of further evaluation on the high quality shale were established. With the criteria, gas demonstration zone could be divided into two subzones.

Acknowledgments

The work presented in this paper was supported by the National Science and Technology Special (2016ZX05047004-001) and the National Natural Science Foundation of China (No. 41302092).

References

[1] Jarvie D M, Hill R J, Pollastro R M. Assessment of the gas potential and yields from shales: The Barnett Shale model [J]. Oklahoma Geol. Surv. Circular., 2004, 110: 37-50.

[2] Wignall P B, Newton R. Black shales on the basin margin: a model based on examples from the Upper Jurassic of the Boulonnais, northern France [J]. Sediment Geol., 2001, 144 (3): 335-356.

[3] Loucks R G, Ruppel S C. Mississippian Barnett Shale: Lithofacies and depositional setting of a deep-water shale-gas succession in the Fort Worth Basin, Texas [J]. AAPG Bull, 2007, 91 (4): 579-601.

[4] Aplin A C, Macquaker J H. Mudstone diversity: Origin and implications for source, seal, and reservoir properties in petroleum systems [J]. AAPG Bull, 2011, 95 (12): 2031-2059.

[5] Slatt R M, O'Brien N R. Pore types in the Barnett and Woodford gas shales: Contribution to understanding gas storage and migration pathways in fine-grained rocks [J]. AAPG Bull, 2011, 95 (12): 2017-2030.

[6] Abouelresh M O, Slatt R M. Lithofacies and sequence stratigraphy of the Barnett Shale in east-central Fort Worth Basin, Texas [J]. AAPG Bull, 2012, 96 (1): 1-22.

[7] Loucks R G, Reed R M, Ruppel S C, et al. Spectrum of pore types and networks in mudrocks and a descriptive classification for matrix-related mudrock pores [J]. AAPG Bull, 2012, 96 (6): 1071-1098.

[8] Zhang S C, Zhu G Y. Gas accumulation characteristics and exploration potential of marine sediments in Sichuan Basin [J]. Acta Pet. Sin., 2006, 27 (5): 1-8.

[9] Zhang J C, Xu B, Nei H K et al. Two essential gas accumulations for natural gas exploration in China [J]. Nat. Gas Ind., 2007, 27 (11): 1-6.

[10] Zhang J C, Xu B, Nie H K, et al. Exploration Potential Of Shale Gas Resources in China [J]. Nat. Gas Ind., 2008, 28 (6): 136-140.

[11] Pu B L, Jiang Y L, Wang Y, et al. Reservoir-forming conditions and favorable exploration zones of shale gas in Lower Silurian Longmaxi Formation of Sichuan Basin [J]. Acta Petrol. Sin., 2010, 31 (2): 225-230.

[12] Zou C N, Dong D Z, Wang S J, et al. Geological characteristics, formation mechanism and resource

potential of shale gas in China [J] . Pet. Explor. Dev., 2010, 37 (6) : 641-653.

[13] Huang S, Wang G Z, Zou B, et al. Preferred targets of the shale has from Silurian Longmaxi Formation in middle-upper Yangtze of China [J] . J. Chengdu Univ. Technol. (Sci. Technol. Ed.) ., 2012, 39 (2): 190-197.

[14] Li Y J, Liu H, Zhang L H, et al. Lower limits of evaluation parameters for the lower Paleozoic Longmaxi shale gas in southern Sichuan Province [J] . Science China : Earth Sci., 2013, 56: 710-717.

[15] Chen S B, Zhu Y M, Wang H Y, et al. Characteristics and significance of mineral compositions of Lower Silurian Longmaxi Formation shale gas reservoir in the southern margin of Sichuan Basin [J] . Acta Petrol. Sin., 2011, 32 (5) : 775-782.

[16] Chen S B, Zhu Y M, Wang H Y, et al. Shale gas reservoir characterisation : A typical case in the southern Sichuan Basin of China [J] . Energy, 2011, 36: 6609-6616.

[17] Chen W L, Zhou W, Luo P, et al. Analysis of the shale gas reservoir in the Lower Silurian Longmaxi Formation, Changxin 1 well, Southeast Sichuan Basin, China [J] . Acta Petrol. Sin., 2013, 29 (3): 1073-1086

[18] Guo X J, Shen Y H, He S L, et al. Quantitative pore characterization and the relationship between pore distributions and organic matter in shale based on Nano-CT image analysis : A case study for a lacustrine shale reservoir in the Triassic Chang 7 Member, Ordos Basin, China [J] . J. Nat. Gas Sci. & Eng., 2015, 27: 1630-1640.

[19] Zhai G M. Petroleum geology of China [M] .Beijing : Petroleum Industry Press, 1989.

[20] Wang Z C, Zhao W Z, Peng H Y. Charaeteristies of multi-source Petroleum systems in Sichuan basin [J] . Pet. Explor. Dev., 2002, 29 (02) : 26-28.

[21] Shen C B, Mei L F, Xu Z P, et al. Architecture and tectonic evolution of composite basin-mountain system in Sichuan Basin and its adjacent areas [J] . Geotect Metallog, 2007, 31 (3) : 288-299.

[22] Su W B, Li Z M, Ettensohn F R, et al. Distribution of black shale in the Wufeng—Longmaxi Formations (Ordovician-Silurian), South China : major controlling factors and implications [J] . Earth Sci-J. China Univ. Geosci., 2007, 32: 819-827.

[23] Hakimi M H, Abdullah W H. Organic geochemical characteristics and oil generating potential of the Upper Jurassic Safer shale sediments in the Marib-Shabowah Basin, western Yemen [J] . Organic Geochem., 2013, 54: 115-124.

[24] Cao J, Wang X L, Sun P A, et al. Geochemistry and origins of natural gases in the central Junggar Basin, northwest China. Organic Geochem, 2012, 53: 166-176.

[25] Jacob H. Disperse solid bitumens as an indicator for migration and maturity in prospecting for oil and gas [J] . Erdol & Kohle Erdgas Petrochemie, 1985, 38 (8) : 365-365.

[26] Sun M D, Yu B S, Hu Q H, et al. Nanoscale pore characteristics of the Lower Cambrian Niutitang Formation Shale : A case study from Well Yuke# 1 in the Southeast of Chongqing, China[J]. International J. Coal Geol., 2016, 154: 16-29.

[27] Wang Y, Zhu Y M, Chen S B, et al. Characteristics of the nanoscale pore structure in Northwestern Hunan shale gas reservoirs using field emission scanning electron microscopy, high-pressure mercury

intrusion, and gas adsorption [J] . Energy & Fuels, 2014, 28 (2) : 945-955.

[28] Liang L X, Xiong J, Liu X J. Mineralogical, microstructural and physiochemical characteristics of organic-rich shales in the Sichuan Basin, China [J] . J. Nat. Gas Sci. & Eng., 2015, 26: 1200-1212.

[29] Dai H M, Huang D, Liu X N, et al. Characteristics and Evaluation of Marine Source Rock in Southwestern Shunan [J] . Nat. Gas Geosci., 2008, 19 (4) : 503-508.

[30] Ross D J, Bustin R M. Shale gas potential of the lower Jurassic Gordondale member, northeastern British Columbia, Canada [J] . Bull. Canadian Pet. Geol., 2007, 55 (1) : 51-75.

[31] Curtis M E, Sondergeld C H, Ambrose R J, et al. Microstructural investigation of gas shales in two and three dimensions using nanometer-scale resolution imaging. AAPG Bull, 2012, 96 (4) : 665-677.

[32] Huang J L, Zou C N, Li J Z, et al. Shale gas accumulation conditions and favorable zones of Silurian Longmaxi Formation in south Sichuan Basin, China [J] . J. China Coal Soc., 2012, 37 (5) : 782-787.

[33] Jarvie D M, Hill R J, Ruble T E, et al. Unconventional shale-gas systems : The Mississippian Barnett Shale of north-central Texas as one model for thermogenic shale-gas assessment [J] . AAPG Bull, 2007, 91 (4) : 475-499.

[34] Rouquerol J, Avnir D, Fairbridge C W, et al. Recommendations for the characterization of porous solids (Technical Report) [J] . Pure Appl. Chem., 1994, 66 (8) : 1739-1758.

蜀南地区富有机质页岩孔隙结构及超临界甲烷吸附能力

朱汉卿[1]，贾爱林[1]，位云生[1]，贾成业[1]，袁　贺[1]，刘　畅[2]

（1. 中国石油勘探开发研究院；
2. 中国石油大学（北京））

摘　要：以蜀南地区龙马溪组下部富有机质页岩为研究对象，通过场发射扫描电镜（FE-SEM）、低压氩气吸附实验和重力法高压甲烷吸附实验，研究页岩孔隙结构特征及超临界状态下页岩储层的甲烷吸附能力，并讨论了页岩孔隙结构对甲烷吸附能力的影响。研究表明，蜀南地区龙马溪组富有机质页岩主要发育有机质孔隙，页岩孔隙结构非均质性强，比表面积为16.846～63.738m^2/g，孔体积为0.050～0.092cm^3/g，微孔和介孔贡献页岩90%以上的比表面积，介孔和宏孔贡献页岩90%以上的孔体积。甲烷在地层条件下处于超临界状态，过剩吸附曲线在12MPa左右时出现极大值，随后开始下降。使用修正过的四元Langmuir–Freundlich（L–F）方程拟合高温甲烷过剩吸附曲线，拟合效果较好，相关系数大于0.997。页岩饱和吸附量为0.0670～0.2202mmol/g，不同页岩样品吸附能力差异明显。海相富有机质页岩中，随着有机质含量的增大，有机质孔隙数量增多，且页岩中微孔比例增大，微孔的吸附能力远大于介孔和宏孔，故页岩吸附能力增强。有机质含量是影响蜀南地区海相富有机质页岩孔隙结构和甲烷吸附能力的主要因素。

关键词：龙马溪组；页岩；孔隙结构；低压Ar吸附；高压甲烷吸附；过剩吸附；四元L–F方程

近年来，伴随着水平井技术及分段压裂技术的日趋成熟，页岩气这种非常规资源得到了长足的发展[1, 2]。根据美国能源信息署（EIA）2016年的统计，美国2015年页岩气年产量达到$4.29\times10^{12}m^3$，占其天然气总产量的46.1%。页岩储层致密，孔渗极低，主要发育纳米级孔隙[3, 4]，常规储层表征技术手段难以有效表征页岩纳米级的孔隙结构形态及分布。Loucks等[5]通过氩离子抛光技术观察到了页岩中纳米级孔隙的形态，并对孔隙类型进行了系统分类。气体吸附法作为一种常用的表征吸附材料孔隙结构的实验方法，越来越广泛地应用于煤层及页岩储层孔隙结构表征[6-9]。其中，最常见的吸附质为氮气，但是由于四极距作用，氮气分子会和吸附剂表面的功能团及离子发生相互作用，从而影响探测的准确性，尤其是对于微孔的探测存在一定的局限性[10, 11]，而微孔对于页岩的甲烷吸附能力有较大的贡献[12-14]。根据国际纯粹与应用化学学会（IUPAC）的建议，对于含微孔和介孔的材料，推荐使用没有四极距作用的单原子分子氩作为吸附质进行实验。

页岩气主要以吸附态和游离态两种形式赋存于纳米级孔隙中[15]，吸附气的含量不仅影响地质储量的评价，更是气井后期稳产的关键[16, 17]。目前对于页岩吸附能力的评价主要依赖于高压等温吸附实验。受限于实验设备的精度，前人的甲烷吸附实验多低于15MPa，温度也达不到地层条件[13, 14, 18]，而温度和压力等环境因素对页岩的吸附性能有较大的影响[19–21]。且甲烷在地层条件下处于超临界状态[22–24]，超临界状态下甲烷的吸附量会出现极值，即过剩吸附现象[25, 26]。经典的 Langmuir 方程拟合过剩吸附曲线会低估页岩的实际吸附能力[16, 27, 28]。

在利用场发射扫描电镜观察镜下微观孔隙形态及类型的基础上，设计 –185.5℃低压氩气吸附实验定量表征页岩纳米级孔隙结构特征，使用重量法等温吸附仪进行高温（90℃）高压（30MPa）下富有机质页岩的甲烷吸附实验，使用四元 Langmuir–Freundlich（L–F）模型拟合过剩吸附曲线，得到页岩吸附能力参数，并探讨孔隙结构对页岩吸附能力的影响。

1 实验样品及方法

1.1 样品

页岩岩心样品取自蜀南长宁示范区的一口页岩气评价井，9 个样品均取自下志留统龙马溪组下部开发层段（图 1），样品深度为 2355.0～2393.7m，取样间距为 3.4～6.9m。样品基础地球化学及矿物组分数据见表 1。其中，样品 TOC 值介于 0.82%～4.47%，平均值为 2.38%。矿物组分中，石英含量为 29.6%～39.7%，平均值为 36.1%，黏土矿物含量为 29.2%～47.3%，平均值为 39.4%，碳酸盐矿物含量为 10.5%～25.7%，平均值为 17.2%，另外还有少量长石（2.5%～10.9%）及黄铁矿（0.7%～3.2%）。

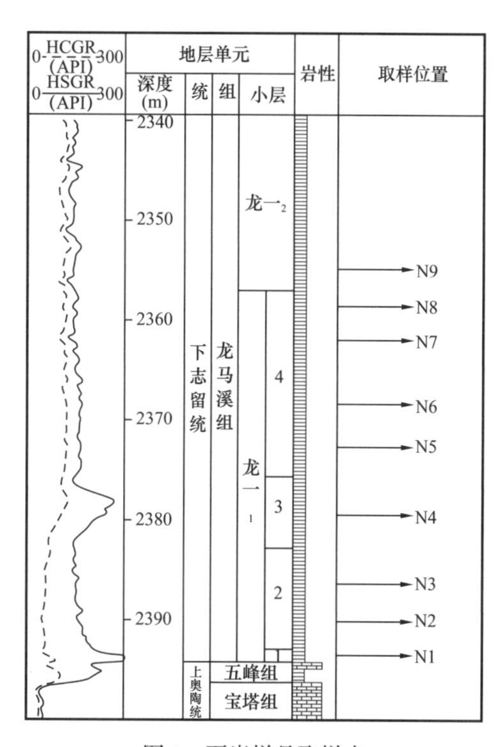

图 1　页岩样品取样点

1.2 扫描电镜实验

为了更加直观地观察页岩孔隙的类型和形貌特征，将页岩制成 5mm × 5mm × 3mm 的样块，待观察的表面使用 IM4000 离子抛光机进行氩离子抛光处理，并在表面镀上一层金膜，以增强岩石的导电性，从而形成清晰的图像，然后将 9 个样品用导电胶粘在样品台上进行观察。实验仪器为 FEI Quanta 200F 高分辨率场发射扫描电镜，实验在北京理化中心完成。

1.3 低压氩气吸附实验

二氧化碳、氮气及氩气是吸附材料领域应用较为广泛的 3 种探针介质，3 种气体分子在分子属性、实验条件、探测范围等方面存在一定的差别（表 2）。二氧化碳分子主要适用于孔径小于 1nm 的窄微孔的探测。氮气分子过去一直作为探测纳米级孔径（小于

100nm）的理想吸附质分子，但是近年来研究表明，其在探测微孔方面存在明显不足，主要是在极低相对压力段（10^{-7}），吸附等温线很难达到平衡[10]。另外，由于四极距作用的存在，氮气分子会和吸附剂表面的功能团发生相互作用，同样影响了探测结果的准确性。选取的氩气为单原子分子，不存在四极距作用，是理想的测量含微孔和介孔材料的吸附质。

表 1　页岩 TOC 及 XRD 分析结果

样品号	样品深度（m）	TOC（%）	矿物含量（%）					
			石英	长石	方解石	白云石	黄铁矿	黏土矿物
N1	2393.7	3.06	37.9	2.5	10.6	14.5	2.5	32.0
N2	2390.3	3.18	39.6	3.4	15.2	10.5	2.1	29.2
N3	2386.5	3.19	34.8	2.5	12.4	5.9	1.8	42.6
N4	2379.6	4.47	39.7	3.6	10.4	7.6	2.9	36.8
N5	2372.7	2.34	36.9	4.5	6.8	3.7	3.2	44.9
N6	2368.5	2.38	29.6	5.7	8.9	5.6	2.9	47.3
N7	2362.1	1.10	35.9	6.4	8.2	3.5	1.5	44.5
N8	2358.7	0.85	32.4	10.9	12.4	3.1	1.2	40.0
N9	2355.0	0.82	37.9	8.9	10.2	5.1	0.7	37.2

表 2　常用吸附质分子特征参数

吸附质	分子动力学直径（nm）	实验温度（℃）	优势探测范围（nm）	相对压力范围	四极距作用
CO_2	0.33	0	0.4～1	0～0.03	有
N_2	0.36	-196	2～100	0～1	有
Ar	0.34	-185.5	0.5～100	0～1	无

低压氩气吸附实验使用美国康塔公司生产的 Autosorb IQ 比表面和孔径分析仪。实验前将样品制成 60～80 目的粉样，并经过 8 小时 110℃ 的高温抽真空处理，以去除样品表面的杂质。以 99.99% 以上纯度的氩气为吸附质，实验温度 -185.5℃，相对压力范围为 4×10^{-7}～0.995，测量平衡蒸气压下页岩粉样的氩气吸附量和脱附量，并根据密度泛函原理（DFT）方法计算页岩比表面积、孔体积以及孔径分布[29, 30]。

1.4　高温高压甲烷等温吸附实验

高温高压甲烷等温吸附实验采用重量法，实验仪器为荷兰安米德 Rubotherm 高温高压吸附仪，最高测试压力为 35MPa，最高测试温度为 150℃，仪器采用循环油浴加热方式，温度的波动范围控制在 0.2℃以内，保证测量的准确性。仪器核心部件为高精度磁悬浮天平，精度可达 10μg。实验温度为 90℃，与页岩样品对应的地层温度相当。实验前将页岩样品制成60～80目的粉样，样品质量约为15g，经过8小时110℃的高温抽真空处理，以充分去除样品表面的水分以及杂质。

吸附实验过程包括空白实验、浮力实验及吸附实验3个部分[31]。空白实验获得样品框的质量和体积，浮力实验获得样品的质量和体积，吸附实验以纯度为99.99%的甲烷为吸附质，设定实验压力最高为30MPa，共设计10个压力测试点。磁悬浮天平的读数是样品框质量和体积、样品质量和体积、吸附甲烷质量和体积共同作用的结果[31, 32]。关系式为

$$\Delta m = m_{cs} + m_{abs} - \left(V_{cs} + V_{abs}\right)\rho_g \tag{1}$$

超临界状态下，吸附相体积 V_{abs} 是不断变化的，且不能忽略。根据过剩吸附量和绝对吸附量的含义，两者之间存在如下关系：

$$m_{ex} = m_{abs} - \rho_g V_{abs} \tag{2}$$

将式（2）代入式（1），可得

$$\Delta m = m_{cs} + m_{ex} - V_{cs}\rho_g \tag{3}$$

按照式（3）即可求出过剩吸附量 m_{ex} 的值。由此可见，与体积法一样，重量法也不能直接测得绝对吸附量的值，实验测得的吸附量一定是过剩吸附量，这是超临界条件下的必然结果[15]。所以在进行页岩吸附能力评价时，需要对实验数据进行修正，将过剩吸附量转化为绝对吸附量。

2 实验结果

2.1 微观孔隙结构镜下特征

根据氩离子抛光后的场发射扫描电镜观察结果（图2），对研究区龙马溪组页岩微观孔隙的形态特征及分布特征进行研究，并参考Loucks等[5]对页岩孔隙类型的分类，将研究区龙马溪组的纳米级孔隙分为有机质孔隙、粒内孔隙、粒间孔隙3类。

有机质孔隙是发育在有机质颗粒内部的孔隙，随着成岩演化的进行，固体干酪根向烃类流体转化，从而在干酪根内部形成大量次生孔隙[33]。研究区龙马溪组富有机质页岩中广泛发育有机质孔（图2），孔隙呈蜂窝状、圆形及椭圆形（图2a、b、d、h）。有机质孔隙大小差异较大，微观非均质性较强，有些孔隙孔径相对较大（图2b、h），有些孔隙的孔径则非常小，肉眼难以识别（图2c）。孔隙边缘较光滑，说明有机质孔隙受后期压实作用的影响较小，有机质孔隙形成于干酪根热裂解时期，此时固结成岩作用已经完成，加上有机质颗粒周边脆性矿物对孔隙起到了一定的支撑作用，使得有机质孔隙得以保存。有机质孔隙构成了研究区富有机质页岩主要的孔隙连通网络，为页岩气提供了大量的吸附和储存空间。

粒内孔隙在研究区富有机质页岩中主要呈现两种类型：一类是碳酸盐岩颗粒内部的次生溶蚀孔隙（图2d、f、g、i）。有机质生烃过程中形成的有机酸对方解石这类易溶矿物产生溶蚀作用，形成溶蚀孔隙。这类孔隙多呈椭圆形及不规则形等，孔隙之间的连通性较差；另一类是草莓状黄铁矿的晶间孔（图2a、d、h）。草莓状黄铁矿形成于缺氧的沉积

环境，直径一般在 3μm 左右，由许多小的黄铁矿晶体组合而成，晶间孔隙多呈不规则状，连通性较差，部分较大的孔隙被有机质充填，并在有机质内部形成有机质孔隙（图 2d）。

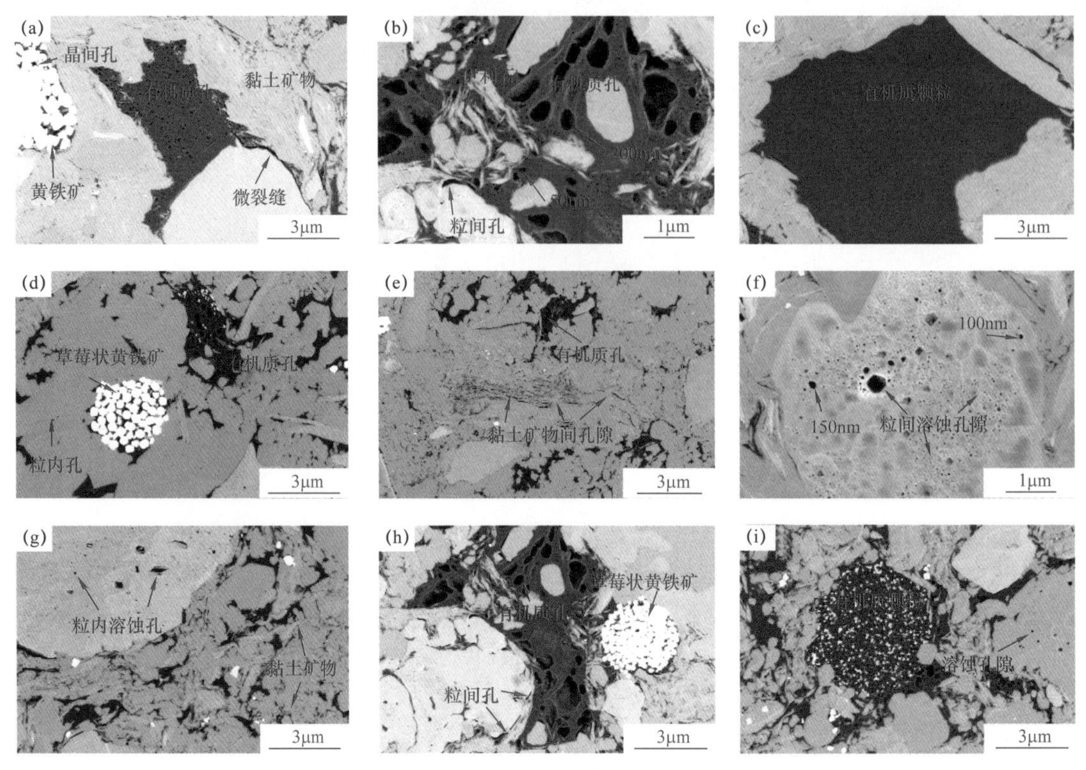

图 2　龙马溪组富有机质页岩 FE-SEM 镜下微观孔隙结构特征

（a）N1 号样，有机质孔，黄铁矿晶间孔，微裂隙；（b）N2 号样，有机质孔，粒间孔，视域内有机质孔隙大小差别较大，大孔隙呈蜂窝状；（c）N3 号样，有机质颗粒中发育孔径极小的孔隙；（d）N4 号样，黄铁矿晶间被有机质充填，并发育有机质孔，视域内发育碳酸盐颗粒内溶孔；（e）N8 号样，发育有机质孔及黏土矿物间狭缝状孔隙；（f）N9 号样，粒间溶蚀孔隙，非均质性较强，孔隙截面呈圆形；（g）N7 号样，不规则状粒内溶蚀孔隙，黏土矿物间不发育孔隙；（h）N4 号样，蜂窝状有机质孔，草莓状黄铁矿；（i）N5 号样，整颗草莓状黄铁矿被有机质充填，并发育有机质孔隙

粒间孔隙也主要呈两种类型：一类是脆性矿物边缘的粒间孔（图 2b、h），这类孔隙主要是由于脆性矿物和塑性矿物的差异压实造成的，脆性矿物的存在使得塑性矿物发生弯曲，也阻止了塑性矿物的进一步压实，从而在脆性矿物的边缘形成这类狭缝型或月牙形粒间孔，脆性矿物颗粒较大时则有可能形成微裂缝（图 2a），对页岩气的渗流起到促进作用；另一类是黏土矿物间的孔隙（图 2e），这类孔隙在页岩样品中较少见，通常情况下由于压实作用，黏土矿物间的孔隙消失殆尽（图 2a、g）。孔隙的形成与黏土矿物在成岩过程中的转化有关，研究区龙马溪组页岩中的黏土矿物主要是伊利石和伊 / 蒙混层，在蒙皂石向伊利石转化的过程中，矿物体积缩小，形成狭缝型孔隙。

2.2　低压氩气吸附曲线特征

9 个页岩样品的低压氩气吸附—脱附曲线如图 3 所示。可以看出，9 个页岩样品的等温吸附线具有很高的一致性，仅在吸附量上存在差别，整体上都呈反 S 形。根据经典的 BDDT 分类[34]，等温吸附线属于 II 型。在极低压力时（P/P_0<0.01）发生微孔充填，吸附

曲线初始段明显上升，该阶段可以用来表征微孔的分布，在极低压力段对压力点进行了加密测试，以期求准页岩纳米级孔隙中微孔的分布；紧接着是单层吸附，在微孔被充满之后，氩气分子开始覆盖整个页岩孔隙表面，对应于吸附等温线膝盖式弯曲的部分（$0.01 < P/P_0 < 0.05$）；单层吸附铺满后，多层吸附发生，吸附曲线进入相对平台区（$0.05 < P/P_0 < 0.4$），经典的 BET 理论就是在利用该阶段等温吸附线计算孔隙的比表面积[35]；当相对压力大于 0.4 时，发生毛细管凝聚作用，孔道中的吸附气体转化为液体，可以通过开尔文方程描述这一过程，通过该方程量化平衡压力与毛细管尺寸的关系，从而可以计算孔隙分布[36]。

由于毛细管凝聚作用的发生，通常会发生吸附曲线与脱附曲线不重合的情况，形成回滞环。回滞环的形态可以用来判断孔隙的形态[10]，但是通过扫描电镜的观察，页岩纳米级孔隙形态各异，是多种孔隙结构的混合，故通过吸附回滞环来判断页岩孔隙形态会带来误判；当相对压力接近 1 时，吸附曲线急剧上升，说明页岩中存在大孔，且孔径分布不均匀，这一点也在扫描电镜的观察中得到了证实。

2.3 超临界甲烷等温吸附曲线特征

9 个页岩样品的高温（90℃）高压（30MPa）甲烷等温吸附数据如图 4 所示。可以看出，所有等温吸附线都具有相似的形状，差别主要体现在甲烷吸附量上。所有的过剩吸附量在压力为 12MPa 左右时出现极大值，随后随着压力的增大，过剩吸附量开始降低。页岩气的主体成分为甲烷，甲烷的临界温度为 -82.54℃，临界压力为 4.599MPa，地层条件下甲烷气体为超临界状态。处于超临界状态的气体，其吸附等温线均表现出相同特征：在压力较低（吸附量低）时表现为 I 型等温线，当压力增大到一定程度时，吸附量出现极大值，然后随着压力的增大，吸附量降低[32, 37, 38]。

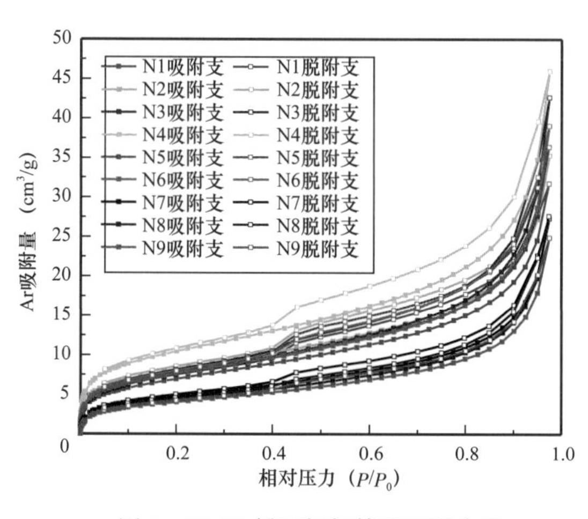

图 3　87.5K 低压氩气等温吸附曲线

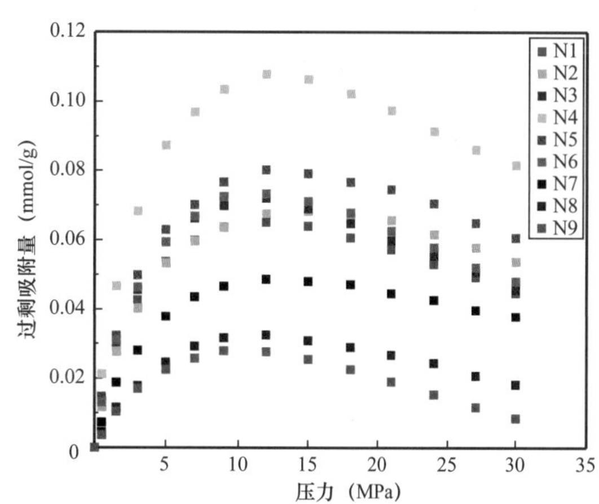

图 4　重力法高压甲烷等温吸附线测试结果

根据 Gibbs 对过剩吸附量的定义，过剩吸附量指的是吸附相中超过气相密度的过剩量[39]。绝对吸附量和过剩吸附量的关系［式（2）］可以改写为

$$m_{\mathrm{ex}} = m_{\mathrm{abs}}\left(1-\frac{\rho_{\mathrm{g}}}{\rho_{\mathrm{a}}}\right) \tag{4}$$

在亚临界条件下，自由气体密度 ρ_{g} 远小于吸附相密度 ρ_{a}，过剩吸附量与绝对吸附量

近似相等；然而在超临界条件下，随着压力的增大，自由气体密度增长很快，当自由气体的增长速度超过吸附相的增长速度时，过剩吸附等温线必然会出现极大值，随后实验测得的吸附量随着自由气体密度的继续增大而下降。前人的高温甲烷吸附实验中大多没有出现极大值的现象[13, 18, 23]，除了忽略了超临界条件下吸附相体积外，也与实验设备的最大测试压力有关[32]。在较低的压力条件下，自由气体密度很小，过剩吸附量与绝对吸附量相差不大，等温吸附线必然呈上升趋势。

3 讨论

3.1 基于低压氩气吸附实验的孔隙结构特征

由于页岩中存在微孔，且微孔孔壁之间的距离较介孔和宏孔更近，孔壁的 van der Waals 势发生重叠，吸附势强度变大，微孔的吸附能力远大于介孔和宏孔，而传统的 BET 方法计算的是介孔的比表面积，忽略了微孔的存在。对于含微孔的材料，IUPAC 推荐使用密度泛函原理（DFT）来计算孔径分布及孔隙结构参数[15]。使用 DFT 方法计算页岩孔隙结构参数，假设吸附等温线是由无数单一的单孔吸附等温线乘以孔径的分布范围得到的，使用 Autosorb IQ 设备自带的氩气在 -185.5℃下的模型，通过快速非负数最小二乘法解方程得到孔径分布曲线，并且给出吸附材料的孔体积和比表面积。

从孔径分布图（图 5）上可以看出，通过 DFT 方法可以计算得到页岩孔隙 0.5～100nm 范围内的孔径分布。甲烷的分子动力学直径为 0.38nm，0.5～100nm 的孔隙都能够作为甲烷气体的储存空间。9 个页岩样品具有相似的孔径分布特征，孔体积密度分布呈现多峰特征（图 5a），微孔主峰位于 0.8nm 和 1.8nm 两个位置，介孔主峰位于 5nm 的位置；样品比表面积则主要由 10nm 以下的孔径贡献（图 5b），且微孔（＜2nm）占据绝对优势。需要指出的是，dV（d）表示孔体积对孔径的微分，dS（d）表示比表面积对孔径的微分，这种计算方法有利于放大微孔在所有孔径中的分布特征，而缩小介孔及宏孔的分布特征。

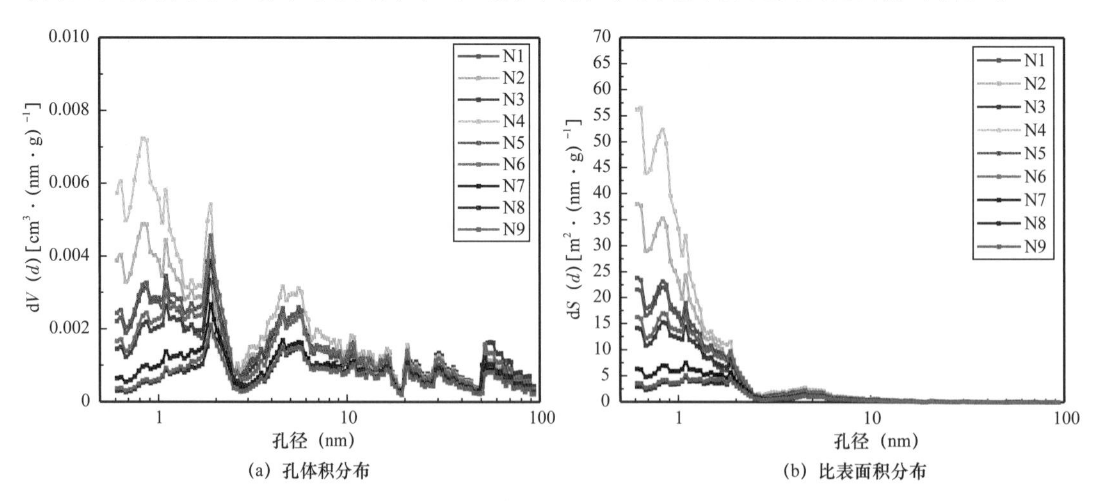

图 5 基于 DFT 方法的页岩孔径分布特征

根据 DFT 的解释结果（表 3），蜀南地区龙马溪组页岩样品总孔体积为 0.050～0.092cm^3/g，平均值为 0.072cm^3/g，总比表面积为 16.846～63.738m^2/g，平均值为 34.920m^2/g。

表 3　页岩样品孔隙结构参数

样品号	总孔体积（cm^3/g）	总比表面积（m^2/g）	微孔		介孔		宏孔	
			孔体积（cm^3/g）	比表面积（m^2/g）	孔体积（cm^3/g）	比表面积（m^2/g）	孔体积（cm^3/g）	比表面积（m^2/g）
N1	0.067	35.916	0.004	20.679	0.035	13.534	0.028	1.703
N2	0.073	47.773	0.006	31.712	0.038	14.280	0.029	1.781
N3	0.092	35.201	0.003	16.761	0.047	15.940	0.042	2.500
N4	0.091	63.738	0.007	43.962	0.049	17.691	0.035	2.085
N5	0.071	39.883	0.005	23.622	0.040	14.673	0.026	1.588
N6	0.082	36.343	0.004	19.260	0.044	14.982	0.035	2.101
N7	0.056	20.993	0.002	9.129	0.031	10.492	0.023	1.372
N8	0.061	17.588	0.001	5.691	0.033	10.283	0.027	1.614
N9	0.050	16.846	0.002	6.326	0.028	9.277	0.021	1.244

实验得到的孔体积和比表面积值明显大于前人对同一地区龙马溪组页岩运用低压氮气吸附实验得到的结果[3]。究其原因，主要是因为低压氮气吸附实验主要针对的是介孔的测量，尤其是在使用 BET 模型计算比表面积时，无法得到微孔对比表面积的贡献，而微孔对页岩吸附能力的贡献不可忽略。

将 9 个页岩样品的孔体积和比表面按照微孔、介孔及宏孔进行分类统计（图 6，表 3）。可以发现，介孔和宏孔贡献了龙马溪组页岩主要的孔体积，两者孔体积占样品总孔体积的 91.9%～97.7%，平均值为 94.9%，微孔对页岩孔体积的贡献在 10% 以内；而比表面积统计表明，龙马溪组页岩主体的比表面积由微孔和介孔贡献，其中，微孔比表面积为 5.69～43.962m^2/g，平均值为 19.682m^2/g，微孔贡献了 32.4%～69.0% 的比表面积，介孔比表面积为 9.277～17.691m^2/g，平均为 13.461m^2/g，介孔贡献了 27.8%～58.5% 的比表面积。对于 TOC 大于 2% 以上的样品（N1～N6），微孔比表面积的贡献均在 45% 以上，尤其是 TOC 最高的 N4 号样品，微孔比表面积贡献了 69% 的比表面。

3.2　超临界吸附模型

Langmuir 方程是应用最为广泛的描述甲烷吸附的等温吸附模型[40]，其假设吸附剂表面均匀，分子之间没有相互作用。然而这种理想状态并不存在，且从扫描电镜和低压吸附实验可知，页岩孔隙结构非均质性较强。选用考虑吸附剂表面非均质性及分子间作用力等复杂因素的 Langmuir–Freundlich（L–F）模型[38]，其公式如下：

$$m_{\rm abs}=\frac{m_{\rm o}kP^{m}}{1+kP^{m}} \tag{5}$$

式中 m 为校正系数，当一个吸附位只吸附一个吸附质分子时，m=1，此时的方程就是传统的 Langmuir 方程。结合过剩吸附的定义［式（4）］与 L–F 模型，得到超临界状态下过剩吸附的拟合模型：

$$m_{ex}=\frac{m_o kP^m}{1+kP^m}\left(1-\frac{\rho_g}{\rho_a}\right) \tag{6}$$

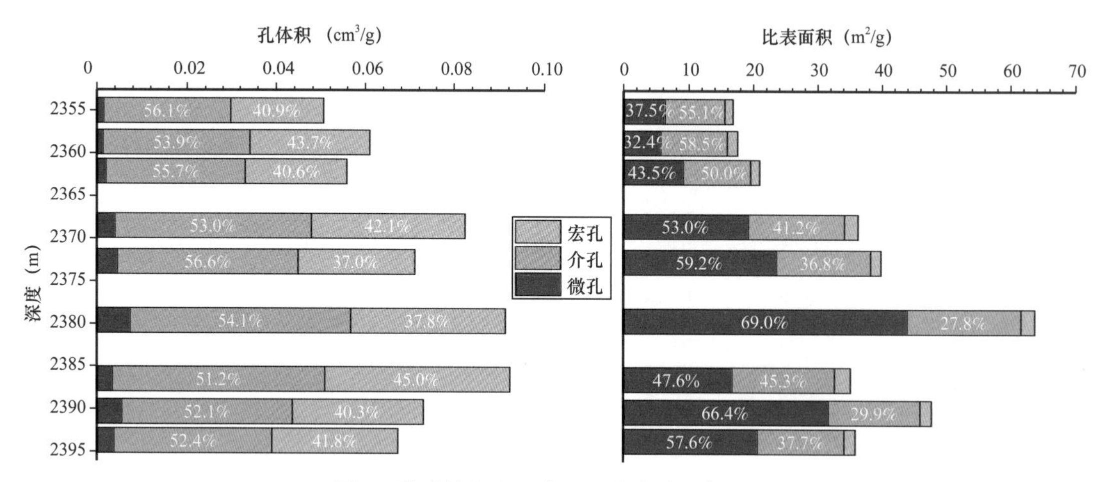

图 6　孔隙体积和比表面积分布堆积条形图

甲烷气体 90℃温度下的密度使用 SRK 状态方程求得，由于气体状态方程较为复杂，不利于后期方程的拟合，使用状态方程求得的甲烷气体密度值回归成与压力相关的多项式函数：

$$\rho_g=a_0+a_1P+a_2P^2 \tag{7}$$

a_0，a_1，a_2 值见图 7。

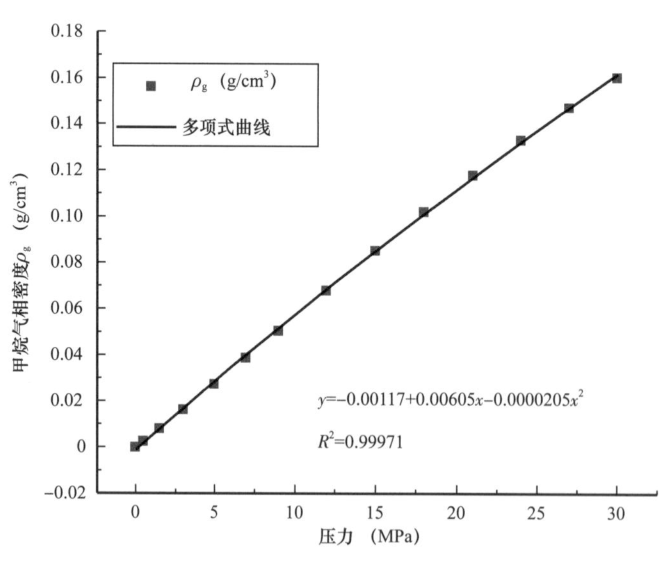

图 7　90℃甲烷气相密度对压力的回归曲线

超临界状态下甲烷吸附相密度的确定是将过剩吸附量转化为绝对吸附量的关键，然而吸附相的密度不能通过实验直接测得。目前确定吸附相密度的方法主要有经验取值法[23]、等温线斜率法[41]及拟合法[38, 41]，为了提高模型拟合结果的准确性，将吸附相密度作为未知参数，并将实验得到的过剩吸附数据通过 Langmuir 模型和修正的 L–F 模型分别进行拟合，对比其结果的差异性（表 4）。结果表明，由 Langmuir 模型拟合得到的吸附相密度介于 0.1873～0.3423g/cm^3，L–F 模型拟合得到的吸附相密度介于 0.1890～0.3081g/cm^3，所有值都处于甲烷临界密度（0.162g/cm^3）与常压沸点液体甲烷密度（0.423g/cm^3）之间，说明两个模型都较合理，但是对比相关系数，L–F 模型的相关系数都比 Langmuir 模型高，说明 L–F 模型的拟合效果优于 Langmuir 模型。因此，选用修正的 L–F 模型拟合所有过剩吸附数据。

表 4　Langmuir 模型和 L–F 模型拟合结果对比表

样品编号	Langmuir 模型				L–F 模型				
	最大吸附量（mmol/g）	Langmuir 常数	吸附相密度（g/cm^3）	相关系数	最大吸附量（mmol/g）	Langmuir 常数	吸附相密度（g/cm^3）	修正系数	相关系数
N1	0.1161	0.2189	0.2919	0.9981	0.1415	0.1876	0.2713	0.8534	0.9996
N2	0.1232	0.1805	0.3423	0.9976	0.1740	0.1356	0.2960	0.8120	0.9995
N3	0.1333	0.2032	0.2677	0.9979	0.1492	0.1878	0.2581	0.9087	0.9985
N4	0.1843	0.2226	0.3272	0.9982	0.2202	0.1935	0.3007	0.8667	0.9993
N5	0.1437	0.1918	0.3239	0.9981	0.1847	0.1571	0.2916	0.8421	0.9995
N6	0.1413	0.1848	0.2670	0.9984	0.1566	0.1723	0.2585	0.9213	0.9988
N7	0.0928	0.1648	0.3129	0.9990	0.0960	0.1613	0.3081	0.9750	0.9991
N8	0.0783	0.1156	0.2283	0.9971	0.0760	0.1174	0.2295	1.0204	0.9976
N9	0.0733	0.1160	0.1873	0.9977	0.0670	0.1204	0.1890	1.0743	0.9981

用 L–F 模型拟合的结果见图 8 中蓝色曲线。表示吸附能力的最大吸附量 m_0 介于 0.0670～0.2202mmol/g 之间，平均值为 0.1406mmol/g。在拟合得到参数 m_0、k 及 m 后，将参数值代入到公式（5）中求取页岩不同压力条件下的绝对吸附量数据，结果如图 8 中红色曲线。从过剩吸附等温线和绝对吸附等温线的对比可以看出，在气体压力较小时（P＜5MPa），过剩吸附量和绝对吸附量相当，而随着压力的增大，绝对吸附量总是大于过剩吸附量，且绝对吸附量与过剩吸附量的差距逐渐增大。故在地层压力和温度条件下，如果不对高温甲烷吸附曲线进行校正，将低估地层实际的甲烷吸附能力，同时对地质储量的评价产生影响。

3.3　孔隙结构特征对甲烷吸附能力的影响

页岩吸附性能除了与温度、压力、湿度等外在因素有关外，还与页岩本身的孔隙结构特征密切相关[13, 22, 42, 43]。从扫描电镜的观察可知，研究区龙马溪组页岩中发育大量的有机质孔隙（图 2），连通性好，构成了页岩主要的孔隙网络，而有机质颗粒本身有较强的吸附性能，故有机质孔表面是页岩气主要的吸附场所。从 TOC 含量、页岩孔隙比表面积及最大甲烷吸附量（m_0）的相互关系可以看出，3 个参数之间互为正相关关系（图 9）。随着页岩中有机质含量的增大，发育在有机质颗粒中的有机质孔隙增多，因此能够为甲烷吸附提供位置的页岩比表面积增大，从而使得页岩的吸附能力增强。与此同时，随着页岩中TOC含量的增多，页岩中微孔提供的比表面积占比增多，而介孔和宏孔提供的比表面积占比减少（图10）。根据上文所述，微孔由于孔壁之间的距离更近，其能够提供的吸附势能远大于介孔和宏孔，所以TOC含量的增大有利于页岩中微孔的发育，从而使得页岩的吸附能力增强。

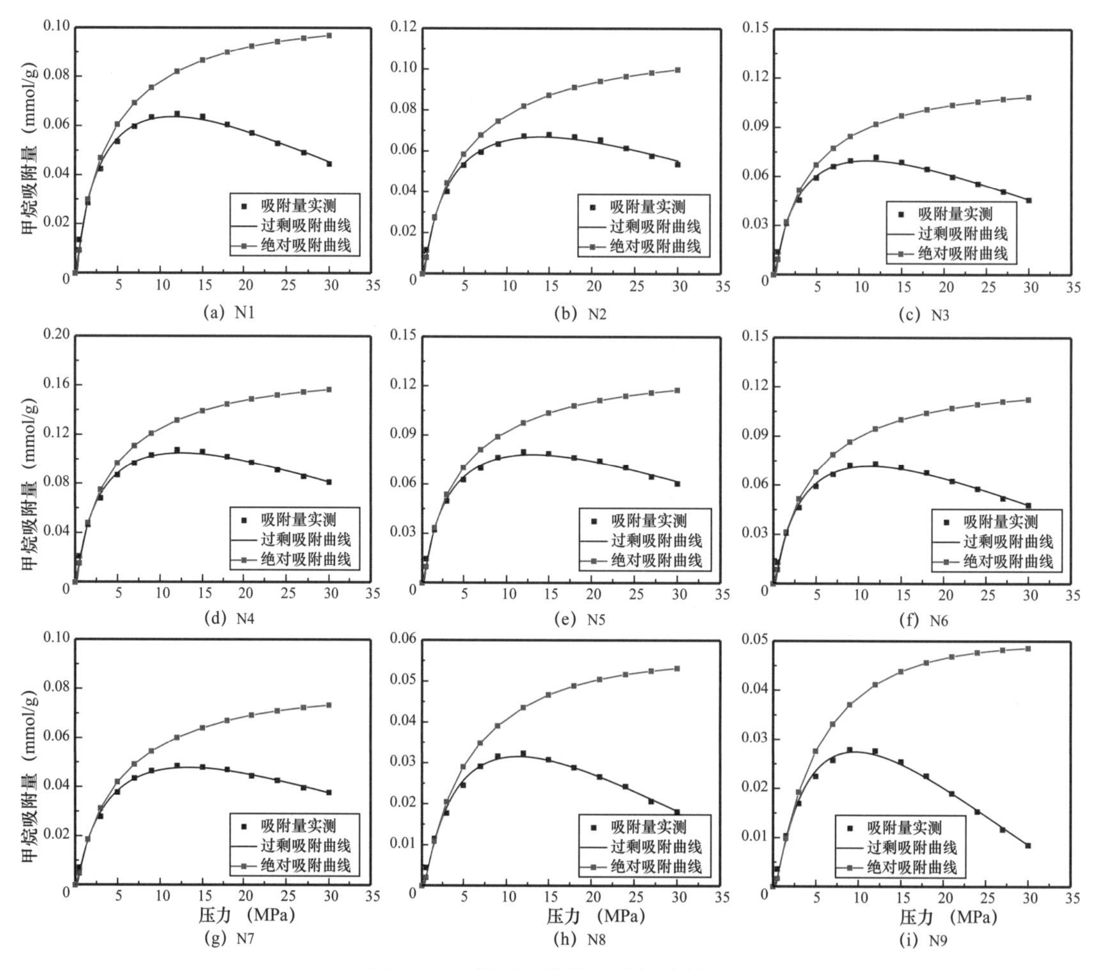

图8 L-F模型吸附等温线拟合结果

页岩中除了有机质外，黏土矿物也能提供一定的比表面积[44]。研究表明，各类黏土矿物中蒙皂石的比表面积最大，其甲烷吸附能力也最强[45]，然而在页岩成岩过程中，大量的蒙皂石已经转化为伊利石，其比表面积大大缩小。从扫描电镜图像上可以看出，研究区龙马溪组页岩中黏土矿物多致密，不发育纳米级孔隙。为了讨论黏土矿物对研究区龙马溪组页岩甲烷吸附能力的影响，将TOC对吸附能力的影响进行归一化处理，然后对黏土矿物含量做交会图（图11），可以看出，归一化之后的页岩最大吸附量与黏土矿物含量之间相关性较差，说明富有机质海相页岩中黏土矿物对页岩甲烷吸附能力的贡献有限。前人研究发现，黏土矿物含量影响陆相页岩的甲烷吸附能力[18]。由于陆相页岩中有机质孔隙有限，多发育无机孔隙，黏土矿物本身具有一定的吸附性能，故能够影响页岩的吸附能力。

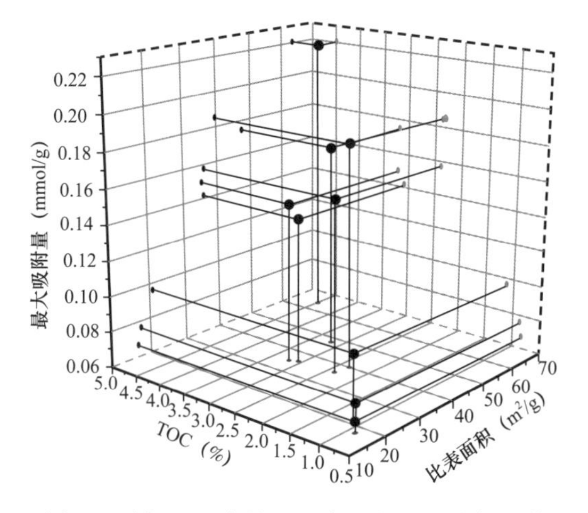

图 9　页岩 TOC 含量、比表面积以及最大吸附量之间的关系

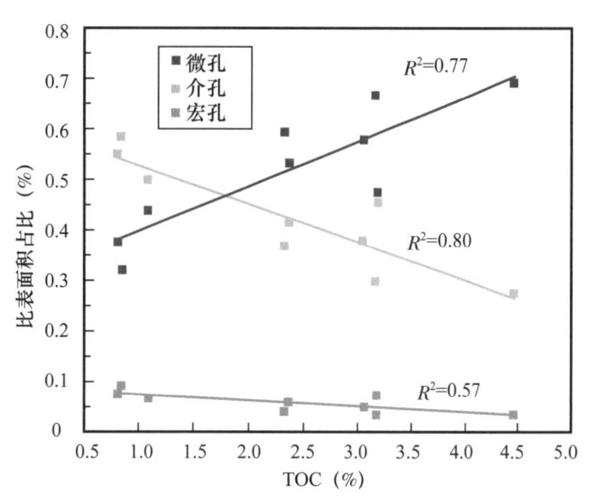

图 10　TOC 与各类孔隙比表面积占比的关系

4　结论

（1）场发射扫描电镜观察表明，蜀南地区龙马溪组下部富有机质页岩发育多种类型纳米级孔隙，以有机质孔为主，孔隙大小不同，形态各异，孔隙结构非均质性强。

（2）低压氮气吸附实验结果表明，研究区龙马溪组页岩样品比表面积为 16.846～63.738m²/g，孔体积为 0.050～0.092cm³/g，微孔和介孔贡献页岩 90% 以上的比表面积，是页岩气主要的吸附场所，介孔和宏孔贡献页岩 90% 以上的孔体积，是页岩气主要的储存场所。

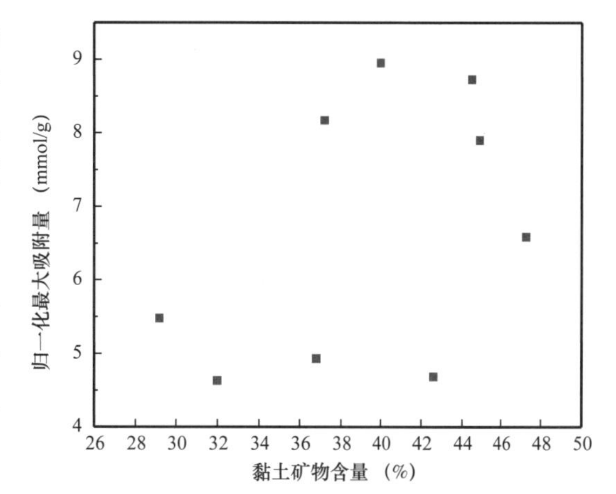

图 11　黏土矿物含量与 TOC 归一化最大吸附量交会图

（3）重力法高温高压甲烷吸附实验表明，超临界状态下过剩吸附曲线在压力为 12MPa 左右时出现极大值，随后开始下降，这是超临界状态下实验测等温吸附线的必然结果。使用修正过的四元 Langmuir-Freundlich（L-F）模型能够较好地拟合过剩吸附曲线，相关系数大于 0.997。

（4）由四元 L-F 模型得到的吸附性能参数 m_0 为 0.0670～0.2202 mmol/g，不同页岩样品甲烷吸附能力存在显著差异。TOC 含量为影响蜀南地区龙马溪组海相页岩甲烷吸附能力的主要因素，随着 TOC 含量的增大，有机质孔隙增多，微孔占比上升，页岩吸附能力增强。研究区富有机质页岩中的黏土矿物对页岩的吸附能力贡献有限。

符号注释：

Δm—天平读数，g；m_{cs}—样品框和样品的总质量，g；m_{abs}—吸附气体的绝对吸附量，g；V_{cs}—样品框和样品的总体积，cm^3；V_{abs}—吸附相体积，cm^3；ρ_g—不同压力点对应的甲烷气体密度，g/cm^3；m_{ex}—吸附气体的过剩吸附量，g；ρ_a—吸附相密度，g/cm^3；m_o—最大吸附量，mmol/g；k—Langmuir 常数，反映了吸附速率与脱附速率的比值；m—校正系数；P—测试压力，MPa；a_0，a_1，a_2—多项式的回归系数。

参考文献

［1］Curtis J B. Fractured Shale-gas Systems［J］. AAPG Bulletin，2002，86（11）：1921-1938.

［2］贾爱林，位云生，金亦秋. 中国海相页岩气开发评价关键技术进展［J］. 石油勘探与开发，2016，43（6）：1-7.

［3］陈尚斌，朱炎铭，王红岩，等. 川南龙马溪组页岩气储层纳米孔隙结构特征及其成藏意义［J］. 煤炭学报，2012，37（3）：438-444.

［4］李贤庆，王哲，郭曼，等. 黔北地区下古生界页岩气储层孔隙结构特征［J］. 中国矿业大学学报，2016，45（6）：1172-1183.

［5］Loucks R G，Reed R M，Ruppel S C，et al. Morphology，Genesis，and Distribution of Nanometer-Scale Pores in Siliceous Mudstones of the Mississippian Barnett Shale［J］. Journal of Sedimentary Research，2009，79（12）：848-861.

［6］Mastalerz M，Schimmelmann A，Drobniak A，et al. Porosity of Devonian and Mississippian New Albany Shale across a maturation gradient：Insights from organic petrology，gas adsorption，and mercury intrusion［J］. AAPG Bulletin，2012，97（10）：1621-1643.

［7］田华，张水昌，柳少波，等. 压汞法和气体吸附法研究富有机质页岩孔隙特征［J］. 石油学报，2012，33（3）：419-427.

［8］Clarkson C R，Solano N，Bustin R M，et al. Pore structure characterization of North American shale gas reservoirs using USANS/SANS，gas adsorption，and mercury intrusion［J］. Fuel，2013，103：606-616.

［9］武瑾，梁峰，吝文，等. 渝东北地区龙马溪组页岩储层微观孔隙结构特征［J］. 成都理工大学学报（自然科学版），2016，43（3）：308-319.

［10］Thommes M，Kaneko K，Neimark A V，et al. Physisorption of gases，with special reference to the evaluation of surface area and pore size distribution（IUPAC Technical Report）［J］. Pure Applied Chemistry，2015，87（9）：1-19.

［11］Piter B，Kevin S，Helge S，et al. On the Use and Abuse of N2 Physisorption for the Characterization of the Pore Structure of Shales［J］. The Clay Minerals Society Workshop Lectures Series，2016，21：151-161.

［12］Chalmers G R L，Bustin R M. The Organic Matter Distribution and Methane Capacity of the Lower Cretaceous Strata of Northeastern British Columbia，Canada［J］. International Journal of Coal Geology，2007，70（3）：223-239.

［13］Ross D J K，Bustin R M. The Importance of Shale Composition and Pore Structure upon Gas Storage

Potential of Shale Gas Reservoirs [J] . Marine and Petroleum Geology，2009，26（6）：916–927.

[14] Wang Y，Zhu Y M，Liu S M，et al. Pore Characterization and Its Impact on Methane Adsorption Capacity for Organic-rich Marine Shales [J] . Fuel，2016，181：227–237.

[15] Gasparik M，Ghanizadeh A，Bertier P，et al. High-pressure Methane Sorption Isotherms of Black Shales from the Netherlands [J] . Energy Fuel，2012，26（8）：4995–5004.

[16] Ambrose R J，Hartman R C，Diaz-Campos M，et al. Shale Gas-in-Place Calculations Part I：New Pore-scale Considerations [J] . SPE Journal，2012，17（1）：219–229.

[17] Zhang T W，Ellis G S，Ruppel S C，et al. Effect of Organic-matter Type and Thermal Maturity on Methane Adsorption in Shale-gas Systems [J] . Organic Geochemistry，2012，47（6）：120–131.

[18] 侯宇光，何生，易积正，等．页岩孔隙结构对甲烷吸附能力的影响 [J] . 石油勘探与开发，2014，41（2）：248–256.

[19] Chalmers G R L，Bustin R M. Lower Cretaceous Gas Shales in Northeastern British Columbia，Part I：Geological Controls on Methane SorptionCapacity [J] . Bulletin Canada Petroleum Geological，2008，56：1–21.

[20] Hao F，Zou H，Lu Y. Mechanisms of Shale Gas Storage：Implications for Shale Gas Exploration in China [J] . AAPG Bulletin，2013，97（8）：1325–1346.

[21] 李笑天，潘仁芳，鄢杰，等，四川盆地长宁—威远页岩气示范区下志留统龙马溪组泥页岩吸附特征及其影响因素分析 [J] . 海相油气地质，2016，21（4）：60–66.

[22] Gasparik M，Bertier P，Gensterblum Y，et al. Geological Controls on the Methane Storage Capacity in Organic-rich Shales [J] . International Journal of Coal Geology，2014，123：34–51.

[23] 刘圣鑫，钟建华，马寅生，等．页岩中气体的超临界等温吸附研究 [J] . 煤田地质与勘探，2015，43（3）：45–50.

[24] 侯晓伟，王猛，刘宇，等．页岩气超临界状态吸附模型及其地质意义 [J]，中国矿业大学学报，2016，45（1）：111–118.

[25] Do D D，Do H D. Adsorption of Supercritical Fluids in Non-porous and Porous Carbons：Analysis of Adsorbed Phase Volume and Density [J] . Carbon，2003，41：1777–1791.

[26] Grimm R P，Eriksson K A，Ripepi N，et al. Seal Evaluation and Confinement Screening Criteria for Beneficial Carbon Dioxide Storage with Enhanced Coal Bed Methane Recovery in the Pocahontas Basin，Virginia [J] . International of Coal Geology，2012，90–91：110–125.

[27] 张晓明，石万忠，舒志国，等．涪陵地区页岩含气量计算模型及应用 [J] . 地球科学,2017,42（7）：1157–1168.

[28] Zhou S W，Xue H Q，Ning Y. Experimental Study of Supercritical Methane Adsorption in Longmaxi Shale：Insights into the Density of Adsorbed Methane [J] . Fuel，2018，211：140–148.

[29] Lowell S，Shields J，Thomas M A，et al. Characterization of Porous Solids and Powders：Surface area，Pore size and Density [M] . Netherlands：Springer 2006.

[30] Zhang Z，Yang Z. Theoretical and Practical Discussion of Measurement Accuracy for Physisorption with Micro-and Mesoporous Materials [J] . Chinese Journal of Catalysis，2013，34（10）：1797–1810.

[31] 俞凌杰，范明，陈红宇，等．富有机质页岩高温高压重力法等温吸附实验 [J] . 石油学报，2015，

36（5）：557–563.
［32］周尚文，王红岩，薛华庆，等．基于 One-Kondo 格子模型的页岩气超临界吸附机理探讨［J］．地球科学，2017，42（8）：1421–1430.
［33］何建华，丁文龙，付景龙．页岩微观孔隙成因类型研究［J］．岩性油气藏，2014，26（5）：30–35.
［34］Brunauer S，Deming L S，Deming W E，et al. On a Theory of the Van Der Waals Adsorption of Gases［J］. Journal of American Chemistry Society，1940，62（7）：1723–1739.
［35］Brunauer S，Emmett P H，Teller E. Adsoprtion of Gases in Multimolecular Layers［J］. Journal of American Chemical Society，1938，60（2）：309–319.
［36］Barrett E P，Joyner L G，Halenda P P. The Determination of Pore Volume and Area Distributions in Porous Substances. I. Computations from Nitrogen Isotherms［J］. Journal of the American Chemical Society，1951，73（1）：372–380.
［37］Menon P G. Adsorption of Carbon Monoxide on Alumina at High Pressures［J］. Journal of America Chemical Society，1965，87：3057–3060.
［38］胡涛，马正飞，姚虎卿．甲烷超临界高压吸附等温线研究［J］．天然气化工，2002，27：36–40.
［39］Zhou L，Zhou Y，Li M，et al. Experimental and Modeling Study of the Adsorption of Supercritical Methane on a High Surface Activated Carbon［J］. Langmuir，2000，16（14）：5955–5959.
［40］Langmuir I. The Adsorption of Gases on Plane Surfaces of Glass，Mica and Platinum［J］. Journal of American Chemical Society，1918，40（9）：1403–1460.
［41］周尚文，王红岩，薛华庆，等．页岩过剩吸附量与绝对吸附量的差异及页岩气储量计算新方法［J］．天然气工业，2016，36（11）：12–20.
［42］Tan J J，Weniger P，Krooss B，et al. Shale Gas Potential of the Major Shale Formations in the Upper Yangtze Platform，South China，Part II：Methane sorption capacity［J］. Fuel，2014，129：204–218.
［43］张烈辉，唐洪明，陈果，等．川南下志留统龙马溪组页岩吸附特征及控制因素［J］．天然气工业，2014，34（12）：63–69.
［44］Ji L M，Zhang T，Milliken K，et al. Experimental Investigation of Main Controls to Methane Adsorption in Clay-rich Rocks［J］. Applied Geochemistry，2012，27（12）：2533–2545.
［45］吉利明，邱军利，夏燕青，等．常见黏土矿物电镜扫描微孔隙特征与甲烷吸附性［J］．石油学报，2012，33（2）：249–256.

页岩岩相表征及页理缝三维离散网络模型

欧成华[1, 2]，李朝纯[1]

（1. 西南石油大学；
2. 油气藏地质及开发工程国家重点实验室）

摘　要：以四川盆地焦石坝地区上奥陶统五峰组—下志留统龙马溪组页岩气田为例，开展页岩岩相表征及页理缝三维离散网络模型研究。页理缝的发育受控于页岩的岩相类型，不同岩相内页理缝发育强度存在显著差异，据此，研制出基于岩相表征的页理缝三维离散网络模型建模方法。该方法通过页岩储层岩相分析和页理缝描述，建立岩相与页理缝特征模式，依靠页岩岩相三维模型，建立页理缝发育指数三维模型和页理缝发育强度三维模型，最终建立页理缝三维离散网络模型。通过四川盆地焦石坝地区五峰—龙马溪组页岩气田储层页理缝离散三维模型的建立，不仅在三维空间充分展示了气田主力产层内页理缝的分布位置、发育规模及每条页理缝的倾角、方位角的信息，为后续的生产模拟提供了页理缝地质参数场，也为类似页岩气田页理缝建模提供了借鉴。

关键词：岩相表征；页理缝发育指数；页理缝发育强度；页理缝三维离散网络模型；四川盆地；上奥陶统五峰组；下志留统龙马溪组

页岩储层孔喉为纳米级，孔渗性极差[1]，没有页理缝的沟通难以大规模产出石油和天然气。研究者通过野外露头观察[2, 3]及各类室内实验[4, 5]，证实了页岩储层中页理缝的客观存在。当页岩储层中压力较大，如达到超高孔隙压力时，就会形成大量页理缝[6–9]，成为页岩油气渗流的有效通道，使得致密的页岩储层也能成为有效产层[1, 10–12]，甚至还能出现较高的油气产能[13–17]，因此，页理缝对于页岩油气的开采意义重大。目前有关页理缝的认识仅局限于岩心观察或各类电子显微镜观测[4, 7–9]，仅能描述页理缝的密度和基本产状，无法实现在三维空间对页理缝进行定量刻画，难以为页岩油气生产提供定量的页理缝三维离散网络模型。针对上述问题，提出基于页岩岩相表征的页理缝三维离散网络建模新方法，并以四川盆地东部焦石坝地区上奥陶统五峰组—下志留统龙马溪组页岩气田为例，采用该方法建立了页理缝三维离散网络模型，实现了对页理缝分布位置、发育规模及延伸方位的三维可视化。

1　研究思路

富油气页岩都是在静水条件下、还原环境中形成的[18]。极其缓慢的沉积速度[19, 20]，加上受沉积时水体能量、温度、盐度及含有物等周期性变化的影响[21]，使得这些富油气页岩中相互平行的纹层状页理极其发育[4, 17–21]，从而将页岩与砂岩、碳酸盐岩、煤系等其他油气储层显著区别开来。

众所周知，页岩通过成岩压实会大量排出原生地层水[22]，由细小颗粒支撑形成的页岩孔隙半径极其细小[1, 11, 12]，地下水在其中难以自由通过，因而在页理面间很难获得由地层水溶解带来的胶结物质。缺少胶结的页理纹层面成为脆弱面，一旦出现构造应力集中[23-26]或页岩中生成的大量烃类物质体积膨胀[6-9]，极易诱发形成页理纹层面间的裂缝，即页理缝。

由此可见，页理是页理缝形成的物质基础，页理的形成与页岩沉积环境及在该环境下发育的沉积物息息相关[18, 20, 21]，而页岩岩相的概念同时涵盖了页岩沉积期古环境及该环境下发育的沉积物两种属性，因此，页理缝与页岩岩相间具有内在联系。页岩岩相分析是页岩储层表征的常用手段[27-29]，研究者先后发现不同页岩岩相中页理缝的发育程度明显不同[6-17, 30]，笔者在研究区五峰—龙马溪组页岩共识别出 8 种页岩岩相，每种岩相的页理缝发育强度具有显著差异。

因此，通过页岩岩相分析，观察不同类型页岩岩相页理缝发育的数量和产状特征，能建立页岩岩相与页理缝发育空间范围，以及发育强度之间的关系，利用这种关系，即可通过建立的页岩岩相三维定量地质模型，得出页理缝发育指数和发育强度三维模型，从而达到依靠页岩岩相来约束页理缝建模的目的。

2　基于岩相表征的页理缝三维离散网络模型建立方法

首先，开展页岩储层岩相分析和页理缝描述，建立页岩储层岩相与页理缝特征关系；其次，利用离散型变量储层建模方法，建立页岩岩相三维模型，利用页岩岩相与页理缝之间关系，建立页理缝发育指数三维模型，圈定页理缝建模的空间范围；然后，依靠页岩岩相三维模型的空间约束，利用页理缝特征模型，建立页理缝发育强度三维模型，揭示页理缝发育强度的空间分布；最后，以页理缝发育指数模型为空间约束，利用页理缝发育强度三维模型，建立页理缝三维离散网络模型（图 1）。

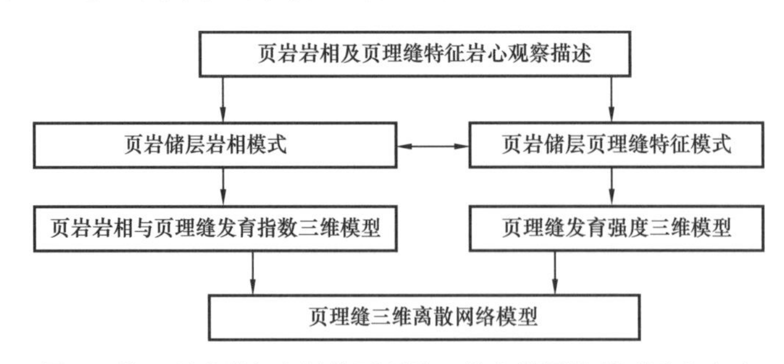

图 1　基于页岩岩相表征的页理缝三维离散网络模型建立方法

2.1　页岩储层岩相与页理缝特征模式的提取

首先开展取心井岩心观察描述，详细记录每块岩心的岩性、颜色、组成等特征，精确测量页理缝的长度、宽度、密度、倾角、方位角等参数，构建页岩储层页理缝特征模式。采用岩相古地理方法，分析页岩岩相类型及特征模式，统计每种页岩岩相页理缝发育强度，揭示页理缝发育与页岩岩相类型及其纹层结构间的内在关系。筛选出页理缝易于发育的页岩岩相类型，作为建立页理缝发育指数三维模型的基本依据。

2.2 页岩岩相与页理缝发育指数三维模型的建立

依据研究获得的页岩储集层岩相模式，开展单井页岩岩相分析，识别出每口井在目的层深度范围内各个深度位置处的页岩岩相类型，建立单井页岩岩相变化剖面。编码不同的页岩岩相类型，构成单井剖面岩相数据，输入商业化建模软件，采用序贯指示模拟算法[31-34]，即可建立起研究区内的页岩岩相三维模型，从而在三维空间定量表征研究区内不同页岩岩相的分布特征。

依靠建立的页岩岩相三维模型，将页理缝发育的页岩岩相定义为 1，其余岩相定义为 0，建立起页理缝发育指数模型 F，该模型中页理缝只发育在值为 1 的网格内。

2.3 页理缝发育强度三维模型的建立

对比分析依靠岩心描述构建的页岩储层页理缝特征模式与测井曲线特征，开展沿井深剖面的页理缝宽度与密度的定量描述。考虑到页理缝宽度和密度都对流体渗流具控制作用，将沿井深剖面每个深度位置处的页理缝宽度和密度的乘积作为页理缝强度属性，构建单井页理缝发育强度数据集 E。将 E 输入到商业化储层建模软件中，以页理缝发育指数模型 F 为约束条件，采用序贯高斯模拟算法[31-34]，即可建立页理缝发育强度三维模型，在三维空间表征页理缝发育强度。

一般而言，构造应力越强，诱导产生的页理缝就越发育[23-26]。反之，页理缝发育强度值越大，页理缝越发育，构造应力也就越强。

2.4 页理缝三维离散网络模型的建立

以页理缝发育指数三维模型为页理缝建模的三维空间约束，以建立的页理缝发育强度三维模型为主输入，页理缝的倾角、方位角为辅助输入，利用储层建模商业软件[31-34]的裂缝离散网络建模模块即可建立页理缝三维离散网络模型，从而在三维空间再现页岩储层页理缝的分布及其属性特征。

3 焦石坝地区五峰组—龙马溪组页岩页理缝三维离散网络模型的建立与应用

3.1 基础地质研究

前人研究表明[15]，位于涪陵地区的焦石坝页岩气田（以下简称实例页岩气田）在构造上处于万县复向斜与斗山背斜带之间（图 2），其五峰组—龙马溪组页岩为深水—半深水陆棚沉积环境。依靠单井岩心观察及室内分析化验，共识别出 4 类沉积微相[28]，分别是硅质陆棚微相、砂质陆棚微相、混积陆棚微相和泥质陆棚微相，各沉积微相类型及特征如表 1 所示。实例页岩气田发育从下到上，水深由深到浅的沉积序列（表 1，图 3），最下部为深水强还原沉积环境的硅质陆棚沉积，是气田当前实际投入开发的主力产层，往上均为半深水弱还原沉积环境，依次为砂质陆棚、混积陆棚和泥质陆棚沉积，成为气田的接替产层、次接替产层和难采储层[35-38]。

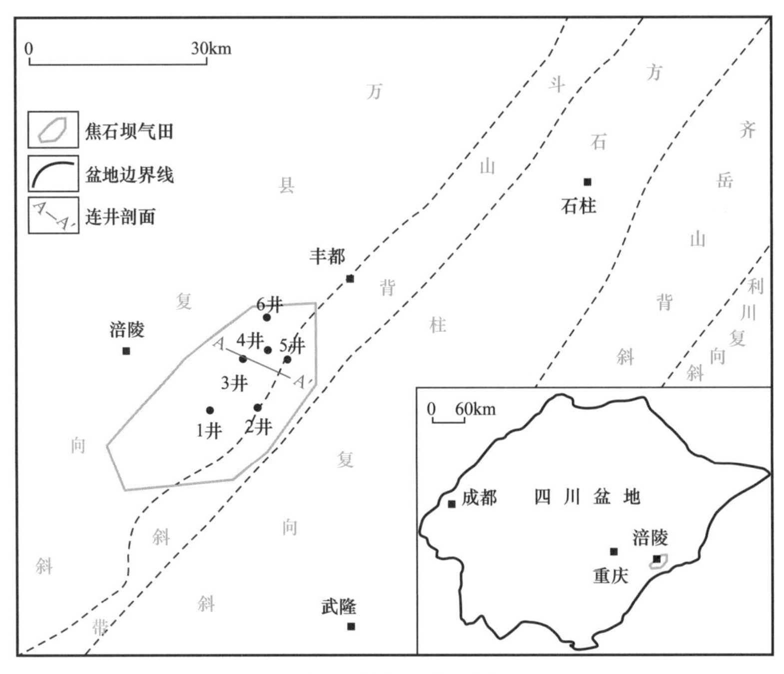

图 2 研究区位置图

表 1 实例页岩气田沉积微相类型、特征及其对生产的作用

沉积相类型	沉积环境	岩性特征	含有物特征	位置及作用
泥质陆棚	半深水、弱还原环境	泥质页岩，少量含粉砂页岩	笔石化石较丰富，黄铁矿含量少，TOC 值较低	上部，难采储层
混积陆棚	半深水、弱还原环境	碳质、粉砂质页岩，灰云质页岩	笔石化石较丰富，黄铁矿含量少，TOC 值低—中等	中上部，次接替产层
砂质陆棚	半深水、弱还原环境	碳质、粉砂页岩，碳质、粉砂质页岩	笔石化石丰富，黄铁矿含量较少，TOC 值中等—高，局部 TOC 值较低	中下部，接替产层
硅质陆棚	深水、强还原环境	灰黑色、黑色碳质、硅质页岩	黄铁矿发育，笔石化石极发育，TOC 值高—极高	下部，主力产层

在沉积环境与沉积相研究的基础上，前人进一步开展了岩相类型及其特征分析，并制定了岩相划分标准[27–29, 35–37]。该标准采用黏土、硅质 / 长英质、碳酸盐 3 端元组分，按常规岩石组成比例划分岩石类型，同时按 TOC 的高低添加富碳（含量大于 4%）、高碳（含量 3%～4%）、中碳（含量 2%～3%）和低碳（含量 1%～2%）前缀。基于上述方案，依靠 6 口取心直井和 163 口水平井的测井解释数据，在实例页岩气田五峰—龙马溪组识别出 8 种岩相类型，各岩相类型的参数特征详如表 2 所示。

以岩相类型及其特征模式为基础，进一步开展了单井岩相分析，同时将岩心观察获得的页理缝宽度、密度及通过两者计算的页理缝发育强度数据同步到单井岩相分析成果中，编制了实例页岩气田单井页岩岩相及页理缝分布剖面（图 3）及统计表（表 2）。结果发现：（1）页理缝的延伸方向与页岩层面基本平行，实例页岩气田主体部位的页岩层倾角通常小于 20°，页理缝的倾角也基本小于 20°，而页理缝方位角的指向性则不明显，基本在 360° 范围内变化；（2）页理缝的发育与页岩岩相类型具有较好的相关性，即硅质页岩页理缝的发育好于砂质页岩，砂质页岩页理缝的发育好于黏土页岩；（3）硅质页岩通常具有极其发育且分布均匀的页理纹层，成为硅质页岩页理缝极其发育且均匀分布的物质基础，砂质页岩常常出现二元或三元互层的纹层结构，造成砂质页岩的页理缝也常常出现二元或三元互层结构，黏土页岩常常出现块状层理，也会出现不规则的纹层结构，使得黏土页岩的页理缝相对欠发育；（4）在同类页岩岩石中，随着 TOC 值的增大，页理缝发育的强度值也有增大的趋势。

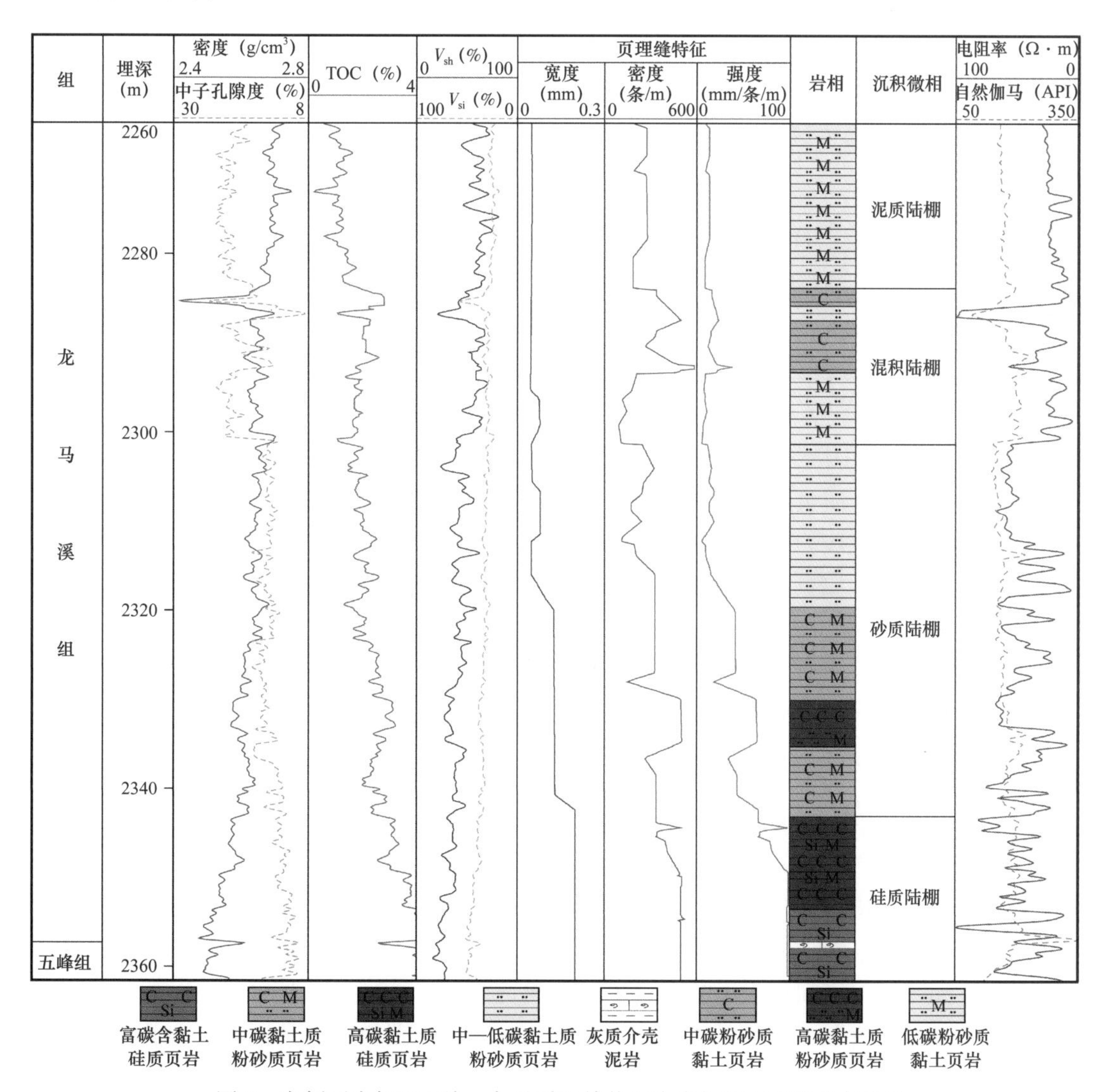

图 3 实例页岩气田五峰—龙马溪组单井页岩岩相及页理缝分布特征

（V_{sh}—泥质矿物含量，V_{si}—硅质矿物含量）

由此可见，硅质矿物和 TOC 值的大小控制了页理缝的发育强度。随着页岩中硅质矿物的增多和 TOC 值的增大，页理缝的发育强度增大，反之则减小。硅质矿物含量增多，反映页岩的脆性增加，成为页理缝更发育的内因；而TOC值增大，则意味着页岩具有更强大的生气能力，生成的气体在页岩孔缝中形成的内压也就更大，导致裂缝更加发育，形成页理缝发育的外因。内、外因共同起作用，以不同页岩岩相类型页理缝发育程度不同的规律表现出来，形成依靠页岩岩相开展页理缝建模的理论基础。

表2 实例页岩气田五峰—龙马溪组页岩岩相类型及特征

岩相类型	TOC（%）	V_{sh}（%）	V_{si}（%）	沉积相	页理缝发育强度（mm/条/m）
富碳含黏土硅质页岩	>4	<25	50～75	硅质陆棚	98～493（139）
高碳黏土质硅质页岩	3～4	25～50	50～75	硅质陆棚	66～98（84）
灰质介壳泥岩	3～4	>25	25～50	硅质陆棚	98
高碳黏土质粉砂质页岩	3～4	25～50	25～50	砂质陆棚	60～65（64）
中碳黏土质粉砂质页岩	2～3	25～50	25～50	砂质陆棚	20～66（44）
中—低碳黏土质粉砂质页岩	1～3	25～50	25～50	砂质陆棚/混积陆棚	5～38（16）
中碳粉砂质黏土页岩	2～3	>50	25～50	混积陆棚	12～38（17）
低碳粉砂质黏土页岩	1～2	>50	25～50	泥质陆棚	0～14（8）

注：括号内数值为样品数。

在实例页岩气田中，若按照页理缝发育强度由高到低排序，则有：富碳含黏土硅质页岩、高碳黏土质硅质页岩和灰质介壳泥岩、高碳黏土质粉砂质页岩、中碳黏土质粉砂质页岩、中—低碳黏土质粉砂质页岩和中碳粉砂质黏土页岩、低碳粉砂质黏土页岩，其中的中—低碳黏土质粉砂质页岩、中碳粉砂质黏土页岩和低碳粉砂质黏土页岩3类页岩岩相的页理缝发育强度较低，可以忽略。

依据上述规律，不仅可以利用页岩岩相来控制实例页岩气田页理缝发育的空间范围，还可以借助页岩岩相估计页理缝发育的强度。

3.2 模型建立

以实例页岩气田单井页岩岩相类型分析成果为输入数据，利用序贯指示建模方法[31-34]，建立了实例页岩气田五峰—龙马溪组页岩岩相三维模型（图4）。结果表明，受沉积环境影响，实例页岩气田的岩相类型随页岩层位的变化而依次变化；在同一层位，常常是以某一类页岩岩相为主，间杂分布1种或2～3种其他页岩岩相；同层页岩沉积稳定，仅局部存在相变，而不同层之间相变迅速，这也是海相页岩沉积的普遍特点。

由于实例页岩气田的中—低碳黏土质粉砂质页岩、中碳粉砂质黏土页岩和低碳粉砂质黏土页岩的页理缝发育强度较低，可以忽略，因此，将这3种页岩岩相定义为0，其他页岩岩相定义为1，建立实例页岩气田页理缝发育指数三维模型（图5）。建模结果限制了实例页岩气田页理缝发育的空间范围，即页理缝只能在上述3类页岩岩相分布的空间之外发育。

以实例页岩气田单井页理缝发育强度数据为输入数据，以页理缝发育指数三维模型为趋势约束，利用序贯高斯方法[31-34]，建立了页岩气田五峰—龙马溪组页理缝发育强度三维模型（图6）。结果表明，页理缝的发育强度呈现下强上弱的特征，从下到上，不同层位间差异巨大，同一层位仅局部有些变化、总体变化不大，与岩相类型的空间分布特征保持一致。

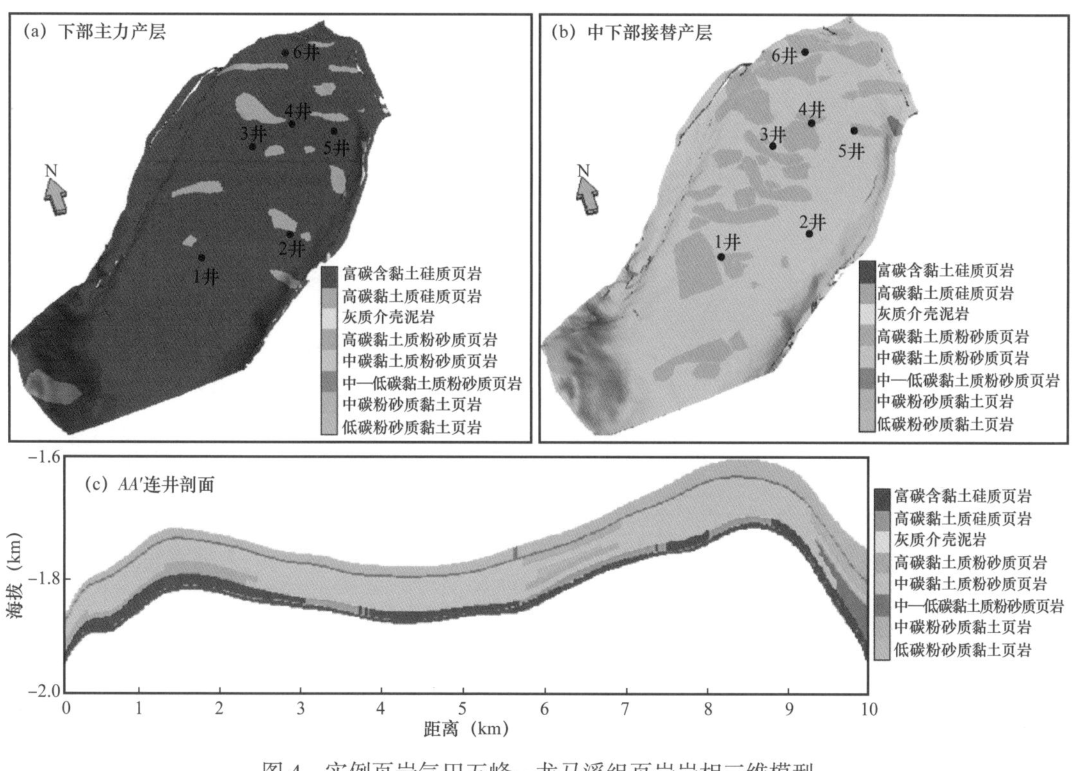

图4　实例页岩气田五峰—龙马溪组页岩岩相三维模型

（图a、b层位位置见表1，图c剖面位置见图2）

以实例页岩气田页理缝发育强度三维模型为页理缝建模的主输入，以岩心观察描述获得的页理缝倾角、方位角等限定页理缝属性特征，以页理缝发育指数三维模型为页理缝建模的空间约束，采用商业软件[31-34]的裂缝离散网络建模模块，建立了页岩气田五峰组—龙马溪组页理缝三维离散网络模型（图7），在三维空间展示了实例页岩气田页理缝的分布特征。图7a—d分别为以页理缝的倾角、方位角、孔隙度和渗透率属性作色的页理缝三维离散网络模型，显示了页理缝的倾角、方位角、孔隙度和渗透率等属性在三维空间的分布特征。图7e—f则详细统计了页岩气田页理缝的倾角、方位角、孔隙度和渗透率属性值的分布频率，统计结果与实例页岩气田页理缝基础地质研究结果保持一致，表明建模结果反映了页岩气田页理缝发育的实际特征。

3.3　模型应用

在开展页理缝三维离散网络模型研究之前，页岩气田生产模拟依靠的是常规方法建立的地质参数模型，该模型的主输入为测井解释获得的基础物性数据。由于测井解释没有提供页理缝的物性参数解释，建立的地质参数模型只能反映页岩储层的基质物性，难以反映页理缝的物性特征，致使模拟结果严重偏离实例页岩气田实际生产动态，只得反过来通过生产历史拟合人为调整地质参数模型。虽然最后的模拟结果总体上吻合了实际生产动态，但调整后的地质参数模型与调整前相比已面目全非，再也无法反映实例页岩气田的实际地质特征，给采用该模拟结果预测气田未来生产动态带来了极大误差。

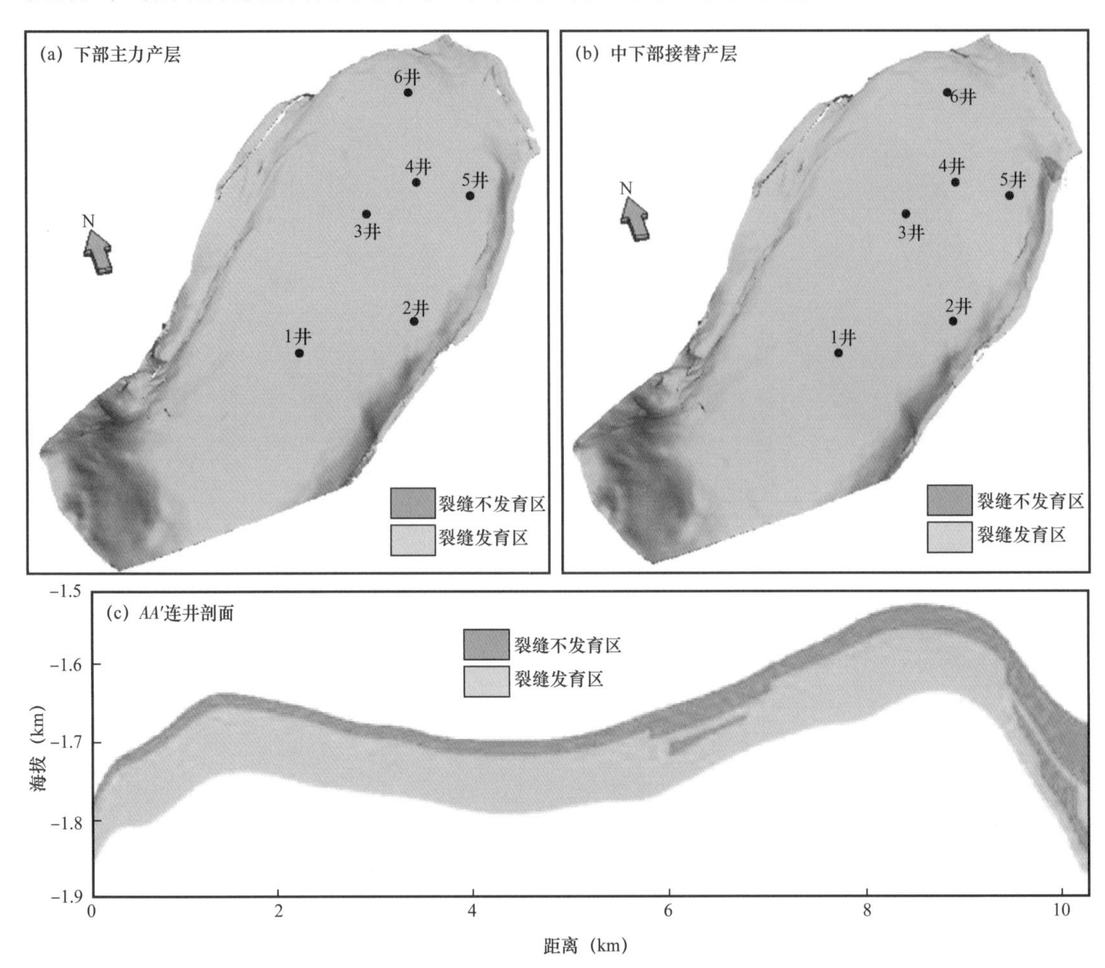

图 5　页岩气田五峰—龙马溪组页理缝发育指数三维模型

（图 a、b 层位位置见表 1，图 c 剖面位置见图 2）

针对上述问题，建立了实例页岩气田储层页理缝三维离散网络模型，该模型详细表征了实例页岩气田一期产能建设区主力生产层位页理缝的分布位置、发育规模及延伸方位，刻画了其中每条页理缝的空间位置、几何尺寸，以及倾角、方位角、孔隙度和渗透率等属性特征。通过将本模型与原地质参数模型叠加使用，未经大的调整，模拟结果即吻合了实际生产动态。更重要的是，采用本模型能定量表征不同平面位置、不同剖面层位的页理缝发育特征，给实例页岩气田后续的生产调整指明了方向。

4　结论

理论分析及实例研究表明，页岩岩相类型与页理缝发育特征具有一一对应关系，这种关系形成了依靠页岩岩相有效约束页理缝分布建模的理论及实践基础。具体表现在两个方面：（1）可以按照页岩岩相类型与页理缝发育特征间的对应关系调整和修正页理缝发育强度三维模型；（2）可以依靠岩相三维模型建立页理缝发育指数三维模型，限制页理缝插值的空间范围，实现对页理缝空间分布位置的有效界定。

基于页岩岩相表征的页理缝三维离散网络建模需要解决 4 个关键技术环节：（1）通过页岩储层岩相分析和页理缝描述，建立岩相与页理缝特征模式；（2）依靠页岩岩相三维模型，建立页理缝发育指数三维模型；（3）提取页理缝特征模式，在页岩岩相三维模型的约束下建立页理缝发育强度三维模型；（4）依靠页理缝发育强度三维模型，在页理缝发育指数三维模型的约束下建立页理缝三维离散网络模型。

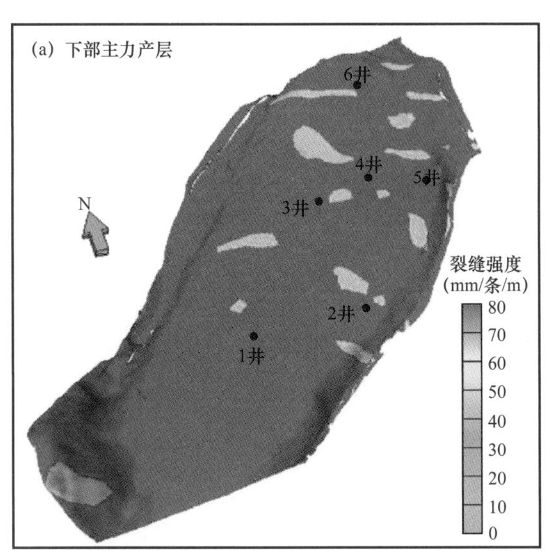

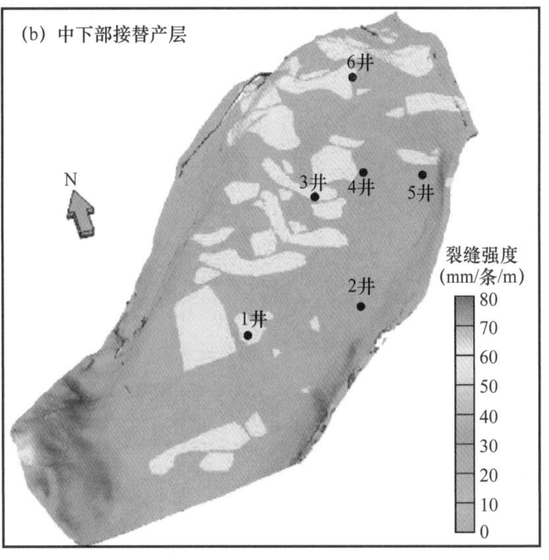

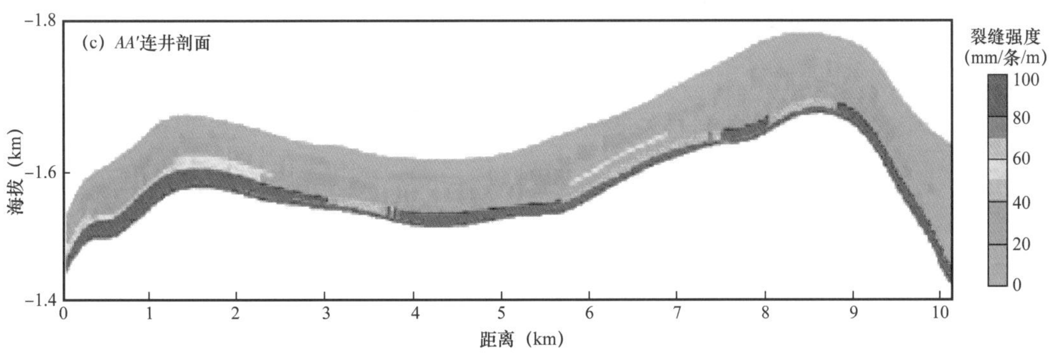

图 6 页岩气田五峰—龙马溪组页理缝强度三维模型

（图 a、b 层位位置见表 1，图 c 剖面位置见图 2）

四川盆地东部焦石坝地区五峰—龙马溪组页岩气田储层页理缝三维离散网络模型研究表明：（1）在实例页岩气田中识别出 8 类页岩岩相，每类页岩岩相页理缝发育的程度均不相同，其中有 3 类页岩岩相页理缝发育程度较差，可以忽略；（2）依靠页岩岩相三维模型为空间约束，建立的页理缝发育指数三维模型和页理缝发育强度三维模型与实例页岩气田的实际地质特征吻合一致；（3）页理缝三维离散网络模型在与实例页岩气田页理缝岩心观察描述特征保持一致的前提下，在三维空间充分展示了实例区内页理缝的分布位置和发育规模及倾角、方位角的变化，为实例页岩气田后续的生产模拟提供了页理缝地质参数；（4）模型在实例页岩气田的成功应用充分证实了基于岩相表征的页理缝三维离散网络建模方法的适用性和可靠性。

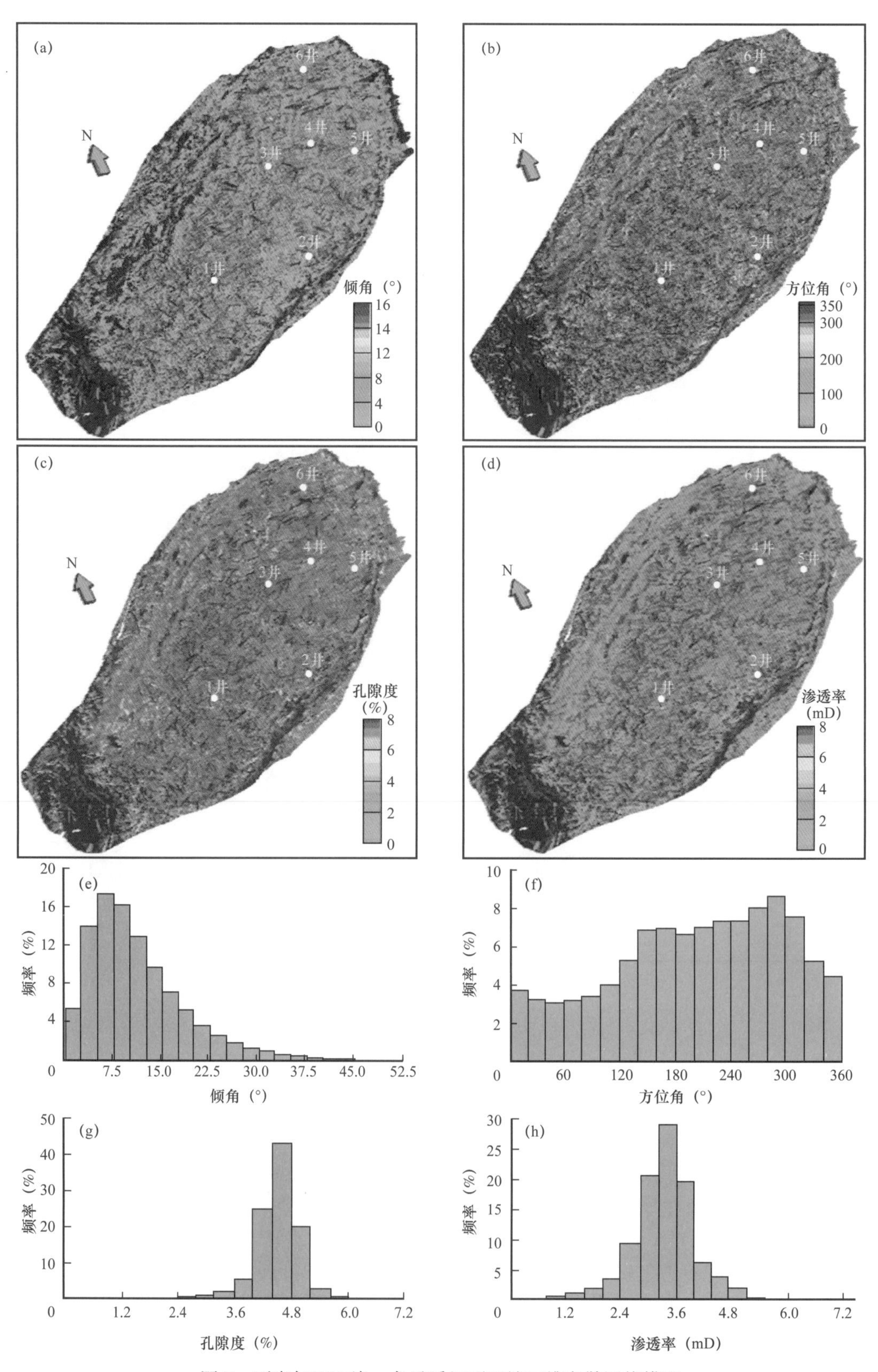

图 7　页岩气田五峰—龙马溪组页理缝三维离散网络模型

参考文献

［1］Torsaeter M，Vullum P E，Nes O M. Nanostructure vs. Macroscopic properties of mancos shale［C］. SPE 162737，2012.

［2］Fjær E，Nes O M. The impact of heterogeneity on the anisotropic strength of an outcrop shale［J］. Rock Mechanics and Rock Engineering，2014，47（5）：1603–1611.

［3］Kim J，Gomaa A M，Zhang H，et al. Novel evaluation method of fracturing fluid additives on Barnett and Marcellus outcrop shale cores using NMR technique［R］. SPE 179122，2016.

［4］Dokhani V，Yu M，Miska S Z. The effect of bedding plane orientation on pore pressure in shale formations：Laboratory testing and mathematical modeling［C］// 47th US rock mechanics/geomechanics symposium. San Francisco，California，USA：American Rock Mechanics Association，2013.

［5］Meier T，Rybacki E，Backers T，et al. Influence of bedding angle on borehole stability：A laboratory investigation of transverse isotropic oil shale［J］. Rock Mechanics and Rock Engineering，2015，48（4）：1535–1546.

［6］Ding W，Li C，Li C，et al. Fracture development in shale and its relationship to gas accumulation［J］. Geoscience Frontiers，2012，3（1）：97–105.

［7］Ding W，Zhu D，Cai J，et al. Analysis of the developmental characteristics and major regulating factors of fractures in marine-continental transitional shale-gas reservoirs：A case study of the Carboniferous–Permian strata in the southeastern Ordos Basin，central China［J］. Marine and Petroleum Geology，2013，45（4）：121–133.

［8］Cho Y，Ozkan E，Apaydin O G. Pressure-dependent natural-fracture permeability in shale and its effect on shale-gas well production［J］. SPE Reservoir Evaluation and Engineering，2013，16（2）：216–228.

［9］Zeng L，Lyu W，Li J，et al. Natural fractures and their influence on shale gas enrichment in Sichuan Basin，China［J］. Journal of Natural Gas Science and Engineering，2016，30：1–9.

［10］Soeder. Porosity and permeability of eastern Devonian gas shale［J］. SPE Formation Evaluation，1988，3（1）：116–124.

［11］Katsube T J，Williamson M A. Effects of diagenesis on shale nano-pore structure and implications for sealing capacity［J］. Clay Minerals，1994，29（4）：451–472.

［12］Zhang P，Li J，Xie L，et al. The quantitative characterization of heterogeneity using nanometer CT technique on the shale reservoir［J］. Acta Geologica Sinica，2015，89（S1）：134–135.

［13］Martineau D F. History of the Newark East field and the Barnett shale as a gas reservoir［J］. AAPG Bulletin，2007，91（4）：399–403.

［14］Sondergeld C H，Newsham K E，Comisky J T，et al. Petrophysical considerations in evaluating and producing shale gas resources［R］. SPE 131768，2010.

［15］Guo X，Hu D，Li Y，et al. Geological features and reservoiring mode of shale gas reservoirs in Longmaxi Formation of the Jiaoshiba Area［J］. Acta Geologica Sinica，2014，88（6）：1811–1821.

［16］Wang Y，Zhai G，Bao S，et al. Latest progress and trend forecast of China's shale gas exploration and development［J］. Acta Geologica Sinica，2015，89（1）：211–213.

［17］袁玉松，周雁，邱登峰，等. 泥页岩非构造裂缝形成机制及特征［J］. 现代地质，2016，30（1）：

155–162.

[18] Könitzer S F, Stephenson M H, Davies S J, et al. Significance of sedimentary organic matter input for shale gas generation potential of Mississippian Mudstones, Widmerpool Gulf, UK [J] . Review of Palaeobotany and Palynology, 2016, 224: 146–168.

[19] Schulz, H M, Biermann S, Van Berk W, et al. From shale oil to biogenic shale gas : Retracing organic–inorganic interactions in the Alum Shale (Furongian–Lower Ordovician) in southern Sweden [J] . AAPG Bulletin, 2015, 99 (5) : 927–956.

[20] 邹才能，董大忠，王玉满，等 . 中国页岩气特征、挑战及前景（一）[J] . 石油勘探与开发，2015，42（6）: 689–701.

[21] Arthur M A, Sageman B B. Marine shale : Depositional mechanisms and environments of ancient deposits [J] . Earth and Planetary Sciences, 1994, 22 (22) : 499–551.

[22] Dræge A, Jakobsen M, Johansen T A. Rock physics modelling of shale diagenesis [J] . Petroleum Geoscience, 2006, 12 (1) : 49–57.

[23] Ou C H, Chen W, Li C C, et al. Structural geometrical analysis and simulation of decollement growth folds in piedmont Fauqi Anticline of Zagros Mountains, Iraq [J] . SCIENCE CHINA Earth Sciences, 2016, 59 (9) : 1885–1898.

[24] Ou C H. Technique improves exploration, exploitation offshore Myanmar [J] . Oil & Gas Journal, 2016, 114 (7) : 56–61

[25] Ou C H, Chen W, Ma Z. Quantitative identification and analysis of sub-seismic extensional structure system : Technique schemes and processes [J] . Journal of Geophysics and Engineering, 2015, 12 (3): 502–514.

[26] Ou C H, Chen W, Li C C. Using structure restoration maps to comprehensively identify potential faults and fractures in compressional structures [J] . Journal of Central South University, 2016, 23 (3) : 677–684.

[27] Abouelresh M O, Slatt R M. Lithofacies and sequence stratigraphy of the Barnett Shale in east-central Fort Worth Basin, Texas [J] . AAPG bulletin, 2012, 96 (1) : 1–22.

[28] Han C, Jiang Z, Han M, et al. The lithofacies and reservoir characteristics of the Upper Ordovician and Lower Silurian black shale in the Southern Sichuan Basin and its periphery, China [J] . Marine and Petroleum Geology, 2016, 75: 181–191.

[29] 吴蓝宇，胡东风，陆永潮，等 . 四川盆地涪陵气田五峰组—龙马溪组页岩优势岩相 [J] . 石油勘探与开发，2016，43（2）: 189–197.

[30] Guo L, Jiang Z X, Guo F. Mineralogy and fracture development characteristics of marine shale-gas reservoirs : A case study of Lower Silurian strata in southeastern margin of Sichuan Basin, China [J] . Journal of Central South University, 2015, 22 (5) : 1847–1858.

[31] Ripley B D. Stochastic simulation (Vol. 316) [M] . New York : John Wiley and Sons, 2009.

[32] Ou C H, Wang X L, Li C C, et al. Three-dimensional modelling of a multi-layer sandstone reservoir : The Sebei Gas Field, China [J] . Acta Geologica Sinica, 2016, 90 (1) : 801–840.

[33] Ou C H, Li C C, Ma Z. 3D modeling of gas/water distribution in water-bearing carbonate gas reservoirs :

The Longwangmiao Gas Field，China［J］. Journal of Geophysics and Engineering，2016，13：745–757.

［34］Zhou F，Yao G，Tyson S. Impact of geological modeling processes on spatial coal bed methane resource estimation［J］. International Journal of Coal Geology，2015，146：14–27.

［35］马乔 . 川东地区页岩岩相及其控气性特征研究［D］. 成都：西南石油大学，2015：41–63.

［36］Tang X，Jiang Z，Huang H，et al. Lithofacies characteristics and its effect on gas storage of the Silurian Longmaxi marine shale in the southeast Sichuan Basin，China［J］. Journal of Natural Gas Science and Engineering，2016，28：338–346.

［37］Ou C H，Ray R，Li C C，et al. Multi-index and two-level evaluation of shale gas reserve quality［J］. Journal of Natural Gas Science and Engineering，2016，35：1139–1145.

［38］Ou C H. Fluid typing extends production in Chinese gas reservoir［J］. Oil & Gas Journal, 2015, 13（2）：54–61.

页岩气水平井地质信息解析与三维构造建模

乔　辉，贾爱林，位云生

（中国石油勘探开发研究院）

摘　要：页岩气开发具有开发评价期短，评价井少（直井）的特点，直井密度低增加了储层精细描述和地质建模的技术难度。因此，建立研究区地层岩性、物性和电性特征的足量识别标准，利用该标准对水平井水平段开展精细小层划分，并对实际钻遇地层厚度进行权正，将水平井水平段等效为多口平面分布的直井。在此基础上，以水平井段解析的各小层数据资料作为控制点，建立研究区精细构造模型。水平井多点地质信息解析，将水平段等效为多口评价井获取相关地质参数资料，可有效地弥补评价井资料少的不足。

关键词：优质页岩；多点地质信息解析；三维地质建模；龙马溪组；威远区块

随着全球能源需求的增长及勘探开发技术水平的提升，页岩气等非常规资源开发发展迅速，已成为油气勘探开发的热点领域。中国页岩气资源丰富，具备良好的资源基础和开发潜力。目前，中国石油在四川威远、长宁和昭通等地区的页岩气区块已达到工业开发水平，并已建成产能[1，2]，中国石化则在涪陵建成了中国第一大页岩气田[3]。

页岩气开发具有大规模采用水平井工厂化作业，直井/评价井数少、密度低的特点，如何充分利用水平井资料对页岩气区储层研究显得尤为重要。在水平井资料运用方面，随着水平井技术广泛地应用于薄层油藏、致密油气、页岩气等复杂油气藏的开发，以及老区调整和剩余储量挖潜[4-7]，国内外学者已逐渐意识到水平井地质信息提取对储层研究的重要性，开展了相关的研究工作[8-11]，如赵国良等[4]以阿曼DL油田为例，提出充分利用多分支水平井资料，采用水平井轨迹对构造层面进行约束和校正，提高了构造模型的精度。郝建明等[7]在高尚堡油田实现了利用水平井资料进行变差函数分析，建立单一目标砂体高精度地质模型。吴胜和等[10]应用坪桥区块水平井资料和露头资料，有效提取三维建模所必需的地质统计学参数，特别是砂体侧向变化参数，弥补了砂体统计学特征受直井井距的限制，建立的沉积微相三维模型更符合地质实际。水平井实钻储层地质信息提取和利用技术是页岩储层描述的重要发展方向之一。同时，中国页岩气区块多具有纵向上优质页岩层薄，最佳巷道位置厚度薄，优质页岩储层钻遇率不高等问题。页岩气井产能影响因素分析表明页岩气井中的长水平段优质页岩储层的钻遇率和井眼轨迹控制是影响页岩气井产能的重要因素[12-18]。

前人研究认为采用随钻地质导向钻井在一定程度上可以提高水平井段钻探优质储层的精度[12-18]，但地质导向判断参数少、不直观，标志点选取难度大，且龙马溪组优质页岩段发育多个小层和旋回，部分小层伽马特征具有一定相似性，特别容易混淆。因此，如何利用水平井资料建立精细的三维构造模型，对水平井井轨迹的优化设计及提高水平井段优质页岩储层的钻遇率具有重要意义。

1 研究区地质背景

震旦纪以来，受多期构造作用的影响，形成了四川盆地现今格局，其中威远区块位于四川盆地西南部，横跨四川省内江市、资中县、自贡市荣县境内。构造上处于四川盆地川中古隆起区的川西南低陡褶皱带（图 1）。

目前，四川盆地主要勘探开发目的层位为奥陶系五峰组至志留系龙马溪组一段 1 亚段，埋深 1500～3500m。龙马溪组页岩沉积时期处于加里东构造运动期，构造运动活跃，海平面升降频繁，发育多期沉积旋回，页岩纵向上尤其是底部的龙马溪组一段 1 亚段层理发育、非均质性强。龙一$_1$亚段自下至上可依次划分为 4 个小层：龙一$_1^1$、龙一$_1^2$、龙一$_1^3$及龙一$_1^4$（图 1）。从沉积角度，五峰组沉织期及龙马溪组沉积早期的龙一$_1$亚段沉积主体为半局限浅海相的深水陆棚，发育一套黑色碳质、硅质页岩和黑色页岩组合沉积，有机质丰度，笔石发育，主体为优质页岩发育地层，厚度约 36～48m。龙马溪组沉积晚期，水体变浅，沉积环境变为浅水陆棚，主要岩性为深灰色泥页岩、灰色泥岩和粉砂质泥岩，笔石含量明显降低，有机质相对不发育。龙一$_2$亚段为大段砂泥质互层或夹层岩性组合，笔石数量少，地层厚度在 105～200m。目前威远区块 W 井区已建产能，主要以水平井方式开发，探明含气面积 54.86km^2，探明地质储量 312.63 × 10^{8}m^3，技术可采储量 78.16 × 10^{8}m^3[16]。

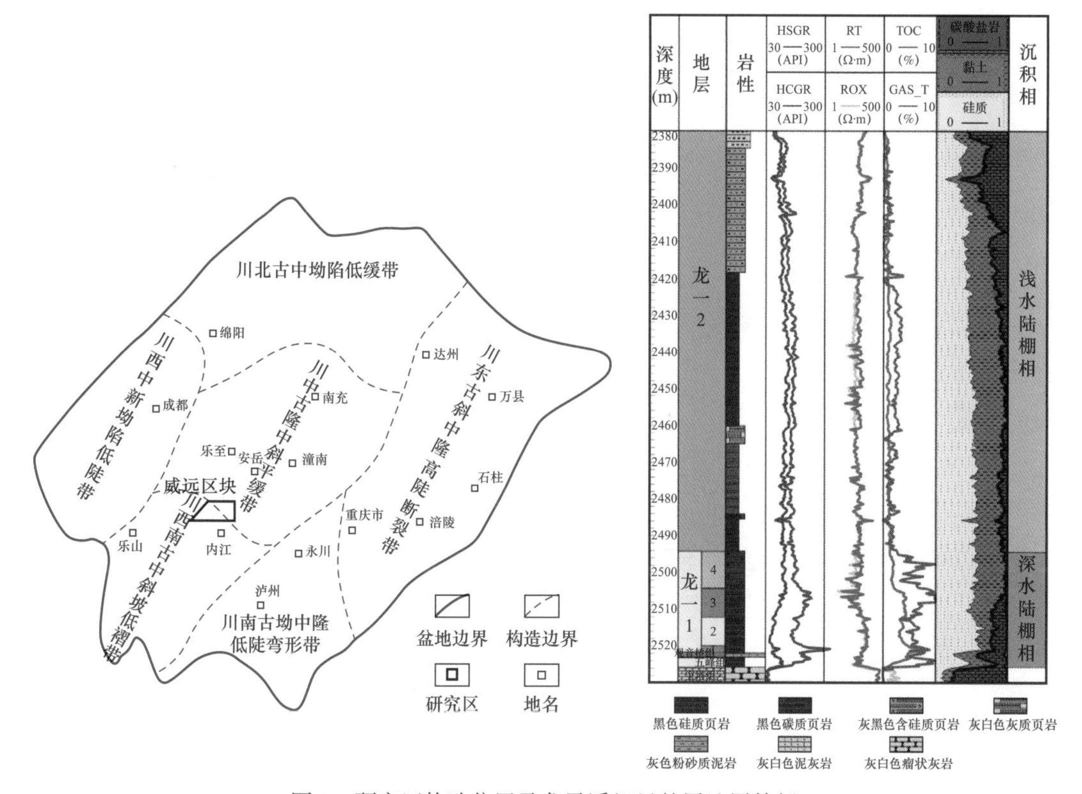

图 1 研究区构造位置及龙马溪组目的层地层特征

2 优质页岩分布特征

四川盆地龙马溪组优质页岩地层为海相沉积，优质页岩厚度受深水陆棚沉积相的整体控制，整体上厚度变化幅度不大，约为10～25m。垂向上，志留纪早期经历了两期海进—海退的变化，发育多期沉积旋回，垂向上小层间非均质性强[16, 19, 20]。在威远地区，靠近龙马溪组底部的龙一$_1$亚段是目前页岩气区块主要目的产层，岩相主要为黑色碳质页岩和硅质页岩相，脆性矿物发育，含气量高[16, 21]。为了满足页岩气大规模效益开发的需求，通过岩心、薄片、测井及实验分析资料对该亚段进行小层细分，从下往上依次划分为龙一$_1^1$、龙一$_1^2$、龙一^{13}及龙一$_1^4$共4个小层[22]。

根据中国石油制定的南方海相页岩气储层分类评价标准[23]，龙马溪组龙一$_1$亚段的龙一$_1^1$、龙一$_1^2$和龙一$_1^3$小层储层以Ⅰ类和Ⅱ类储层为主，为主要的优质页岩层段，其中龙一$_1^1$小层Ⅰ类储层最发育。通过对比4个小层纵向上的有机质含量、含气量、孔隙度、裂缝密度、古生物发育情况及地层脆性指数，龙一$_1^1$小层是最有利的甜点层段，是水平井钻探的目标靶体位置（图2）。

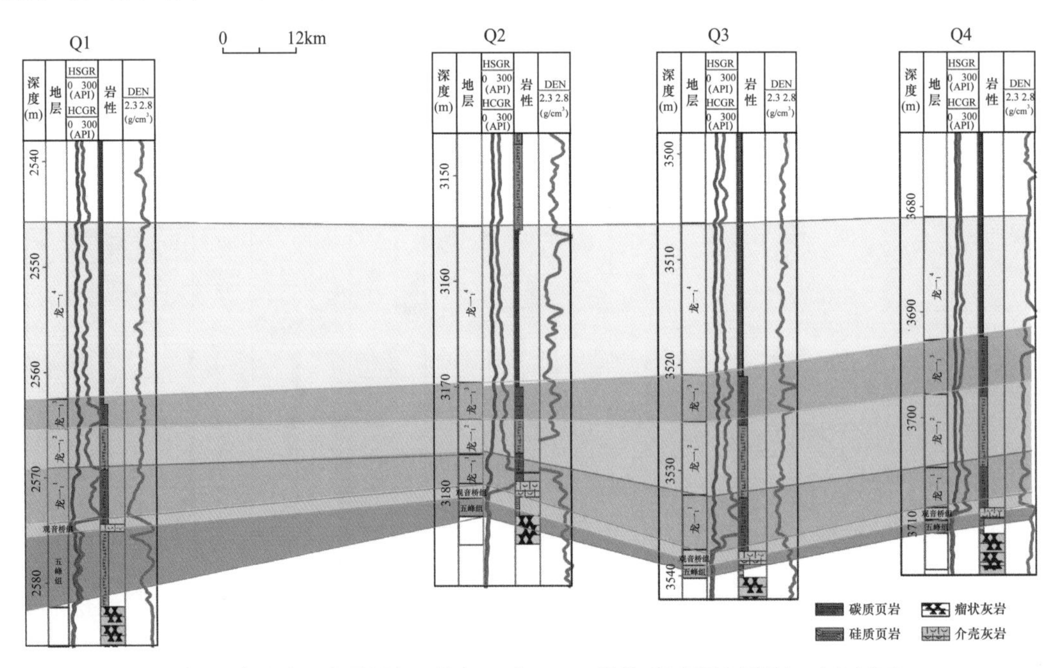

图2 威远地区龙马溪组五峰组—龙一$_1$亚段优质页岩地层精细对比剖面

3 水平井多点地质信息解析

3.1 研究思路

目前中国页岩气区块多采用水平井开发，威远研究区也仅存在少量探井为直井，采用仅有的几口探井开展地质研究显然不能满足储层精细描述及实际气藏开发的需要，因此，充分利用水平井资料进行地质研究尤为重要。

常规油藏中水平井的水平段多在某一特性的目的层穿行，偶见上下穿层现象[7, 10]。与常规油藏不同的是，威远页岩气藏目的层段小层厚度薄，五峰组、龙一$_1^1$、龙一$_1^2$、龙一$_1^3$小层各自厚度多在5m左右；平面上地层分布稳定，具有甜点连续分布特征。水平井钻井过程中受地层厚度薄、地层倾角变化等因素影响，钻井过程中钻头方位不断发生变化，导致水平段不断上下穿行，钻遇不同小层，在长度约1500～2000m的水平段，井轨迹多次与多个小层的构造面顶面相交，可获得比单一直井多很多倍的信息，等效于获取多口直井的地质信息，弥补了评价井少的不足（图3）。

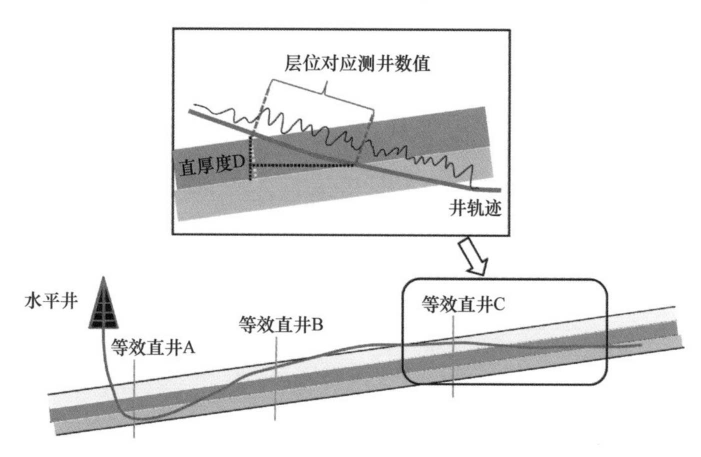

图3　水平井多点地质信息解释示意图

3.2　水平井钻遇层位精细分析

水平井具有难以与邻井进行精细地层对比的特征，而龙马溪组优质页岩段发育多个小层和旋回，部分小层，例如龙一$_1^1$与龙一$_1^3$、龙一$_1^2$与龙一$_1^4$小层的GR特征具有一定相似性，通过测井曲线直观识别小层有一定的难度。因此，研究中首先建立主要钻遇地层，如宝塔组、五峰组、龙一$_1^1$、龙一$_1^2$、龙一$_1^3$及龙一$_1^4$小层的岩性、物性、电性曲线识别标准，划分中主要参考了GR、DEN、AC及TOC4个参数，然后利用建立的小层划分标准对水平井的长水平段进行精细的小层划分，主要钻遇地层识别标准见表1。

为提高测井资料的准确性，在进行水平井分析前，对研究区内不同仪器测得的不同井的测井曲线进行校正。通过对比不同仪器测得的标志层测井曲线的方法，对测井数据进行了极差正规化处理，避免应用过程中出现较大误差[24]。统计发现，威远地区优质页岩层段龙一$_1$亚段的GR＞150API，龙一$_1^1$与龙一$_1^3$小层的GR＞250API；DEN为2.3～2.6g/cm^3，且密度随着有机质含量的增高而降低，龙一$_1^1$小层密度值在此页岩层中最低；AC为60～70μs/s，TOC为3%～8%，龙一$_1^1$小层TOC最高，为6%～8%。

根据水平井井轨迹信息与测井曲线组合特征，结合邻近评价井的小层电性特征，精细分析每口水平井在地层中的轨迹分布。由威远区块W4水平井段小层精细划分剖面（图4）可见，该井在钻井过程中钻头不断调整钻井方向，水平井段钻遇五峰组、龙一$_1^1$、龙一$_1^2$、龙一$_1^3$及龙一$_1^4$小层，水平段长度分别是100m、842m、653m、0、0，各小层钻遇率分别是6.27%、52.79%、40.94%、0及0。

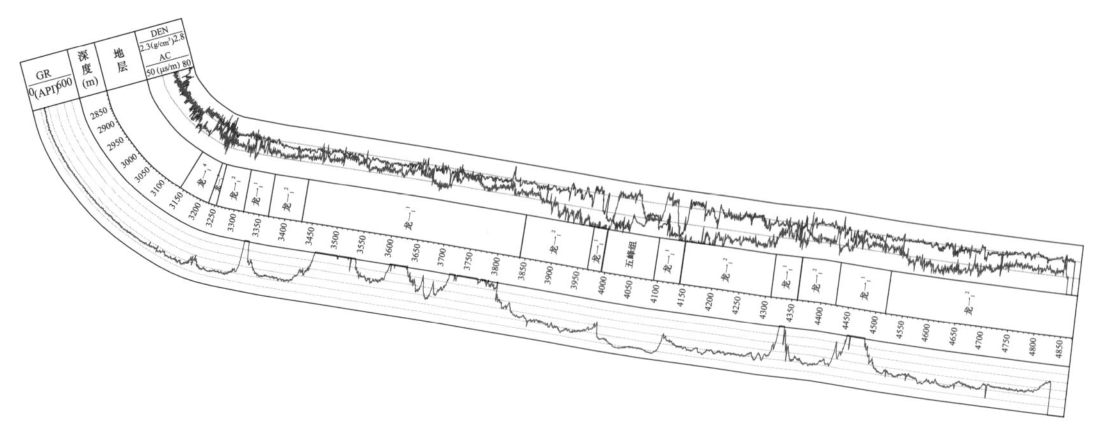

图 4　威远区块 W4-1 水平井段小层精细划分图

表 1　主要钻遇地层识别标准

地层		岩性特征	储层参数特征				厚度（m）	沉积相
			GR（API）	DEN（g/cm^3）	AC（μs/s）	TOC（%）		
龙一$_2$亚段		深灰色页岩	100～150	>2.6	60～70	1～2	100～150	深水陆棚
龙一$_1$亚段	龙一$_1^4$	灰黑色页岩	150～200	2.6	70	3	6～25	
	龙一$_1^3$	黑色碳质、硅质页岩	250～300，部分井>300	2.6	60～70	4～8	3～9	
	龙一$_1^2$	黑色碳质页岩	150～200	2.6	60～70	3～4	4～11	
	龙一$_1^1$	黑色碳质、硅质页岩	>250	2.3～2.6	60～70	6～10	1～4	
五峰组		顶为观音桥段介壳灰岩，主体为碳质、硅质页岩	100	2.55～2.75	70	1	0.5～15	
宝塔组		灰白色瘤状灰岩	50	>2.6	<60	<1	2～100	

3.3　水平井地层厚度校正

地层厚度是了解地层纵向和平面展布规律的重要参数。绘制地层等厚图的关键是确定井点位置的地层厚度，在直井情况下地层厚度通过测井曲线获得并进行地层倾角校正。研究区直井较少时，利用仅有的直井不足以体现地层真实厚度，而利用水平井长水平段多次钻遇各小层这一地质资料解析后可获得多个小层地层视厚度数据，进而对视厚度进行校正即可获得多点地层真厚度数据点[14, 25]（图 5）。

地层真实厚度采用式（1）求取：

$$H = H_{视} \times \cos\alpha \tag{1}$$

当地层倾向与钻井方向相同时，地层视厚度采用式（2）求取：

$$H_{视} = Z_{AB} - \sqrt{a^2 - Z_{AB}^2} \times \tan\alpha \tag{2}$$

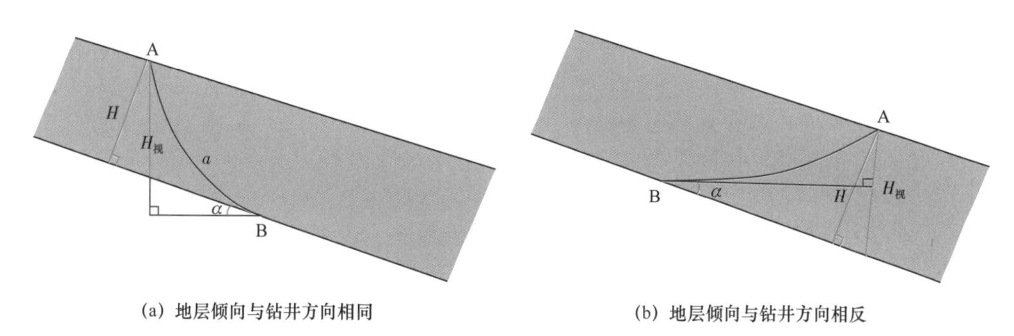

(a) 地层倾向与钻井方向相同　　(b) 地层倾向与钻井方向相反

图 5　地层倾向与钻井方向关系图

当地层倾向与钻井方向相反时，地层视厚度采用式（3）求取：

$$H_{视}=Z_{AB}+\sqrt{a^2-Z_{AB}^2}\times\tan\alpha \tag{3}$$

其中，Z_{AB} 为 AB 点海拔高差；a 为 AB 点距离，$a=\sqrt{\Delta x^2+\Delta z^2}$；$\alpha$ 为地层倾角。

图 6（a）与图 6（b）是威远区块 W 井区分别利用区块仅有的少量探井和利用水平井穿不同小层这一特征经厚度校正求取的龙一$_1^3$小层页岩地层厚度平面分布图。受直井资料少的限制，图 6（a）中 W 井区地层厚度仅为 4～5m，而无法看出井区内部地层精细的厚度变化规律，图 6（b）解析后的地层厚度点，其约束下绘制的厚度图 6（b）比图 6（a）更符合地层实际情况。总体来说，W 井区地层厚度约为 2～7m，中间局部地区较厚，向南北方向减薄，总体上厚度分布稳定。

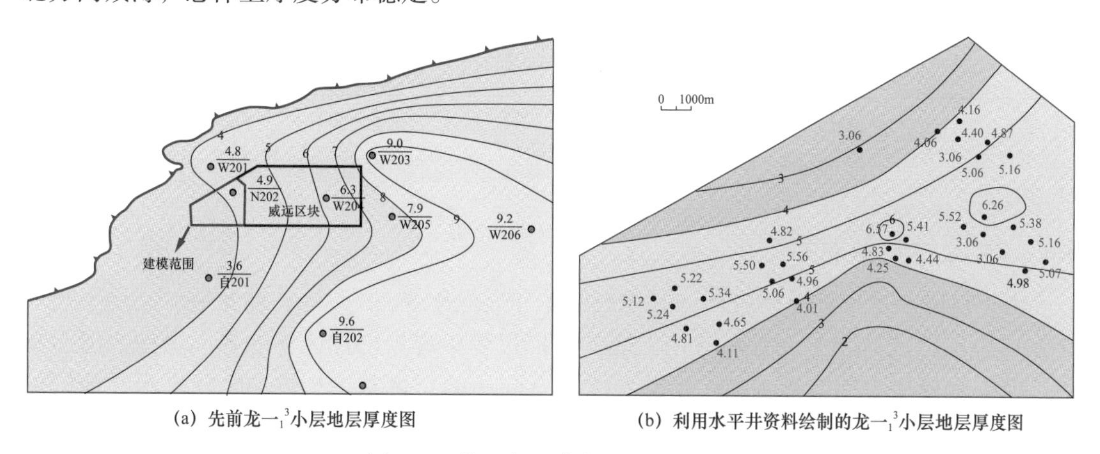

(a) 先前龙一$_1^3$小层地层厚度图　　(b) 利用水平井资料绘制的龙一$_1^3$小层地层厚度图

图 6　W 井区龙一$_1^3$小层地层厚度图

4　利用水平井资料建立三维构造模型

4.1　构造模型的建立

构造模型主要反映地质构造及构造背景下的地层厚度分布、垂向地层之间的接触关系和断裂系统的发育等，且能够更加精确地显示微构造[4, 26]，对水平井优化设计及提高目标储层的钻遇率具有重要意义[16-17]。水平井资料在常规油藏精细描述和建模中的应用主要包括：与其他资料一起识别砂体侧向边界；水平井轨迹对构造建模的层面和断层进行约束或检验；描述储层物性参数在平面上的连续变化特征等[7, 10, 11]。研究区直井少，主要通过水平井资料处理，建立研究区构造模型，用于研究地层三维空间展布、优质页岩储层钻遇并指导水平井钻探。由于钻井过程中钻头方位的不断变化，在部分情况下钻机并未钻穿某层位就不断地向上或向下调整钻头方位，此时若钻头往上覆地层钻，则该小层分层深度代表上覆地层底界，若水平段往下伏地层钻，则该深度代表该地层的底界。

建模时首先建立关键层位的层面模型控制全区的地层格架，以此为约束建立其他小层层面，最终建立全区层面模型[27]。以水平井井点处理后的分层数据作为约束条件应用精细地震解释出的地层构造层面为趋势，井震结合建立宝塔组、五峰组、龙一$_{1}^{1}$、龙一$_{1}^{2}$、龙一$_{1}^{3}$和龙一$_{1}^{4}$小层的构造面（图 7）。此井—震结合的构造面既保证了井点处地层深度的准确性，又展示了井间构造变化趋势，因此较准确反映了目的小层构造特征。从模型上可见，W 井区具有西北高东南低的特点，地层比较平缓，倾角约 8°。

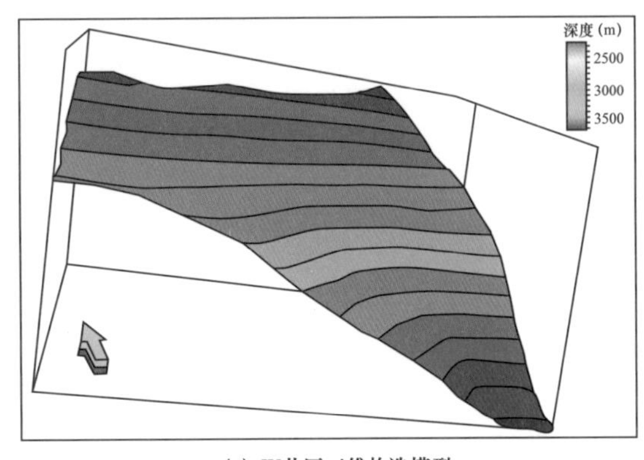

(a) W井区三维构造模型

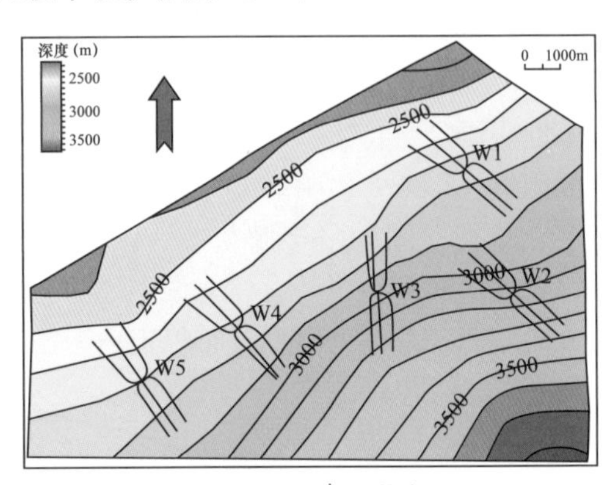

(b) W井区龙一$_{1}^{1}$小层构造图

图 7 W 井区三维构造模型及龙一$_{1}^{1}$小层平面构造图

4.2 模型分析及验证

由于目的小层厚度总体较薄，五峰组、龙一$_{1}^{1}$、龙一$_{1}^{2}$、龙一$_{1}^{3}$和龙一$_{1}^{4}$小层平均厚度分别是 4m、5.1m、5.3m、4m 和 15.5m，建模过程中容易出现建立的各小层构造层面相交，因此在建模中逐井检查各水平井在三维空间中的轨迹是否与该井实钻水平井地层剖面一致。若出现不相符时，及时调整构造层面。通过对比 W3 井实钻地层剖面分析（图 8a）及该井三维空间井轨迹图（图 8b），该井水平段各钻遇的地层与模型完全吻合。因此，本研究认为建立的三维地质模型相对可靠，可用于指导后续水平井钻井及地质属性建模。

4.3 利用水平井资料建立构造模型的重要作用及实践效果

威远区块普遍采用长水平段水平井开发，水平段长度 1500m 左右，而纵向上优质页岩层薄，最佳巷道位置龙一$_{1}^{1}$段厚度仅 5m 左右，井控程度较低，地面和地下条件复杂，地层埋藏深度大、构造复杂，且地质导向判断参数少、不直观，标志点选取难度大。页岩气井中的长水平段优质页岩储层的钻遇率和井眼轨迹控制是影响页岩气井产能的重要因素。通过水平井产能资料分析表明，W 井区水平井产能影响因素分析表明：水平井巷道位置及Ⅰ类储层钻遇长度对水平井产量具有重要影响，测试产量大于 $20\times10^8m^3$ 的井，靶体均位于龙一$_1^1$小层（图 9a）；且单井产量随优质储层钻遇长度的增加而提高（图 9b）。因此，提高优质页岩储层钻遇率是提高页岩气井产量的关键。

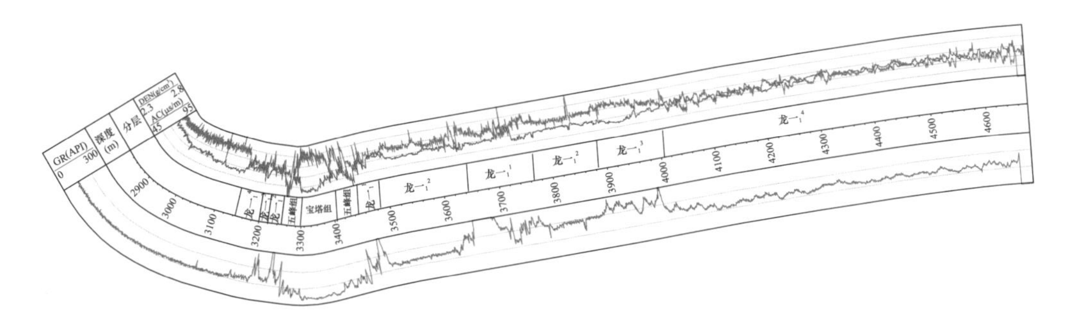

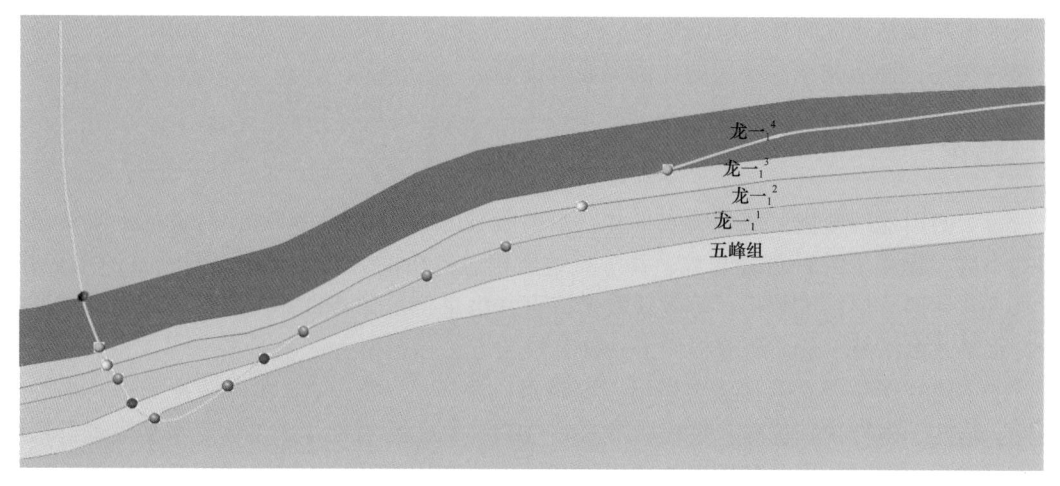

图 8　W3-1 井实钻地层剖面及对应的三维空间轨迹

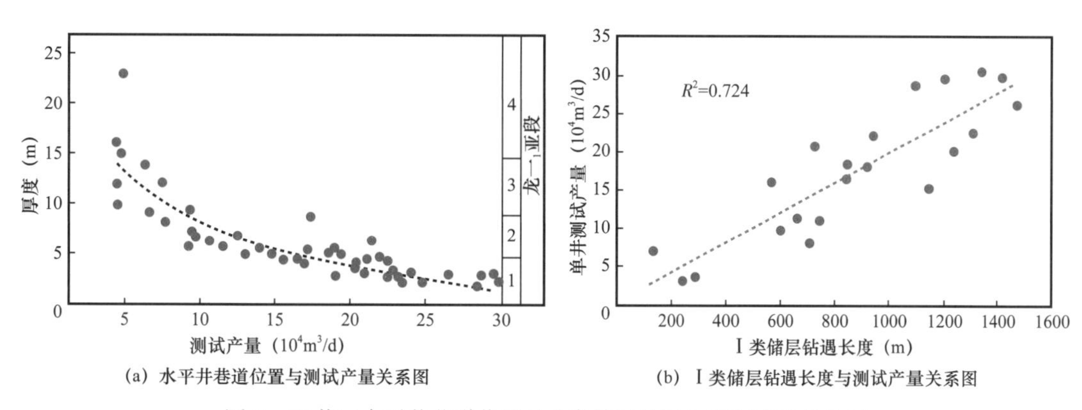

(a) 水平井巷道位置与测试产量关系图　　(b) Ⅰ类储层钻遇长度与测试产量关系图

图 9　W 井区水平井巷道位置及Ⅰ类储层钻遇长度与产量的关系

目前在钻井中，常采用随钻地质导向，该方法可一定程度上提高目标层位的钻遇率[6, 7]。龙马溪组优质页岩段发育多个小层和旋回，部分小层的伽马特征具有一定相似性，特别容易混淆，有时很难判断钻头钻进切线方向和钻头所在地层的实际位置，井眼轨迹容易钻出优质页岩段设计箱体。同时，由于部分井区地层倾角较大，地层倾角变化频繁，沿下倾方向的水平井钻遇地层视厚度明显增加，沿上倾方向的水平井钻遇地层视厚度明显变薄，对跟踪过程中准确预测靶点造成了较大的难度，影响目的层水平井段的钻遇率。因此，在水平井优化设计方面，可以在三维地质模型内进行水平井设计，井轨迹大体平行于地层构造起伏。

在构造模型的基础上对储层参数进行模拟，依据随钻过程中自然伽马测井曲线的特征，模拟自然伽马属性的空间分布，最终完成水平井轨迹的设计与着陆点深度预测。在确保水平井段优质页岩钻遇率方面，将井轨迹随钻测井数据实时加载到精细建立的三维地质模型中进行三维可视化显示，若实钻数据与钻前模型预测数据相差较大，则对模型进行快速更新，预测下一柱至下几柱范围内实时钻井储层监督的储层发育情况，从而指导水平段的钻进，确保水平井段优质页岩的钻遇率，保证产能建设的有效跟踪与快速实施。目前，研究区内采用本方法指导钻井，水平井段优质页岩的钻遇率得到了显著提高。统计表明，区块水平井段钻遇优质页岩层段龙一$_1^1$、龙一$_1^2$、龙一$_1^3$小层的比例由之前的52%提高到现在的93%，钻遇龙一$_1^1$小层的比例也显著提高。

5 结论

（1）四川盆地威远区块多为水平井，水平段长度约1500～2000m，且各水平段钻遇不同的小层。提出了一种页岩气水平井多点地质信息解析的方法，利用当前开发的水平井，对水平井钻遇小层精细解析，等效获取多口直井信息，可弥补直井少的不足，为解决直井少，水平井多情况下地质研究提供了一种有效的方法和思路。

（2）建立了主要钻遇地层宝塔组、五峰组、龙一$_1^1$、龙一$_1^2$、龙一$_1^3$及龙一$_1^4$小层的岩性、物性、电性识别标准，主要参考GR、DEN、AC及TOC 4个参数对水平段进行精细的小层划分，并对水平井各钻遇的地层厚度进行校正，为页岩气储层描述及三维地质建模奠定基础。

（3）研究区目的层段龙一$_1^1$小层Ⅰ类储层最发育，测试产量大于$20\times10^8m^3$井的水平井靶体均位于龙一$_1^1$小层，且统计发现优质页岩储层钻遇长度越长，水平井单井产量越高。因此，以水平井段解析的各分层数据做控制点，建立目的区块精细的构造模型，模型沿水平井轨迹的剖面与实钻水平井剖面一致，表明模型较精确，建立的模型有效指导了水平井优化设计及确保水平井段优质页岩钻遇率。

参考文献

[1] 邹才能，董大忠，王玉满，等. 中国页岩气特征、挑战及前景（一）[J]. 石油勘探与开发，2015，43（2）：689-701.

[2] 邹才能，董大忠，王玉满，等. 中国页岩气特征、挑战及前景（二）[J]. 石油勘探与开发，2016，42（2）：166-178.

[3] 郭旭升，胡东风，魏志红，等. 涪陵页岩气田的发现与勘探认识[J]. 中国石油勘探，2016，21（3）：24-37.

[4] 赵国良，沈平平，穆龙新，等. 薄层碳酸盐岩油藏水平井开发建模策略——以阿曼DL油田为例

［J］. 石油勘探与开发，2009，36（1）：91-96.

［5］马新华，贾爱林，谭健，等. 致密砂岩气开发工程技术与实践［J］. 石油勘探与开发，2012，39（5）：572-579.

［6］贾爱林，位云生，金亦秋. 中国海相页岩气开发评价关键技术进展［J］. 石油勘探与开发，2016，43（6）：949-955.

［7］郝建明，吴健，张宏伟. 应用水平井资料开展精细油藏建模及剩余油分布研究［J］. 石油勘探与开发，2009，36（6）：730-736.

［8］Mu L X，Xu B J，Zhag S C，et al. Integrated 3D Geology Modeling Constrained by Facies and Horizontal Well Data and the Classifying OOIP Calculation for Block M of the Orinoco Heavy Oil Belt［C］// International Thermal Operations and Heavy Oil Symposium. Society of Petroleum Engineers，2008.

［9］朱卫红，周代余，冯积累，等. 油田典型油气藏水平井开发效果评价［J］. 石油勘探与开发，2010，37（6）：716-725.

［10］吴胜和，武军昌，李恕军，等. 水平井区沉积微相三维建模研究［J］. 沉积学报，2003，21（2）：266-271.

［11］王伟. 水平井资料在精细油藏建模中的应用［J］. 岩性油气藏，2012，24（3）：79-82.

［12］刘旭. 页岩气水平井钻井的随钻地质导向方法［J］. 天然气工业，2016，36（5）：69-73.

［13］陈志鹏，梁兴，王高成，等. 旋转地质导向技术在水平井中的应用及体会——以昭通页岩气示范区为例［J］. 天然气工业，2015，35（12）：64-70.

［14］Liang X，Wang L Z，Zhang J H，et al. An integrated approach to ensure horizontal wells 100% in the right positions of the sweet section to achieve optimal stimulation：a shale gas field study in the Sichuan Basin，China［C］// Abu Dhabi International Petroleum Exhibition and Conference. Society of Petroleum Engineers，2015.

［15］Kok J C L，Shim Y H，Tollefsen E M. LWD for Well Placement and Formation Evaluations Towards Completion Optimization in Shale Plays［C］// SPE Annual Technical Conference and Exhibition. Society of Petroleum Engineers，2011.

［16］刘乃震，王国勇. 四川盆地威远区块页岩气甜点厘定与精准导向钻井［J］. 石油勘探与开发，2016，43（6）：1-8.

［17］王理斌，段宪余，钟伟，等. 地质建模在苏丹大位移水平井地质导向中的应用［J］. 岩性油气藏，2012，24（4）：90-92.

［18］吴雪平. 页岩气水平井地质导向钻进中的储层"甜点"评价技术［J］. 天然气工业，2016，36（5）：74-80.

［19］梁兴，王高成，徐政语，等. 中国南方海相复杂山地页岩气储层甜点综合评价技术——以昭通国家级页岩气示范区为例［J］. 天然气工业，2016，36（1）：33-42.

［20］李玉喜，何建华，尹帅，等. 页岩油气储层纵向多重非均质性及其对开发的影响［J］. 地学前缘，2016，23（2）：118-125.

［21］梁超，姜在兴，杨镱婷，等. 四川盆地五峰组—龙马溪组页岩岩相及储集空间特征［J］. 石油勘探与开发，2012，39（6）：691-698.

［22］赵圣贤，杨跃明，张鉴，等. 四川盆地下志留统龙马溪组页岩小层划分与储层精细对比［J］. 天然气地球科学，2016，27（3）：470-487.

[23] 中国石油天然气股份有限公司勘探与生产公司 . 中国石油页岩气测井采集与评价技术管理规定（试行）[R]. 北京：中国石油天然气股份有限公司勘探与生产公司，2014.

[24] 肖佃师，黄文彪，张小刚，等 . 王府凹陷青山口组含油气泥页岩层的测井曲线标准化 [J]. 东北石油大学学报，2014，38（1）：46–53.

[25] Berg C R，Newson A C. Geosteering Using True Stratigraphic Thickness [C] // Unconventional Resources Technology Conference. Society of Exploration Geophysicists，American Association of Petroleum Geologists，Society of Petroleum Engineers，2013：1196–1205.

[26] 胡向阳，熊琦华，吴胜和 . 储层建模方法研究进展 [J]. 石油大学学报：自然科学版,2001,25（1）：107–112.

[27] 穆龙新，贾爱林，陈亮 . 储集层精细研究方法 [M]. 北京：石油工业出版社，2000.

三、气藏工程篇

页岩气井全生命周期物理模拟实验及数值反演

高树生[1, 2]，刘华勋[1, 2]，叶礼友[1, 2]，胡志明[1, 2]，安为国[2]

（1. 中国石油集团科学技术研究院；
2. 中国石油勘探开发研究院渗流流体力学研究所）

摘　要：页岩储层极低的孔隙度、渗透率和复杂的赋存、输运状态导致其特有的 L 形曲线生产特征，页岩气流动机理复杂。对页岩气井衰竭开发全生命周期生产过程进行了全直径岩心物理模拟实验，获取了模拟页岩气井生产过程完整的压力、日产气量等重要的生产数据，解决了页岩气井生产时间短及作业引起的间断性等导致的难以获取完整生产数据的问题。开发模拟实验研究结果表明，模拟实验生产特征与气井相一致；利用模拟实验数据可以准确判断岩心的临界解吸压力（12MPa）、游离气量（3820.8mL）与吸附气量（2152.2mL）及不同时间、地层压力时对应的日产气量中游离气与吸附气比例，废弃压力对应的生产时间和最终采出程度。运用岩心与气井无因次时间的相似性以及试井与相似理论，开展了数值反演计算页岩气井的生产动态曲线，预测气井的开发效果，提出了渗透率与压裂效果（缝网密度）是页岩气藏有效开发的关键，其中渗透率是根本，压裂技术是手段。

关键词：页岩气井；全生命周期；物理模拟实验；数值反演；相似理论；渗透率；压裂效果

岩气藏与常规气藏不同，页岩既是天然气生成的源岩，也是聚集和保存天然气的储层和盖层，页岩气主体上以游离相态存在于黑色泥页岩的裂缝、孔隙及其他储集空间，以吸附状态存在于干酪根、黏土颗粒及孔隙表面，吸附气的比例介于 20%～80%[1]，极少量以溶解状态存在，在烃源岩内就近聚集，表现为典型的原地成藏模式[2]。近年来，中国页岩气勘探开发工作获得重要突破，先后形成了涪陵、长宁—威远和昭通等国家级页岩气示范区，提交探明地质储量 $5441.29\times10^8m^3$，2015 年底落实了 $65\times10^8m^3/a$ 产能建设目标，2016 年底完成了 $85\times10^8m^3/a$ 的产能，实现了页岩气工业化生产[2]。

页岩气储层特征远不同于常规气，发育多类型微纳米级孔隙，而且以纳米级孔隙为主[4]，气体赋存状态复杂，储层孔隙度、渗透率极低，自身基本上没有渗流能力，主要靠扩散和裂缝形成有效导流能力[5-8]，页岩气井大规模整体多段压裂是其有效开发的前提和基础，页岩储层特征与开发方式也决定了其独特的生产特征，初期产量高、递减快，后期低产、稳产时间长，形成典型的 L 形生产特征曲线。如何合理解释页岩气生产特征一直是石油科研工作者们不懈努力的目标。有研究[9-11]认为，页岩储层的基质具有低孔、极低渗的特征，可以用双孔模型描述，页岩气流动可以描述为随着裂缝中自由气体产出，裂缝和基质产生压差，基质表面的吸附气解吸下来后通过裂缝流向井筒，基质内部的气体在浓度差作用下扩散到表面，气体在基质中的流动服从分子扩散，从裂缝流向井筒的过程用达西定律来描述，吸附气体解吸过程用 Langmuir 方程来描述。Schepers 等[12]和 Civan 等[13]提出三孔双渗模型，认为裂缝和基质中的流动都遵循达西定律，基质不仅仅是气体解吸的源项，相比于扩散，基质中的渗流仍然占主导。Swami 等[14, 15]提出了一个考虑微裂缝中自由气，纳米孔隙自由气，孔隙壁面吸附气以及干酪根溶解气的四孔隙气体流动模型，建立了页岩纳米孔隙的流动方程并且进行数值求解，认为在页岩产量预测以及数值模拟过程中应该考虑 Kn 扩散、气体的 Langmuir 解吸、气体滑脱效应及溶解气的扩散。Alharthy 等[16]提出了包含对流、扩散、滑脱流动的二孔隙，以及三孔隙模型并且进行了数值计算，认为三孔隙模型能更好地模拟页岩气藏中气体的流动规律，解吸气体的流动通道不会被介孔所影响。段永刚等[17]提出了考虑吸附气，游离气及溶解气的页岩气双重介质压裂井宏观渗流数学模型，并对页岩气井的不稳定产能进行了计算分析。李治平等[18]考虑了纳米级孔隙气体解吸与扩散效应的影响，认为地层压力低于气体临界解吸压力后，气体解吸使页岩基质收缩变形，气体渗流通道增加，渗透率变好，将气体分子自由程大于孔隙直径（D）的分子所占总的分子量的比例设为 α，小于 D 的分子占 $1-\alpha$，由此建立了气体渗流与扩散方程。王玉普等[19]提出页岩高温高压吸附模型；郭小哲等[20]建立了考虑解吸—吸附、扩散、滑脱及应力敏感的双重介质气水两相渗流模型，朱维耀等[21]利用稳定状态依次替换法，研究了页岩基质储层内压力扰动的传播规律，数值计算分析页岩气井产量递减规律。

文献调研结果表明，目前就页岩气井产能计算与生产动态预测的方法绝大部分都是纯理论研究结果，缺乏实践基础，而且存在明显的缺陷：首先是所有模型都建立在很多假设条件成立的基础上，而实际上大多情况并非如此；其次是计算过程中用到的一些关键参数大多数都无法准确获取；还有就是生产过程中的游离气与吸附气量难以准确界定，根据 Langmuir 模型确定的解析气量存在大的误差[19]。鉴于上述原因，目前的产能计算与生产动态预测方法很难合理解释页岩气井生产曲线的特殊性，当然也不能用于指导页岩气井的有效开发。以页岩气井全生命生产周期的生产动态为研究对象，首次运用全直径页岩岩心模拟气井的整个生产过程，生产条件与气井保持一致，获得的生产特征曲线与气井具有高度的一致性，只要能够合理解释模拟岩心的生产曲线特征，就可以有效地解释气井的生产曲线特征，而室内开展的全直径岩心页岩气全生命周期物理模拟实验可以获取完整的生产数据及计算过程中需要的一些关键参数，从而解决当前页岩气井产能计算与生产动态预测方法难以解决的难题。

1 气井全生命周期物理模拟实验

1.1 实验方案及流程

页岩全直径岩心（直径为 10cm，长度为 15cm）取自川南昭通地区龙马溪组，为了尽量保证岩心的原始特性，随车携带保压取心装置到钻井现场进行第一时间取样，拿到岩心简单清洁后迅速将其置入保压取心装置中，饱和天然气至 10MPa，关闭岩心两端的阀门，运回实验室备用。实验设备是中国石油勘探开发研究院渗流所自主研发的页岩气解吸—扩散—渗流耦合物理模拟实验装置，实验流程如图 1 所示。

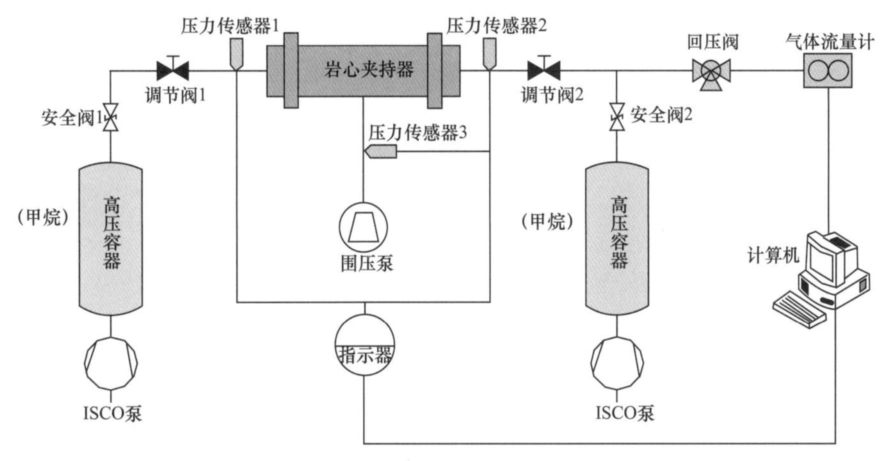

图1　气井全生命周期生产物理模拟实验流程示意图

实验压力传感器量程40MPa，精度0.3%，满足实验压力计量要求，具体实验过程如下：

（1）将保压取心设备的核心装置全直径岩心夹持器取出置入恒温25℃实验室，接入模拟实验装置流程，用高压泵将高压中间容器中的天然气增压并维持在28MPa，打开调节阀1和2，从岩心两端充注甲烷气（储层天然气组分甲烷占比高达99%），对其进行二次饱和增压至模拟地层压力28MPa，同步增加围压到50MPa，由于页岩极其致密与吸附特征，完全恢复到地层状态需要的时间很长，增压饱和持续了200天，直到岩心出、入口高压中间容器的压力不再降低，此时页岩气几乎完全处于原始地层的存储状态，关闭入口调节阀门1与出口调节阀门2，撤掉高压气源，准备开始全生命周期生产模拟实验。

（2）打开出口调节阀门2，实验开始，为了控制初始气量过大，出口保持合理的回压，当流量降低至合适的范围，调节回压为大气压，开始页岩气井全生命生产周期长期生产的动态模拟。

（3）整个模拟实验过程中入口压力、出口压力及出口流量分别由设置在出入口的压力传感器1、2与出口的气体质量流量计通过计算机数据采集软件实时连续记录。

1.2　实验结果

全直径页岩岩心气井全生命周期衰竭开发模拟实验进行了1631天，根据无因次时间相似原理、物理模拟页岩岩心物性参数和物理模拟岩心所在区块龙马溪组地质参数，计算页岩岩心模拟生产时间相当于不同渗透率、不同压裂规模条件下页岩气井生产4.5～550天，达到了模拟页岩气井全生命生产周期的目的。考虑到岩心夹持器出口端存在1mL死体积，早期产气主要来自岩心夹持器出口端死体积，不能反映页岩岩心流动特征，因此，实验数据分析时删除早期异常点，根据死体积大小、初始压力、出口端回压阀压力和气体高压物性参数，早期260mL气主要产自死体积，消除初期由于实验流程缺陷带来的一些不确定性实验数据，模拟实验日产气量曲线特征与页岩气井生产特征几乎完全一致（图2a），前40天日产气量由第一天的135.9mL迅速降至9mL以下，累计气量迅速升至1038.6mL，采出程度约17.4%；之后日产气量下降与累计产气量上升幅度同时变缓，生产至 100 天时，日产气量降至 3mL；随后进入 1500 多天漫长的低产阶段，日产气量由 3mL 缓慢降至 1mL 左右，这一生产过程进行到 444 天时，累计产气量达到 1909.5mL，采出程度 32%，产气量有一个明显的恢复过程，由 1mL/d 增加到 2mL/d 左右，稳产一段时间后又开始缓慢下降直到 1mL/d 左右，当前累计产气量 3634mL，采出程度 60.8%。页岩岩心的模拟生产曲线特征与致密砂岩岩心的模拟结果完全不同，后者的生产曲线特征表现为日产气量与地层视压力是一个连续下降的过程，而且生产时间不会太长。页岩的极低渗流能力与吸附、解吸作用可能是导致这一巨大差异的主要原因，根据页岩气生产压力曲线特征，可以判断地层视压力达到 14.7MPa（地层压力 12MPa）时，岩心中的吸附气开始大量解吸，而且解吸量随地层压力降低而不断增加（图 2b），这可以通过地层视压力略有上升之后下降幅度变缓和日产气量略有增加之后下降幅度变缓的曲线特征得到证明。

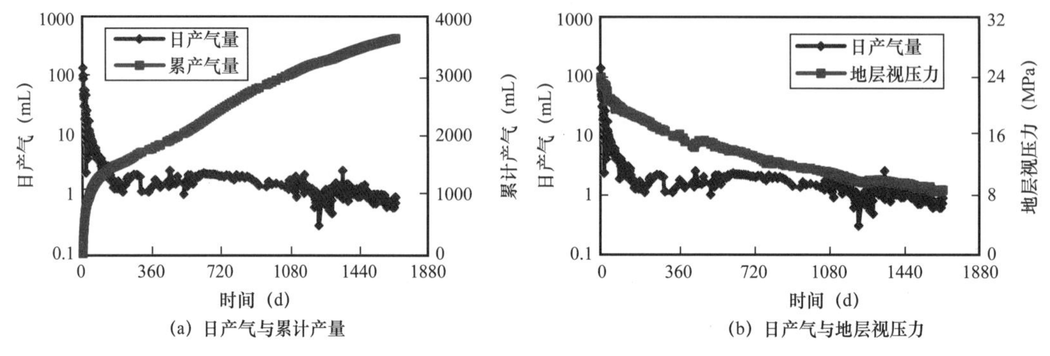

图 2　岩心日产气与累计产气量及地层视压力随时间变化关系

页岩全直径岩心物理模拟实验结果证明，页岩气衰竭开发过程可以分为 3 个主要阶段：前 40 天产气速率迅速降低的高速开发阶段；40～440 天产气速率降低减缓的中速开发阶段；之后进入产气速率稳定、递减极其缓慢的长期低速稳产阶段。由图 2 可以看出，前 40 天地层视压力变化很小，地层视压力大于 20MPa，产出气量基本都是游离气量；随后 400 天地层视压力随累计产气量增加下降幅度较大，且二者呈线性关系，拟合直线斜率的绝对值为 0.0068MPa/mL，地层视压力降至 14.78MPa，产出气量包含有少量的解吸附气量；440～1631 天地层视压力随累计产气量的增加降幅明显变缓，而且二者也基本符合线性关系，拟合直线斜率的绝对值为 0.0038MPa/mL，当前地层视压力降至 9MPa，产出气量中解吸附气的比例越来越大。由此可见，页岩岩心的气井全生命周期物理模拟实验产气动态与气井开发生产动态基本一致，因此，运用物理模拟实验结果可以合理、有效地解释和预测页岩气井的生产动态。

2　实验数据分析

全直径页岩岩心气井全生命周期 1631 天的物理模拟实验过程中，计算机自动采集了围压、入口压力、出口流量随时间变化的所有实验数据，分析、处理实验数据可以很容易得到地层压力下降规律、累计产气量与日产气量随地层压力下降的变化规律，结合气藏工程基础理论，可以推算岩心生产过程中的游离气量、吸附气量及孔隙度和广义渗透率等重要参数值。

2.1 游离气与吸附气量计算

图 3 为页岩岩心模拟气井全生命周期生产过程中，地层视压力与累计产气量关系曲线。对于封闭气藏而言，根据物质平衡方程，地层视压力随累计产气量增加而线性降低。由于生产初期受气井开发或实验条件限制，初期的生产数据不符合理论计算的线性关系，因此，一般使用生产一段时间后的生产或实验数据进行生产动态规律分析，但图 3 岩心衰竭开发模拟实验曲线的中后段显然也不符合这一规律。地层视压力大于 14.7MPa（地层压力 12MPa）时二者是线性关系，小于 14.7MPa 后地层视压力降低趋势变缓，开始偏离上述直线。地层视压力越低偏离越远，表明室温条件下页岩岩心中吸附气的临界解吸压力为 12MPa，与该地区页岩高压吸附实验测试得到的临界解吸压力一致[16]，地层压力低于临界解吸压力后，吸附气开始大量解吸供气，累计产气量开始高于物质平衡计算量，地层压力越低，降至同一地层压力时获得的累计产气量差距越大，也就是说地层压力越低，解吸附气量在产气量中的占比就越大，而且低速稳产开发阶段，累计采气量与地层视压力也基本呈线性关系，只是直线的斜率较中速开发阶段更小，吸附气解吸供给作用更加明显，这也是页岩气井生产后期在低产量下能够长期稳产的主要原因。由此可以把岩心衰竭开发模拟实验过程分为 3 段，受实验条件影响的高速开发稳定高压阶段，中速开发的地层视压力线性下降阶段，低速开发的地层视压力线性下降阶段，其中后两段用来分析气井的生产动态规律。

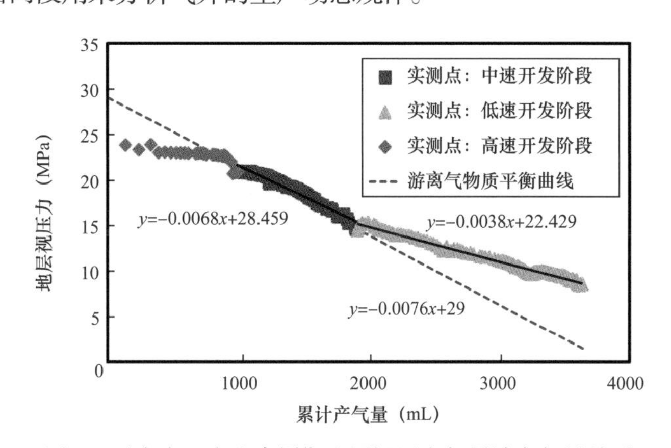

图 3　页岩岩心全生命周期地层视压力与累计产气量关系

图 3 中的红色虚线是根据岩心已知孔隙度值 1.03%（平行样测试结果），利用物质平衡原理，计算得到的封闭气藏游离气物质平衡曲线（红色虚线），其代表了页岩岩心中的游离气生产动态，可以作为基准数据用来区分不同地层压力时采出气量中的游离气与吸附气量，函数表达式：

$$y=-0.0076x+29 \tag{1}$$

根据该式计算的不同地层压力下的累计产气量为游离气量，而实测点对应的气量则为游离气与吸附气量之和，二者的差值即为岩心产气过程中伴随的吸附气量。中速开发阶段的累计产气量与地层视压力关系可以用蓝线拟合函数表示：

$$y=-0.0068x+28.46 \tag{2}$$

低速稳产开发阶段的累计产气量与地层视压力关系可以用黑线拟合函数表示：

$$y=-0.0038x+22.43 \tag{3}$$

为了校正实验误差导致的测试数据偏差，分别运用中速开发阶段（蓝线）与低速稳产开发阶段（黑线）的线性拟合函数来计算对应地层视压力下的累计产气量，之后再减去根据封闭气藏游离气物质平衡曲线函数（红线）计算相同地层视压力下的游离气量，即可得到累计产吸附气量（图 4a），将累计产吸附气量、游离气量对时间进行求导即可得到日产气量（图 4b），计算结果与实测结果几乎完全一致，这样做的优点在于可以消除实验误差引起的实测点异常值，光滑曲线，对于最终处理结果没有任何影响，由此可以实现岩心产气量中吸附气与游离气的准确分离。

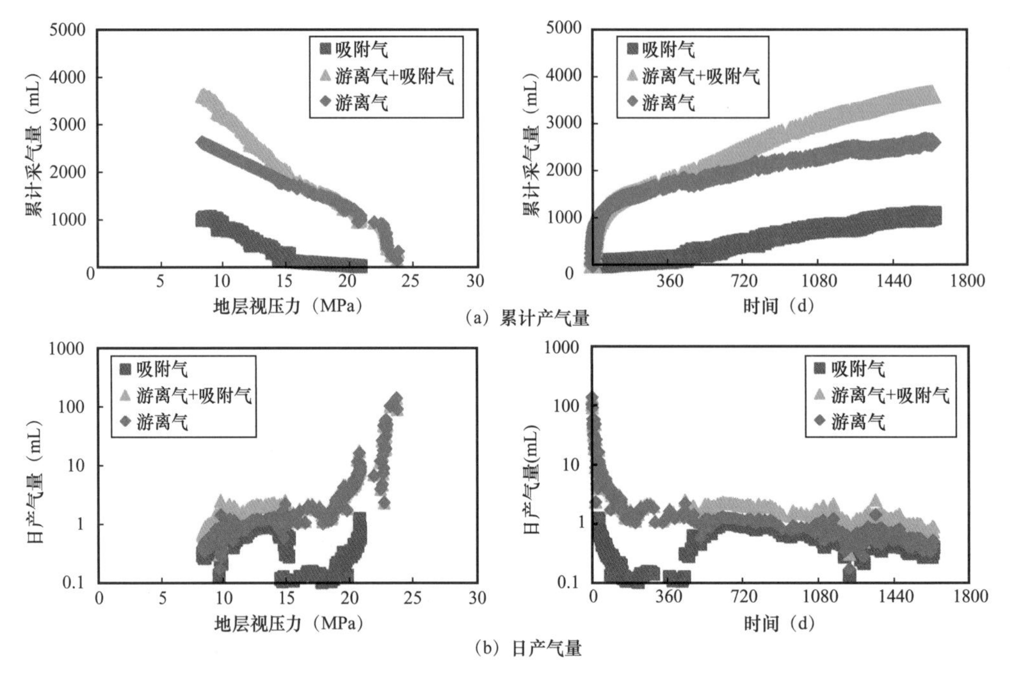

图 4　页岩岩心衰竭开发过程中游离气、吸附气累计日产气量和日产气量动态特征

根据岩心的孔隙度值可以计算初始状态下岩心中游离气的含量为 3820.8mL，根据低速稳产开发阶段实验点的拟合曲线（黑线）可以推算岩心吸附气含量 2151.2mL，故岩心初始状态页岩气总含气量约为 5972mL，其中吸附气量约占总含气量的 36%，吸附气与游离气的比例约为 1∶2，折合页岩储层岩石含气量 $2.8m^3/t$，该值与该地区含气量测试值基本一致，总含气量较低[22]。

2.2　孔隙度

如果处于开发状态下的页岩气井缺失储层孔隙度、渗透率参数时，可以利用气井生产数据来推算，具有一定的可行性。由于页岩气的解吸附特性[19]，高压条件下页岩气产出气主要为游离气，根据封闭气藏物质平衡方程，累计采气量与地层视压力呈线性关系为

$$G_{p1}=G_1\left(1-\frac{z_i p}{zP_i}\right) \tag{4}$$

式（4）表明，早期页岩气累计采气量与地层视压力曲线斜率为 $G_1 z_i/p_i$，截距为游离

气总含量 G_1，因此可根据早期页岩气累计产气量与地层视压力曲线斜率确定游离气总量 G_1 及相应的游离气占据的孔隙体积 V_{p1} 和孔隙度 ϕ：

$$V_{p1}=G_1B_{gi} \tag{5}$$

$$\phi=\frac{V_{p1}}{V} \tag{6}$$

选取中速开发阶段累计产气量与地层视压力的直线段进行分析（图3），曲线斜率为144.9mL/MPa，推算游离气含量4191.7mL，略大于依据岩心孔隙度计算的游离气含量，引入高压物性[19]和岩样形状参数，利用式（5）和式（6）计算得到岩心孔隙体积为13.18mL，孔隙度约为1.13%，略大于平行岩样孔隙度，这是由于中速开发阶段有少量吸附气解吸出来，产出气里面包含部分吸附气，故根据中速开发阶段累计产气曲线确定的页岩孔隙度偏大，但由于此阶段解吸气量相对较少，因此偏差程度较小，故依据中速开发阶段累计产气曲线确定页岩孔隙度是可行的。

2.3 广义渗透率

页岩孔喉细小，多为纳米级孔喉，同时存在连续流、滑移流、过渡流或分子自由流等多种形式复杂流动，目前缺乏有效的技术手段表征页岩气流动能力，通常采用等效的广义渗透率 K 来表征页岩气流动能力的强与弱[23, 24]，计算公式如常规气测渗透率一致：

$$K=\frac{200q\mu zP_{sc}L}{A\left(p_2^2-p_1^2\right)} \tag{7}$$

根据物模岩心入口压力、出口压力和流量，计算不同地层压力下岩心的广义渗透率（图5），可以看出，实验页岩岩心的广义渗透率非常低，数量级约为 10^{-9}mD；而且排除高压阶段少量异常点，不同地层压力下页岩岩心广义渗透率基本一致，可以理解为岩心在整个衰竭开发实验过程中气体的流动能力基本一致，尽管不同阶段产出气的机理可能不同，但是体现在日产气量的宏观层面上却是相同的，因此引入页岩的等效广义渗透率，通过达西流动模型来计算气井的产能是可行的。

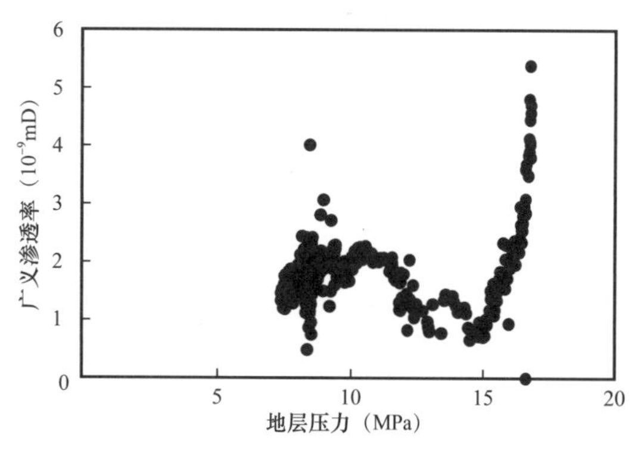

图5　不同地层压力下页岩岩心的广义渗透率曲线

3 相似性分析与数值反演

3.1 物模实验数值反演方法研究

在页岩储层进行体积压裂时，由于页岩特殊物理性质及其内部天然裂缝的影响，会产生一个水力裂缝与天然裂缝相互连通的复杂缝网系统。而根据 Meyer 等[26]基于自相似原理及双重介质模型提出的离散化缝网模型（DFN），体积改造区为椭球体，包含一条主裂缝及多条次生裂缝，主裂缝垂直于最小主应力方向，次生裂缝与主裂缝将基岩切割成规则的小立方体单元（图 6）。

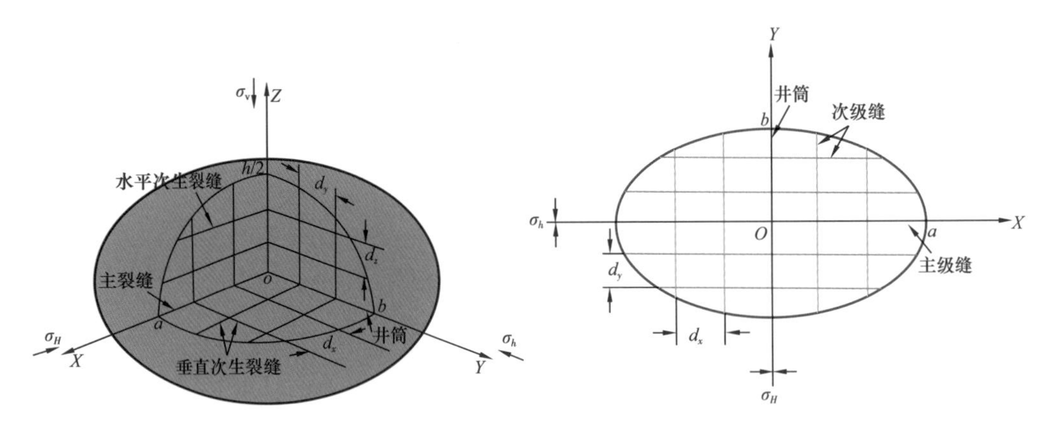

图 6　DFN 几何模型三维与平面俯视图

页岩气藏体积压裂后缝网导流能力在 100mD·m 数量级[21]，相对于页岩气藏，基本可认为无限导流裂缝，即基岩向裂缝供气能力决定着页岩气井产能。

根据试井无因次理论[28]，物模岩心气藏生产时间换算到真实气藏气井生产时间可通过无因次时间 t_D 来实现，其中无因次时间 t_D 表达式如下：

$$t_{\mathrm{D}}=\frac{Kt}{\mu\phi C_{\mathrm{g}}L_{\mathrm{e}}^{2}} \tag{8}$$

页岩气藏基岩块六面泄气，根据文献［29］，对应的基岩块特征长度 L_e 与基岩块边长 a 关系式如下：

$$L_{\mathrm{e}}=\frac{\sqrt{3}}{6}a \tag{9}$$

经物模岩心气藏生产时间换算得到的真实气藏气井生产时间为

$$t_{\mathrm{t}}=\frac{\mu_{\mathrm{t,g}}\phi_{\mathrm{t}}C_{\mathrm{t,g}}a^{2}}{12\mu_{\mathrm{m,g}}\phi_{\mathrm{m}}C_{\mathrm{m,g}}L_{\mathrm{e}}^{2}}\frac{K_{\mathrm{m}}}{K_{\mathrm{t}}}t_{\mathrm{m}} \tag{10}$$

图 7 为依据式（10）计算的矿场气井生产时间与物模岩心产气时间相似关系曲线，其中右图矿场基岩渗透率为模拟岩心的等效广义渗透率 2×10^{-9}mD，图 7 中的 5 条曲线分别对应不同体积压裂效果下（缝网密度，用等效立方体基质岩块边长来表示，分别为 0.55～30m），矿场气井生产时间与物模岩心产气时间的相似对应关系。可以看到当基岩块边长等于 0.55m 时，物模岩心产气时间与矿场气井生产时间正好相等，即二者的生产动态完全一致，也就是说对于极低孔渗的页岩气藏，如果压裂规模足够大，同样能够取得好的开发效果。随着基岩块边长增加，气井生产时间开始大于岩心产气时间，而且基岩块边长越大，差距越大，当边长达到 10m 时，气井的生产时间约是岩心的 300 倍，当边长达到

30m 时，气井的生产时间约是岩心的 3000 倍，可见对于极低渗流能力的页岩气藏，体积压裂效果直接决定了气藏的开发效果。

图 7（a）是以压裂后缝网控制的等效基岩块边长 30m、孔隙度 2.5% 为基础，开展的不同渗透率条件下（渗透率分布从 1×10^{-3}～1×10^{-7}mD），矿场气井生产时间与物模岩心产气时间的相似对应关系。可以看出，当岩心渗透率达到 1×10^{-3}mD 时，矿场气井生产时间约为物模岩心产气时间的 0.006 倍，远小于物模岩心产气时间，尽管基岩块边长达 30m，但是由于渗透率较高、泄流面积大，气藏开发效果很好，随着渗透率降低，矿场气井生产时间开始增加，当渗透率降至 1×10^{-5}mD 时，矿场气井生产时间约为物模岩心产气时间的 0.6 倍，当渗透率降至 1×10^{-7}mD 时，矿场气井生产时间约为物模岩心产气时间的 60 倍。

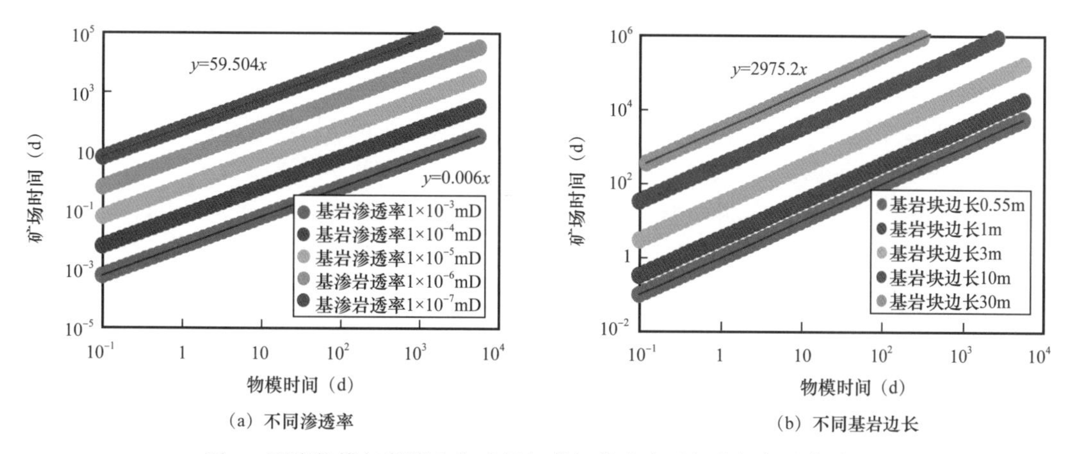

（a）不同渗透率　　（b）不同基岩边长

图 7　页岩物模气藏岩心与矿场气藏气井生产时间的相似对应关系

由此可见，决定页岩气藏开发效果最重要的两个参数是：渗透率和体积压裂效果（缝网密度），如果渗透率大于 1×10^{-5}mD，压裂效果好差对于气藏的开发效果影响不是非常明显；如果渗透率小于 1×10^{-7}mD，气藏的开发效果就决定于压裂形成的缝网密度大小，二者的有机结合才是页岩气藏能否有效开发的前提。

当物模岩心气藏与矿场页岩气藏初始压力、温度一致（初始状态一致），而且边界条件一致（都为封闭边界），根据渗流力学理论和相似理论，相同无因次时间 t_D 下矿场气藏气井采出程度与物模实验岩心采出程度应该一致，即相同无因次时间下物模与矿场气藏状态始终一致，即

$$\frac{G_{t,p}}{G_t}=\frac{G_{m,p}}{G_m} \tag{11}$$

其中，

$$G_t=\frac{V\rho G_c}{Z} \tag{12}$$

式（11）变换可获得真实气藏累计采气量表达式：

$$G_{t,p}=\frac{G_{m,p}}{G_m}G_t \tag{13}$$

因此，当已知真实气藏储层物性参数，可根据式（10）和式（13）及物模实验获得的累计采气量与时间关系曲线，即可获得真实气藏累计采气量与时间关系曲线，相应的气井日产气量为

$$q_t=\frac{\partial G_{t,p}}{\partial t_t} \tag{14}$$

3.2 气井生产动态曲线数值反演与分析

以取自川南昭通地区龙马溪组的实验岩心所代表的气藏为例，进行气井生产动态分析。储层原始地层压力 28MPa，渗透率 2×10^{-9}mD，孔隙度 1.05%，含气量 2.8m^3/t，假设 SRV 区域孔隙体积 $1000\times10^4m^3$，地质储量 $0.44\times10^8m^3$，对于两种不同压裂规模，基岩块等效边长分别为 0.55m 和 10m，开展气井生产动态分析。

图 8（a）是储层渗透率 2×10^{-9}mD、基岩块边长 0.55m 条件下气井生产动态分析曲线。根据图 7 可知，这一生产条件下气井生产时间与物模岩心产气时间比是 1∶1，二者在相同的采出程度下开发时间完全相同。图 8（a）生产动态曲线表明，虽然储层渗透率极低，但是如果压裂规模（缝网密度）足够大、效果足够好，同样能够得到较好的开发效果。可以看到，初期日产气量最高能够达到 $100\times10^4m^3$，之后迅速降低到 $3\times10^4m^3$，这一高速开发过程大约能够持续 100 天，累计产气量达到近 $1000\times10^4m^3$，主要是游离气，解吸附气

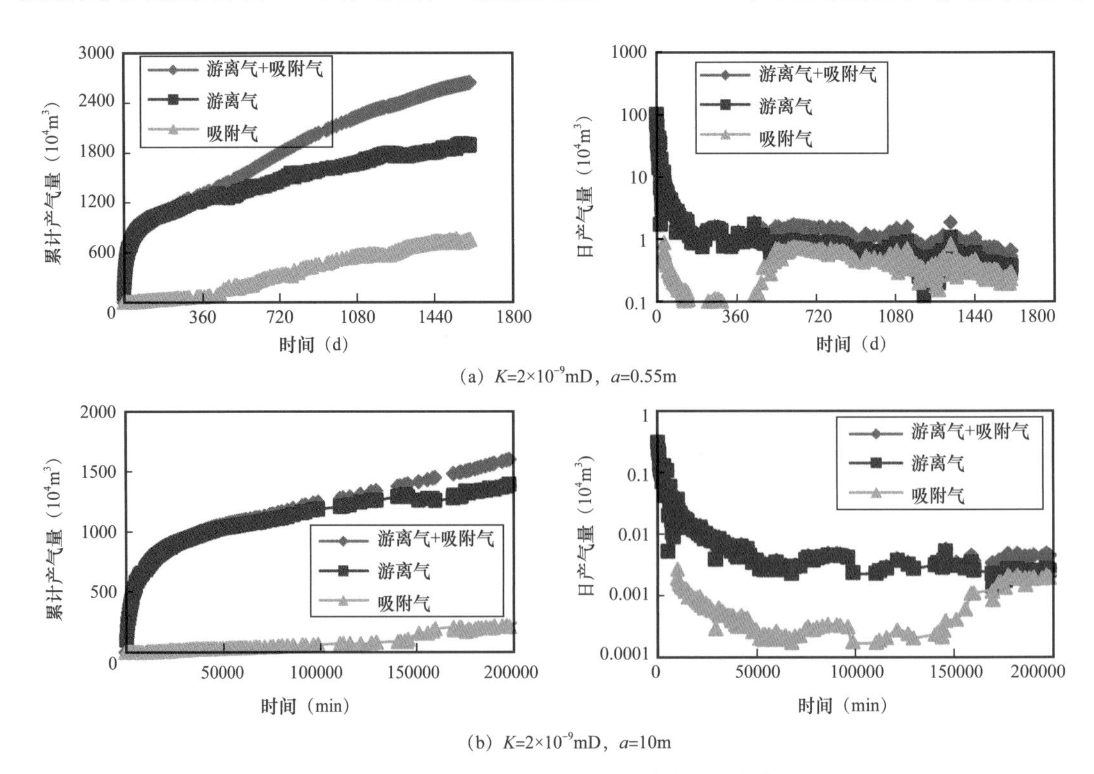

(a) $K=2\times10^{-9}$mD，a=0.55m

(b) $K=2\times10^{-9}$mD，a=10m

图 8　不同压裂规模下日产气量与累计产气量数值反演

量很少，只有约 $26 \times 10^4 m^3$；然后进入中速开发阶段持续约一年，日产气量约 $2 \times 10^4 m^3$，累计产气量达到约 $1300 \times 10^4 m^3$，其中解吸附气量只有约 $85 \times 10^4 m^3$；最后进入长期低速稳产阶段，日产气量约 $1 \times 10^4 m^3$，已经低速稳产了 3 年多，累计产气量达到 $2646 \times 10^4 m^3$，解吸附气量达到了约 $700 \times 10^4 m^3$，约占累计产气量的 25%，证明吸附气解吸是页岩气藏开发后期长期低速稳产的主要供气源，最终采出程度与物模一致，达到了 60%。可见，页岩气藏尽管储层渗流能力极低，但是如果开发压裂工艺技术能够满足要求，体积压裂效果好，缝网密度足够大，一样会得到高效开发。

图 8（b）是储层渗透率 2×10^{-9}mD、基岩块边长 10m 条件下气井生产动态分析曲线。此时气井生产时间与物模岩心产气时间比是 300∶1，二者在相同的采出程度下开发时间差距巨大，与上述生产动态完全不同。可以发现，气井在这一生产条件下，要达到物模实验目前的采出程度需要生产近 550 年，初期最高日产气量只有 $0.3 \times 10^4 m^3$，累计产气量达到 $1000 \times 10^4 m^3$ 需要约 114 年，后期长期稳产日产气量只有约 $0.002 \times 10^4 m^3$，显然不具备工业开采价值。

由此可见，页岩气储层渗透率和体积压裂效果（缝网密度）是决定气藏能否有效开发的关键，其中渗透率是根本，如果渗透率较高，对于体积压裂的要求就会低一些，即缝网密度可以适当小一些，与文献［30］认识基本一致，进一步证明本文物理模拟实验矿场换算结果的可靠性；但是如果渗透率很低，那么有效开发的前提就只有改善体积压裂的效果，尽可能增加裂缝网络密度值，极大地增大限流面积，这也是当前页岩气开发效果较好的关键原因所在，比如礁石坝页岩气藏，由于其孔、渗物性较好，对于压裂工艺的要求相对降低，在当前的压裂技术下开发效果良好；而对于本研究的昭通地区龙马溪组页岩气藏，由于孔、渗物性极差，而且当时的压裂工艺技术达不到要求，所以开发效果很差，一些气井甚至根本不能开发，随着大规模体积压裂技术水平的提高，有效开发孔、渗物性极差的页岩储层正逐渐成为可能。

研究结果表明，运用试井理论和相似理论可以将页岩气井全生命周期衰竭开发物理模拟实验结果合理地转化为气井全生命周期衰竭开发生产动态曲线，有效预测和解释矿场气井的生产动态特征，优化压裂参数，提出高效开发建议。

4 结论

（1）建立了页岩气耦合渗流物理模拟实验流程，运用全直径页岩岩心，首次开展了页岩气井全生命生产周期物理模拟实验，实验结果与气井的开发动态具有很好的一致性，可以用于有效预测气井的产能变化动态、吸附气量、游离气量及最终累计采气量，为页岩气井产能计算与生产动态预测提供了坚实的实验基础。

（2）通过对全直径页岩岩心衰竭开发物理模拟实验 1631 天的数据分析可以看出：模拟页岩储层的临界解吸压力为 12MPa，游离气量 3820.8mL、吸附气量 2151.2mL，合计地质储量 5972mL，吸附气与游离气之比约为 1∶2，含气量 $2.8m^3/t$，孔隙度 1.03%，广义渗透率约 2×10^{-9}mD。通过绘制游离气与吸附气的日产气量和累计产气量随地层压力和生产时间变化关系曲线，地层压力小于 12MPa 后吸附气开始大量解吸供气，地层压力越低吸附气在日产气中所占的比例越大。

（3）物理模拟实验结果表明，页岩气衰竭开发过程可以分为3个主要阶段：早期很短时间内产气速率迅速降低、累计产气量迅速增加的高速开发阶段；中期短时间内产气速率较高、降低减缓的中速开发阶段；后期产气速率低、递减极其缓慢且长期稳定的稳产阶段。

（4）根据试井理论和相似理论，推导了无因次时间 t_D 的数学表达式，实现了物模气藏岩心产气时间与矿场气藏气井生产时间的合理换算，通过数值反演方法，进而得到了页岩气藏气井累计产气量与日产气量计算公式，气井生产动态分析结果表明，渗透率与体积压裂效果（缝网密度）是气井开发效果好差的关键，其中渗透率是根本。

符号注释：

x—物模岩心累计采气量，mL；y—物模岩心地层视压力，MPa；G_{p1}—早期无解吸时累计采气量，mL；G_1—游离气量，mL；p—地层压力，MPa；z—气体压缩因子，无量纲；p_i—原始地层压力，MPa；z_i—地层原始压力下气体压缩因子，无量纲；V_{p1}—游离气占据的岩心孔隙体积，mL；V—岩心表观体积，mL；B_{gi}—原始地层压力下气体体积系数，无量纲；K—广义渗透率，mD；q—岩心流量，mL/min；P_{sc}—标准大气压，0.101MPa；L—岩心长度，cm；A—岩心横截面积，cm^2；μ_g—气体黏度，mPa·s，$\mu_{m,g}$—物模岩心气体黏度，$\mu_{t,g}$—真实气藏天然气黏度；p_1—岩心出口压力，MPa；p_2—岩心入口压力，MPa；t—时间，s，t_m—物模岩心生产时间，s，t_t—真实气藏生产时间；t_D—无因次时间，无量纲；ϕ—孔隙度，f，ϕ_m—物模岩心孔隙度，ϕ_t—真实气藏孔隙度；C_g—气体压缩系数，1/MPa，$C_{t,g}$—真实气藏气体压缩系数，$C_{m,g}$—物模岩心气体压缩系数，g；L_e—特征长度，m；a—压裂后基岩块边长，m；$G_{m,p}$—物模岩心累计采气量，mL；$G_{t,p}$—矿场累计采气量，m^3；G_t—矿场气藏地质储量，m^3；G_m—物模岩心地质储量，mL；q_t—矿场气藏气井产气速度，m^3/s；V—真实气藏SRV区域体积，m^3；ρ—基岩密度，kg/m^3；G_c—含气量，m^3/t。

参 考 文 献

［1］吴奇，胥云，王腾飞，等.增产改造理念的重大变革——体积改造技术概论［J］.天然气工业，2011，31（04）：7-12.

［2］武瑾，梁峰，吝文，等.渝东北地区巫溪2井五峰组—龙马溪组页岩气储层及含气性特征［J］.石油学报，2017，38（5）：512-524.

［3］董大忠，王玉满，李新景，等.中国页岩气勘探开发新突破及发展前景思考［J］.天然气工业，2016，36（1）：19-32.

［4］纪文明，宋岩，姜振学，等.四川盆地东南部龙马溪组页岩微—纳米孔隙结构特征及控制因素［J］.石油学报，2016，37（2）：182-195.

［5］苏玉亮，盛广龙，王文东，等.页岩气藏多重介质耦合流动模型［J］.天然气工业，2016，36（2）：52-59.

［6］温庆志，高金剑，李杨，等.页岩储层SRV影响因素分析［J］.西安石油大学学报（自然科学版），2014，29（06）：58-64.

［7］Carlson E S，Mercer J C.Devonian shale gas production：mechanisms and simple models［J］.Journal of

Petroleum technology, 1991, 43（04）: 476–482.

［8］ Carlson E S.Characterization of Devonian shale gas reservoirs using coordinated single well analytical models［C］//SPE Eastern Regional Meeting.Society of Petroleum Engineers, 1994.

［9］ 纪文明，宋岩，姜振学，等．四川盆地东南部龙马溪组页岩微—纳米孔隙结构特征及控制因素［J］. 石油学报，2016, 37（2）: 182–195.

［10］ Javadpour F.Nanopores and apparent permeability of gas flow in mudrocks（shales and siltstone）［J］. Journal of Canadian Petroleum Technology, 2009, 48（08）: 16–21.

［11］ Javadpour F, Fisher D, Unsworth M. Nanoscale gas flow in shale gas sediments［J］.Journal of Canadian Petroleum Technology, 2007, 46（10）.

［12］ Schepers K C, Gonzalez R J, Koperna G J, et al.Reservoir modeling in support of shale gas exploration［C］// Latin American and Caribbean Petroleum Engineering Conference. Society of Petroleum Engineers, 2009.

［13］ Civan F, Rai C S, Sondergeld C H. Shale–gas permeability and diffusivity inferred by improved formulation ofrelevant retention and transport mechanisms［J］.Transport in Porous Media, 2011, 86（3）: 925–944.

［14］ Swami V, Settari A.A pore scale gas flow model for shale gas reservoir［C］//SPE Americas Unconventional Resources Conference. Society of Petroleum Engineers, 2012.

［15］ Swami V, Settari A T, Javadpour F.A numerical model for multi–mechanism flow in shale gas reservoirs with application to laboratory scale testing［C］//EAGE Annual Conference &Exhibition incorporating SPE Europe.Society of Petroleum Engineers, 2013.

［16］ Alharthy N S, Al Kobaisi M, Kazemi H, et al. Physics and modeling of gas flow in shale reservoirs［C］// Abu Dhabi International Petroleum Conference and Exhibition. Society of Petroleum Engineers, 2012.

［17］ 段永刚，魏明强，李建秋，等．页岩气藏渗流机理及压裂井产能评价［J］. 重庆大学学报，2011, 34（04）: 62–66.

［18］ 李治平，李智锋．页岩气纳米级孔隙渗流动态特征［J］. 天然气工业，2012，32（4）: 50–53.

［19］ 王玉普，左罗，胡志明，等．页岩高温高压吸附实验及吸附模型［J］. 中南大学学报（自然科学版），2015，46（11）: 4129–4135.

［20］ 郭小哲，王晶，刘学锋．页岩气储层压裂水平井气—水两相渗流模型［J］. 石油学报, 2016, 37（9）: 1165–1170.

［21］ 朱维耀, 亓倩, 马千, 等．页岩气不稳定渗流压力传播规律和数学模型［J］. 石油勘探与开发，2016, 43（2）: 261-267.

［22］ 钟光海，谢冰，周肖，等．四川盆地页岩气储层含气量的测井评价方法［J］. 天然气工业，2016, 36（8）: 43–51.

［23］ Sakhaee–Pour A, Bryant S.Gas permeability of shale［J］.SPE Reservoir Evaluation & Engineering, 2012，15（04）: 401–409.

［24］ 王瑞，张宁生，刘晓娟，等．页岩气扩散系数和视渗透率的计算与分析［J］. 西北大学学报，2013, 43（1）: 76–80.

［25］ 杨胜来，魏俊之．油层物理学［M］. 北京：石油工业出版社，2010.

[26] Meyer B R，Bazan L W.A discrete fracture network model for hydraulically induced fractures-theory，parametric and case studies [R] .SPE 140514,2011.

[27] 程远方，李友志，时贤，等.页岩气体积压裂缝网模型分析及应用[J].天然气工业,2013,33(9):53-58.

[28] 孔祥言.高等渗流力学[M].合肥：中国科学技术大学出版社，2010.

[29] 朱维耀，孙玉凯，王世虎，等.特低渗透油藏有效开发渗流理论和方法[M].北京：石油工业出版社，2010.

[30] 高树生，刘华勋，叶礼友，等.页岩气藏SRV区域气体扩散与渗流耦合数学模型[J].天然气工业，2017，37(1):97-104.

页岩高压等温吸附曲线及气井生产动态特征实验

端祥刚[1]，胡志明[1]，高树生[1]，沈　瑞[1]，刘华勋[1]，常　进[1]，王　霖[1,2]

（1. 中国石油勘探开发研究院；2. 中国科学院大学渗流流体力学研究所）

摘　要：选取四川盆地长宁—威远龙马溪组页岩储层样品，采用高压等温吸附仪开展高压等温吸附曲线测试，运用自主研发页岩气流固耦合实验系统开展了单岩心对比和多岩心串联气井衰竭开发物理模拟实验；在总结吸附、解吸规律基础上，建立了高压等温吸附模型，修正了含气量计算方法，明确吸附气动用规律。研究表明，页岩储层高压条件下的等温吸附规律与常规低压下吸附规律不同，高压等温吸附曲线随压力变化存在最大过剩吸附量，对应压力为临界解吸压力。高压等温吸附曲线可用于评价页岩吸附气量及吸附气动用程度；高压等温吸附模型能够拟合和表征页岩高压等温吸附规律；修正后的含气量计算方法，可以更客观地评估含气量与吸附气比例，是储量评估和产量递减分析的理论基础；吸附气动用程度与压力密切相关，储层压力低于临界解吸压力，吸附气才能有效动用；气井生产过程中，近井地带压力下降幅度大，吸附气动用程度高，远离井筒，吸附气动用程度低或不动用。

关键词：页岩；高压等温吸附；过剩吸附量；临界解吸压力；气藏储量；生产动态

页岩储层富含有机质，有机质中发育大量微纳米级孔隙，作为自生自储的非常规气藏，大量页岩气以吸附态赋存于页岩孔隙中，其比例一般超过40%，研究页岩吸附规律对含气量计算、储量评估及产量预测具有重要意义[1, 2]。目前页岩等温吸附规律的研究多沿用煤层气的吸附理论，以室内等温吸附实验为主[3, 4]，测试压力一般在6～15MPa，远低于国内现阶段投入开发页岩储层的压力[5-8]（四川长宁—威远储层温度为70～120℃，压力为30～60MPa），低温低压下的测试方法和理论能否反映真实页岩储层气体的吸附/解吸规律值得商榷。国外研究[9-12]表明，页岩等温吸附曲线在高压下存在先上升后下降的趋势，这与常规吸附规律有所不同，说明目前采用低压测试曲线和Langmuir模型获得储层条件的含气量存在一定的局限性[13-15]。吸附气在含气量中所占比例很大，是气井进入低产、稳产期后的重要气源，明确储层条件下的页岩气吸附/解吸规律，是制订页岩气开发规划和研究产量递减规律的基础。

页岩高压等温吸附方面的研究目前未形成统一的机理认识，吸附/解吸规律不清，将导致含气量计算不准、开发规划预测误差大。因此，采用高压等温吸附仪器（最高测试压力69MPa），选取四川盆地长宁—威远地区志留系龙马溪组页岩样品，开展储层压力条件下的等温吸附测试、吸附气的产出特征、吸附气动用规律等实验，在此基础上建立等温吸附模型并修正页岩含气量计算方法，探索页岩气高效开发的基础理论。

1 高压等温吸附实验设计

1.1 实验样品

实验样品选自四川盆地长宁—威远地区龙马溪组龙一段1亚段，基础参数见表1。实验中将样品分成两部分，一部分烘干后粉碎，筛选0.15～0.25mm（60～100目）页岩样品进行等温吸附测试，另一部分柱状样品用于流动实验。

表1 样品基础参数

地区	样品编号	取心层位	取样深度（m）	总有机碳含量（%）
长宁	N01井1号	龙一段1亚段	2516	4.60
	N03井2号	龙一段1亚段	2391	4.20
威远	V02井1号	龙一段1亚段	2568	3.46
	V03井2号	龙一段1亚段	3177	3.71
	V04井3号	龙一段1亚段	3500	2.93

1.2 实验设备

等温吸附测试实验采用经典的容量法，设备为高压气体等温吸附仪，最大工作压力为69MPa，其压力传感器精度为最大量程的0.05%，恒温油浴最高可达177℃，控制精度为0.1℃。流动实验采用自主设计的页岩气衰竭开发物理模拟实验装置，可实现不同尺度、不同气体和不同岩心的页岩气流动物理模拟实验。

1.3 实验方案

1.3.1 高压等温吸附测试实验

实验装置如图1所示，步骤为：（1）将100g样品放入样品缸，检查气密性，利用基准缸精确测量实验系统的自由空间体积（包括参照缸及连接管线空间体积、样品缸剩余的自由空间及连接管线空间体积和页岩颗粒间空隙），连续测量多次，直至误差小于5%；（2）抽真空后关闭样品缸，向参照缸中充入一定压力的甲烷气体，待压力平稳后打开样品缸阀门，让两缸气体连通，达到压力平衡稳定后，记录平衡压力，计算测试吸附量：

$$n_{\text{test}}=10^{6}\left[\frac{p_{0}V_{\text{c}}}{Z_{0}RT}-\frac{p_{1}\left(V_{\text{c}}+V_{\text{s}}\right)}{Z_{1}RT}\right] \tag{1}$$

（3）关闭样品缸，继续向参照缸充入气体，循环上述平衡过程，直到完成全部实验。

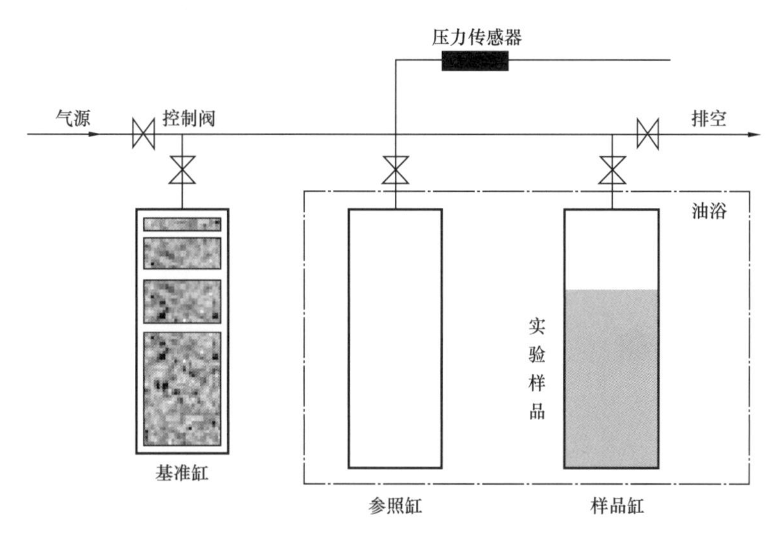

图 1　容积法等温吸附曲线测试装置

1.3.2　页岩气开发特征模拟实验

实验装置如图 2 所示，方案为：（1）采用同层位的柱状页岩样品，干燥后放入驱替系统，饱和甲烷气体至原始地层压力状态后，打开出口端模拟储层条件下的衰竭开发过程；（2）采用惰性气体氦气（忽略吸附作用）开展比对实验，分析吸附作用对产气规律的影响；（3）采用 5 块页岩样品串联的多测压点模拟实验，研究压力传播距离和吸附气动用压力的关系，根据各测点压力的变化，结合页岩物性参数和物质平衡方程计算产气规律和吸附气产出比例。

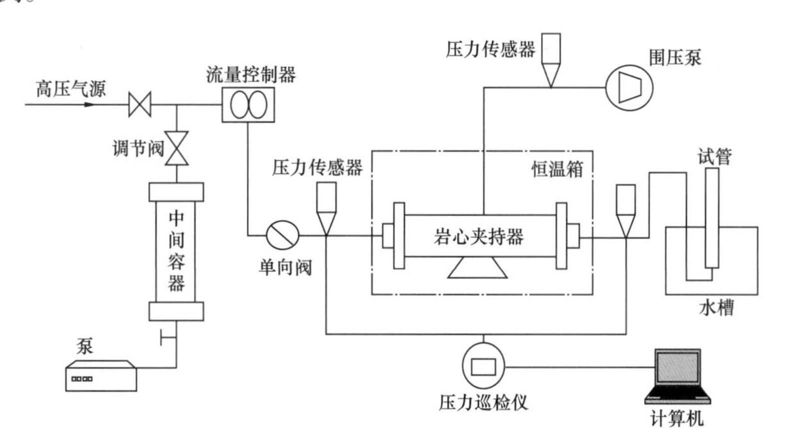

图 2　页岩气开发特征模拟实验装置

2　高压等温吸附特征

2.1　吸附量的定义

吸附是剩余力场使体相组分在相界面处富集的现象，吸附量为界面层溶剂中所含溶质量与体相中相同溶剂中所含溶质量之差，也称之为过剩量[16]。以页岩吸附甲烷气为例

（图 3），孔隙壁面存在吸附力场，吸附层内的甲烷分子密度远大于远离壁面的游离空间的甲烷密度，甲烷的过剩吸附量可表示为

$$G_{ex}=\left(\rho_a-\rho_g\right)V_a\frac{RZ_{sc}T_{sc}}{Mm_0p_{sc}} \tag{2}$$

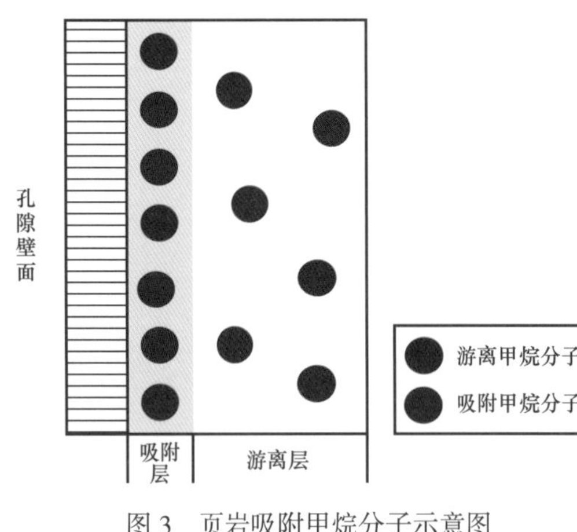

图 3　页岩吸附甲烷分子示意图

当压力较低时，游离相密度 ρ_g 较小，且远小于吸附相密度 ρ_a，因此 ρ_gV_a 项对吸附量的影响较小，一般认为 ρ_aV_a 即为页岩的吸附量。实际上根据 Gibbs 的定义[17]，ρ_aV_a 为绝对吸附量，即吸附空间内所有的甲烷气分子，而页岩吸附量为过剩吸附量，对于温度和压力较高的页岩储层，游离气处于超临界状态，游离相密度较大，如果忽略 ρ_gV_a 会导致测试结果与页岩吸附量的差值很大。因此，必须区别高压和低压下的吸附特征，才能更好地描述高压下页岩的等温吸附曲线。

目前没有直接的方法获取吸附相密度和吸附相体积，容积法和重量法都是通过间接的手段获取过剩吸附量。由式（1）可以看出，测试吸附量为游离气体的减少量，在初始标定样品缸自由体积时，包括了吸附空间和游离空间体积，随着吸附的进行，吸附分子逐渐占据一部分自由体积，并随压力变化而变化，因此吸附量计算需要在不同压力下进行自由体积修正，应该减去吸附相体积，则测试绝对吸附量的计算式应为

$$n_{abs}=10^6\left[\frac{p_0V_c}{Z_0RT}-\frac{p_1\left(V_c+V_s-V_a\right)}{Z_1RT}\right] \tag{3}$$

吸附相体积难以准确获取，目前的测试方法都不进行体积修正，所获取的测试吸附量均为过剩吸附量，而不是绝对吸附量。同样，重量法测量得到的也是过剩吸附量[18]。综上所述，目前测试得到的吸附量均为过剩吸附量，而不是绝对吸附量，过剩吸附量与绝对吸附量有如下关系：

$$G_{ex}=G_{abs}-\rho_gV_a\frac{RZ_{sc}T_{sc}}{Mm_0p_{sc}} \tag{4}$$

需要明确的是，目前尚没有足够精确的测试技术来获得绝对吸附量，通常采用假设吸附相密度或吸附相体积计算绝对吸附量[19, 20]。

2.2　页岩高压等温吸附特征

从不同页岩样品在高压条件下的等温吸附曲线（图 4）可以看出，虽然样品同属于龙马溪组，但不同地区、不同井的最大等温吸附量却存在一定的差异。威远地区的最大过剩吸附量在 1.11～2.16m³/t，长宁地区两口井样品的最大过剩吸附量在 1.45～1.68m³/t。在低压阶段

（小于 10MPa），吸附量随着压力的增加而快速上升，但是超过一定压力（12.0～18.5MPa）之后，吸附量随着压力的增加而降低。

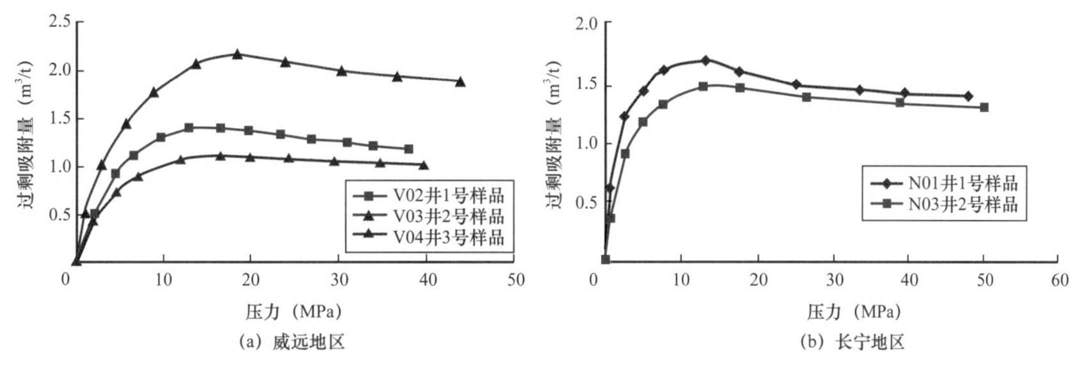

图 4　不同地区页岩等温吸附曲线

图 5 所示，实验结果（N03 井 2 号样品）与低压下的等温吸附曲线（Langmuir 模型拟合）的变化规律不同，高压条件下页岩等温吸附曲线不再是一条单调递增的曲线，而是存在最大过剩吸附量，其物理含义为页岩的最大吸附能力，该值能够为评价不同地区吸附气量提供依据。最大过剩吸附量对应的压力为临界解吸压力，其物理意义为只有当系统压力小于临界解吸压力时，吸附气才开始大量解吸。

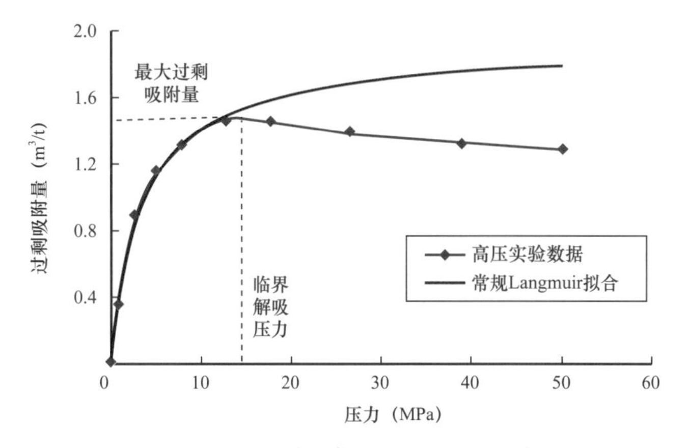

图 5　最大过剩吸附量和临界解吸压力

等温吸附曲线下降是页岩气在储层高压条件下的必然趋势，符合页岩气超临界吸附特征，其原因是测试曲线为过剩吸附量曲线，从定义和式（2）可以看出，过剩吸附量是个相对量，而不是传统意义上的绝对吸附量。吸附分子主要受岩石有机质和黏土矿物等固体分子对甲烷分子的色散力[3]作用，当吸附进入高压阶段后（约大于 15MPa），壁面吸附甲烷分子的吸附力场随压力增加变化不大，吸附分子逐渐增加并达到饱和，在密度曲线上表现为吸附相密度在高压下趋于平稳（图 6）。而游离态分子仅受到气体分子间作用力，随着压力增加，游离分子之间的作用力持续增加，使得游离相密度一直增加。当超过一定压力之后，二者的密度差必然存在一个极值，因此过剩吸附量也必然在对应的压力存在最大值。

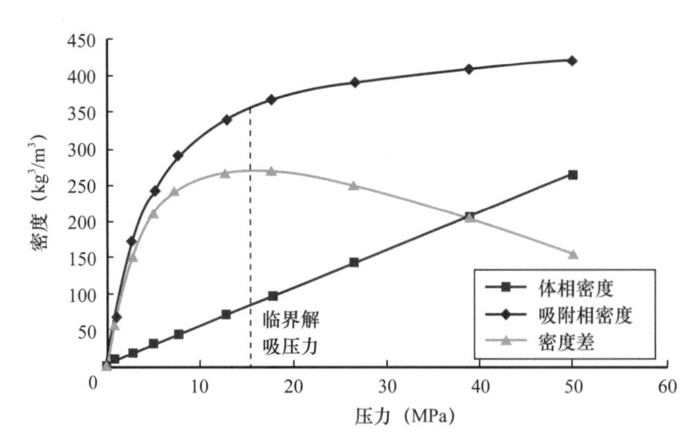

图 6　吸附相和游离相密度与压力关系曲线

需要指出的是，过剩吸附量在临界压力以上随着压力的增加而降低并不表示页岩的吸附能力在降低，实际上吸附空间内的绝对吸附量总是随着压力的增加而增加，类似于图 5 中采用常规 Langmuir 绝对吸附量模型拟合得到的吸附量变化曲线。在页岩高压吸附中，由于游离分子所受作用力持续增加，导致游离气在吸附空间的比例上升，表现为过剩的甲烷分子在减少，因此测试过剩吸附量降低。

3　高压等温吸附模型

过剩吸附量随着压力呈现先增加后降低的趋势，采用描述绝对吸附量的 Langmuir 模型及其他亚临界模型难以描述页岩的高压等温吸附规律[20]。因此，需要建立新的模型，由式（4）和式（2）可知，过剩吸附量和绝对吸附量有如下关系：

$$G_{\mathrm{ex}} = G_{\mathrm{abs}}\left(1-\frac{\rho_{\mathrm{g}}}{\rho_{\mathrm{a}}}\right) \tag{5}$$

上式中需要假设吸附相的密度，不同学者分别以液相密度 423kg/m³、范德瓦尔斯密度 373kg/m³ 及临界密度作为吸附相密度来拟合过剩吸附量曲线[18, 21]。虽然采用各种吸附相密度能够在一定程度上描述过剩吸附量下降的变化趋势，但是存在的问题是，这些方法在不同的压力阶段都采用同样的吸附相密度，而在压力增加过程中吸附甲烷的体积不同，因此吸附相密度是变化的，仅仅在吸附饱和以后，吸附相密度才接近一个定值，因此需要对此方法进行修正。

处于吸附态的分子具有一定体积，随着过剩吸附量的增大，吸附态分子所占据的体积也会不断增大直至吸附饱和。因此，可以假设吸附相体积近似等于吸附相分子所占的总体积，而吸附相分子所占据的总体积可由吸附量对应的分子个数与每个吸附分子所占据的体积相乘得到，那么过剩吸附量与绝对吸附量的关系可由吸附相体积修正得到：

$$G_{\mathrm{ex}} = G_{\mathrm{abs}}\left(1-\rho_{\mathrm{g}}\frac{10^{3}}{M}N_{\mathrm{A}}V_{\mathrm{mole}}\right) \tag{6}$$

采用 Langmuir 单层吸附模型描述页岩超临界条件下的绝对吸附量，其拟合结果与超临界吸附特征相符，具有一定的实用性，但是 Langmuir 模型中一个最主要的假设条件是固体表面是均匀的，这与页岩孔隙壁面的非均质性严重不符，故采用 Freundlich 等温吸附方程来修正固体表面的非均质性，获得 L-F 方程来描述绝对吸附量，进而可获得过剩吸附量模型为

$$G_{ex}=G_{L}\frac{bp^{n}}{1+bp^{n}}\left(1-\rho_{g}\frac{10^{3}}{M}N_{A}V_{mole}\right) \tag{7}$$

假设每个吸附相分子所占据的体积为球形体积，那么特征体积的计算式如下：

$$V_{mole}=\frac{1}{6}\pi {d_{m}}^{3} \tag{8}$$

可以看出，修正关系式的关键在于确定吸附相分子所占的特征直径。由于壁面对分子的作用力远大于气体间分子作用力，而页岩吸附甲烷为单层吸附[17]，因此吸附分子的特征直径可视为与吸附层厚度相等。文献［16，22］认为，单层分子的吸附厚度等于气体分子运动直径，临界解吸压力以上吸附达到饱和，单层吸附层的厚度在 0.5nm 左右，Ambrose 等[23]估算了给定温度和压力下的页岩气体单层厚度，约为 0.7nm。采用上述模型对不同井的实验数据进行拟合（图 7），拟合特征直径在 0.44～0.48nm，与文献中给出的单层吸附层厚度认识基本一致。拟合结果表明，采用高压吸附模型可以很好拟合并预测高压等温吸附特征曲线，同时采用高压吸附模型可以预测地层条件下的过剩吸附量和绝对吸附量，进而为含气量计算提供吸附气数据。

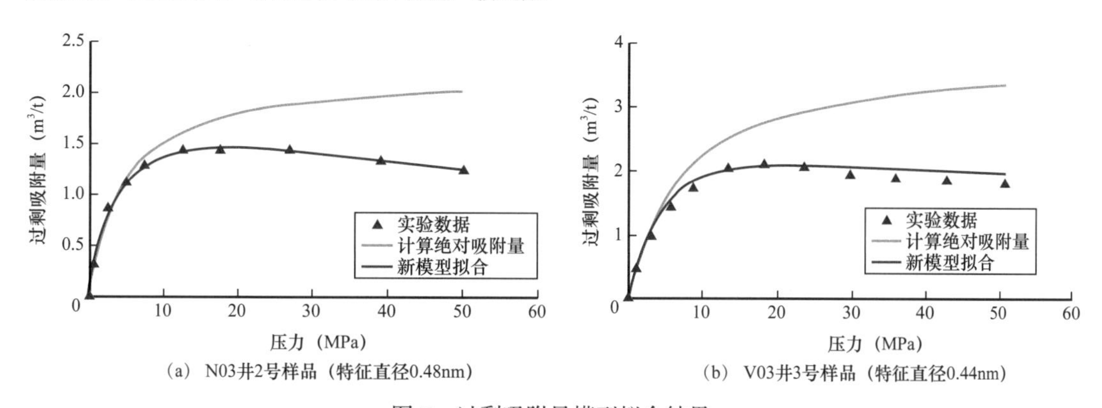

图 7　过剩吸附量模型拟合结果

4　含气量计算方法修正

含气量的准确计算是页岩气藏储量评估及产能预测分析的基础，一般通过实验（取心现场解吸和实验室等温吸附实验）和测井解释等方法获得。页岩气含气量主要考虑吸附气和游离气（忽略溶解气），吸附气量采用 Langmuir 模型计算，包括图 8 中吸附气和准吸附气（其中准吸附气为吸附空间内的游离气体，该部分气体不同于吸附气，其密度与游离气密度一样），游离气量计算仅考虑图 8 中游离相中的游离气体[24]，公式如下：

$$G_{\text{free}} = \frac{10^3 \phi S_g}{\rho_s B_g} \tag{9}$$

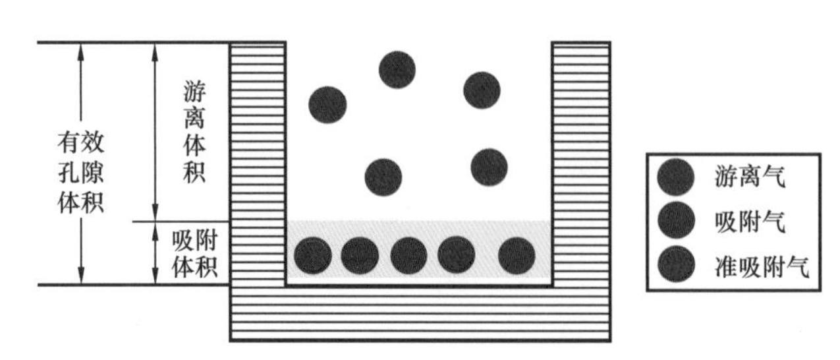

图 8　含气量组成示意图

随着近两年研究的深入，逐渐认识到吸附态甲烷占据一定的孔隙体积，需要对游离气体所占孔隙体积进行修正，假设吸附气所占的孔隙度为 ϕ_s，游离气量可修正为

$$G_{\text{free}} = 10^3 \frac{\phi S_g - \phi_s}{\rho_s B_g} \tag{10}$$

目前方法的局限性在于，一是把测试的过剩吸附量当作绝对吸附量，然后利用 Langmuir 公式外推得到储层温压下的绝对吸附量，这将对含气量的计算造成很大误差。二是游离气量的计算需要估算吸附气所占孔隙体积，该体积只能通过计算获取，无法获得相对准确值[25]。因此下面结合过剩吸附量模型提出含气量的计算方法。

根据图 8 及式（4）过剩吸附量与绝对吸附量关系，可将含气量转换为过剩吸附气量和真实游离气量：

$$G_{\text{total}} = \rho_g V_p \frac{R Z_{sc} T_{sc}}{M m_0 p_{sc}} + G_{ex} \tag{11}$$

由式（11）可知，只需测量过剩吸附量和有效孔隙体积就可以计算含气量，而不必考虑无法测量的绝对吸附量。

将利用高压吸附模型预测的储层条件下的吸附气含量与 Langmuir 模型计算的结果对比（表 2）可知，新模型计算的吸附气量比 Langmuir 模型低了 21.18%～38.56%，这对含气量和储量计算的影响是不可忽视的。

表 2　储层温压条件下不同模型吸附气量计算结果

地区	样品编号	吸附气量（m^3/t）		相对偏差（%）
		Langmuir 模型	新模型	
长宁	N01 井 1 号	1.92	1.45	24.48
	N03 井 2 号	1.70	1.34	21.18
威远	V02 井 1 号	2.25	1.53	32.04
	V03 井 2 号	2.68	1.84	31.33
	V04 井 3 号	1.95	1.20	38.56

5　气井生产动态特征

5.1　吸附气解吸生产动态对比实验

吸附气的有效动用是页岩气井生产后期稳产的保证，在分析吸附气解吸规律时，常规等温吸附方法存在两大难点：（1）难以准确判断临界解吸压力，（2）难以确定最大解吸气量。因此，有必要开展页岩气井生产动态模拟实验，结合高压等温吸附特征，研究吸附气的动用规律。

过剩吸附量特征曲线表明，当系统压力小于临界解吸压力时，吸附气才开始大量解吸，且不同样品的临界解吸压力不同，范围 12.0～18.5MPa；当系统压力大于临界解吸压力时，过剩吸附量随压力的增加而降低。气藏降压开采过程中，地层压力高于临界解吸压力时，吸附气基本不动用，气井主要产出游离气；地层压力小于临界解吸压力时吸附气才会大量解吸并产出。

选取 N03 井 2 号样品，开展有吸附作用的甲烷与无吸附作用的氦气衰竭开发对比实验，累计产气量对比如图 9 所示。分析可知，生产初期（生产时间小于 1900.6 分钟）氦气采气速度较快，大部分的气很快被采出，甲烷初期采气速度明显低于氦气。这主要因为一方面甲烷实验初期生产压力大于临界解吸压力，吸附气未被动用，主要产出游离气；另一方面孔隙内甲烷吸附层占据了一部分流动通道，降低了气体流动能力。生产进入后期（生产时间大于 1900.6 分钟），随着生产时间延长，压力逐步下降，出口端压力低于临界解吸压力，甲烷吸附气开始解吸成为供给气源，因而累计产气量缓慢增加，而氦气由于没有解吸气补充，累计产气量基本不再变化。

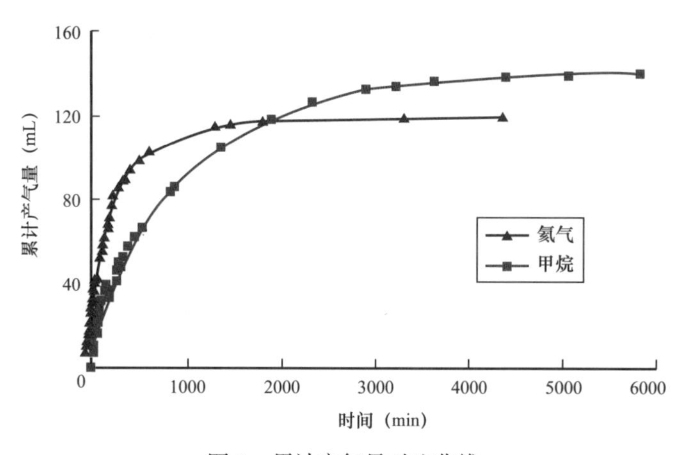

图 9　累计产气量对比曲线

从单位压差产气量指标看（图 10），氦气因无吸附 / 解吸作用，该指标在很长时间内保持在 4mL/MPa 上下波动，变化幅度很小，生产进入末期（系统压力小于约 4MPa），因系统能量衰竭，该指标才迅速下降直至停产。而甲烷在生产前期（系统压力大于约 15.8MPa），系统压力高于临界解吸压力，单位压差产气量随着压力的降低，缓慢上升；压力进一步下降，低于临界解吸压力时，进入吸附气解吸供给阶段，部分吸附气开始产出，压力下降越大，解吸气量越大，气源供给越多，同时释放的流动通道越大，甲烷气体流动能力越强，单位压降产气量明显上升，动用储量快速增加。至生产末期，与氦气实验类似，因能量衰竭指标迅速下降停产。

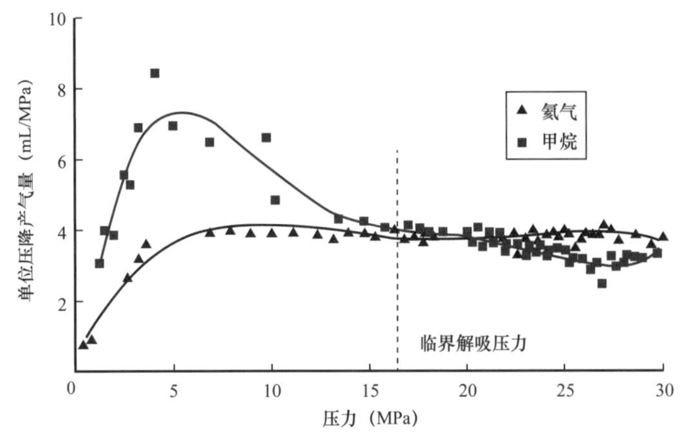

图 10　单位压降产气量对比曲线

5.2　吸附气解吸范围实验

采用单岩心无法获取生产过程中不同泄气半径处的压力分布数据，因而实验中采用 5 块岩心串联的方式进行衰竭开发，获取压力剖面数据，分析吸附气的动用程度与泄气半径的关系。

实验数据如图 11 所示，可以看出，初始产气速度很高，但下降很快，47 分钟后由初期的 107.6mL/h 下降至 10mL/h 以下，呈现典型的 L 形递减规律，随后进入低产阶段，累计生产 650 小时后，产气速度仍为 1.5mL/h。采用物质平衡方程，根据孔隙度、压力及高压等温吸附数据计算得到的累计产气量和实际产气量拟合度很高，证实了新模型的可靠性。至生产后期计算累产气量略高于实测累计产气量，主要原因在于计算使用的吸附气量为页岩粉碎颗粒的吸附气量，其值要高于柱状页岩样品的吸附气量，生产后期吸附气的解吸供给量略高。

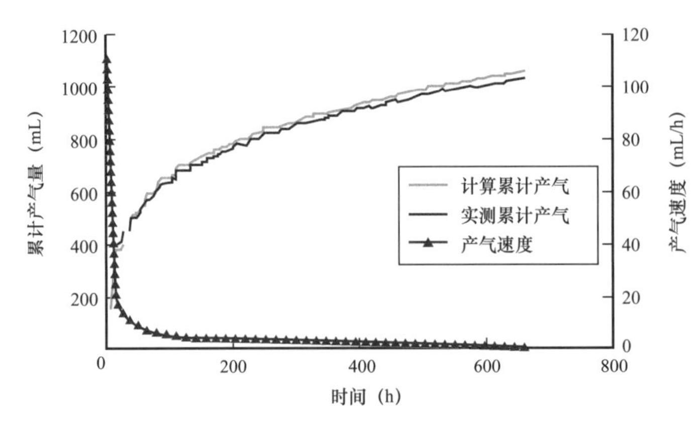

图 11　串联岩心产气速度和累计产气量曲线

计算不同压力阶段的游离气和吸附气量结果如图 12 所示。可以看出，开发初期产出气主要为游离气，累计产气量与平均视压力呈线性关系，这与常规气藏开发规律一致，但由于临界解吸压力附近吸附气开始解吸供给，累计产气量开始偏离游离气产量曲线，至实验结束，吸附气量占累计总产气量的 15%。

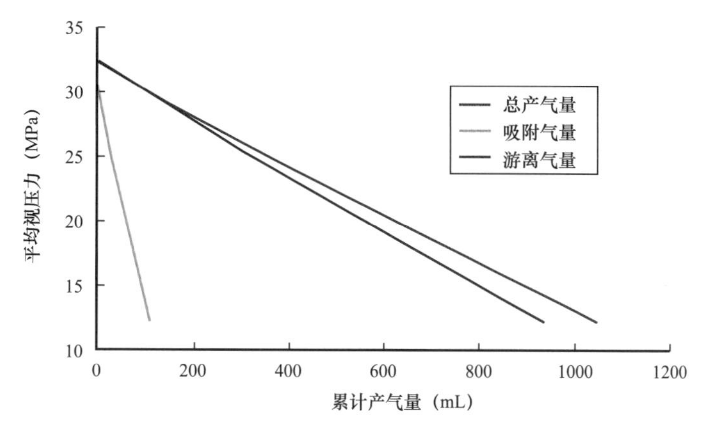

图 12　计算吸附气量、游离气量及累计产气量曲线

将 5 块串联岩心按入口端至产气端方向编号，依次为 1 号、2 号、3 号、4 号、5 号，这样便于单独分析每块岩心的压力变化和吸附气采出量。

图 13 为不同位置的岩心生产过程中的压力变化，可以看出，越接近出口端，岩心的平均压力越低。4 号和 5 号岩心的平均压力很快下降至临界解吸压力（约 15MPa）以下，这两块岩心中的吸附气大量解吸，吸附气产出比例分别达到了 14.4% 和 22.5%（图 14）；2 号、3 号岩心的平均压力在实验期末才降至临界压力以下，吸附气产出量很小，比例低于 5.0%；1 号岩心的平均压力一直大于临界解吸压力，吸附气基本不产出。这说明在气井生产过程中，近井地带压力下降快、幅度大，远低于临界解吸压力，吸附气动用程度高；远离井筒，压力下降程度小，吸附气动用程度低。因此，在生产制度设计时要充分考虑这一因素，建议生产初期主要考虑压裂液返排、排水等，控压生产，尽可能降低出砂、缝网闭合等不利因素的影响，随后放压生产，充分发挥吸附气潜力，提高日产气量与累计产气量。

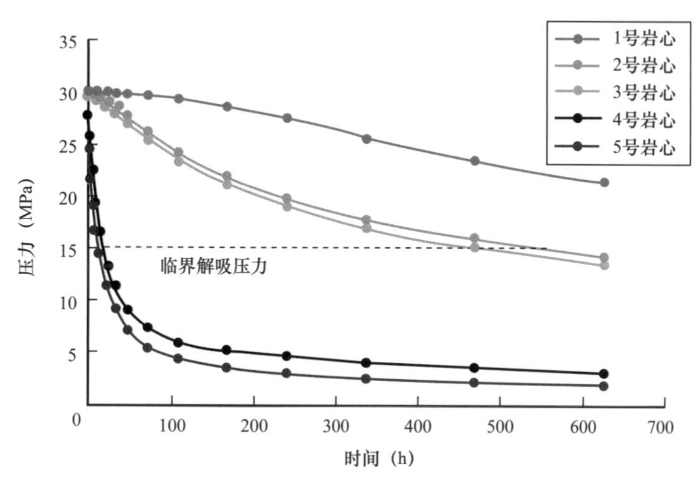

图 13　生产过程中不同位置岩心平均压力变化

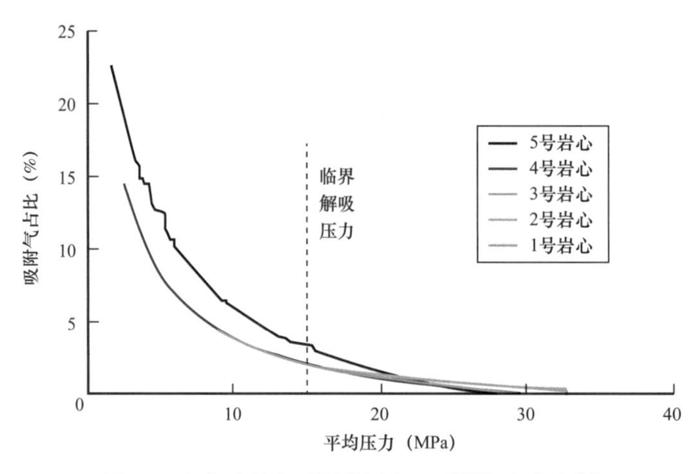

图 14　生产过程中不同位置岩心吸附气产出比例

6　结论

等温吸附实验测得的吸附量为过剩吸附量，页岩高压条件下的等温吸附规律与常规低压下吸附规律不同，高压等温吸附曲线随压力变化存在最大过剩吸附量，对应压力为临界解吸压力，高压等温吸曲线可用于评价页岩吸附气量及吸附气动用程度。

建立的高压等温吸附模型能够拟合和表征页岩高压等温吸附规律；修正的含气量计算方法，可以更客观评估含气量与吸附气比例，是储量评估和产量递减分析的理论基础。

吸附气动用程度与压力密切相关，储层压力低于临界解吸压力，吸附气才能有效动用。气井生产过程中，近井地带压力下降幅度大，吸附气动用程度高，远离井筒，吸附气动用程度低或不动用。建议页岩气开发采用初期控压、后期放压的开发模式，以最大程度提高吸附气的采收率。

符号注释：

b—Langmuir 结合常数，反映吸附与脱附速率的比值；B_g—体积系数，f；d_m—单个吸附相分子所占的特征直径，m；G_{abs}—绝对吸附量，m^3/t；G_{ex}—过剩吸附量，m^3/t；G_{free}—游离气量，m^3/t；G_L—Langmuir 体积，表示最大吸附能力，m^3/t；G_{total}—总含气量，m^3/t；m_0—页岩样品质量，kg；M—甲烷分子摩尔质量，g/mol；n—与吸附分子、吸附剂表面作用强度有关的参数，当 n=1 时就退化成 Langmuir 模型；n_{abs}—测试绝对吸附量，mol；n_{test}—测试吸附量，mol；N_A—阿伏伽德罗常数，$6.02\times10^{23}mol^{-1}$；$p$—气体压力，MPa；$p_0$—平衡前参照缸压力，MPa；$p_1$—平衡后系统压力，MPa；$p_{sc}$—标准状况压力，MPa；$R$—通用气体常数，$8.314m^3\cdot Pa/(mol\cdot K)$；$S_g$—含气饱和度，%；$T$—系统温度，K；$T_{sc}$—标准状况温度，K；$V_a$—吸附相体积，$m^3$；$V_c$—参照缸体积，$m^3$；$V_p$—页岩有效孔隙体积，$m^3$；$V_{mole}$—每个吸附相分子所占的特征体积，$m^3$；$V_s$—样品缸自由体积，$m^3$；$Z_0$—平衡前压缩因子，无因次；$Z_1$—平衡后压缩因子，无因次；$Z_{sc}$—标准状态压缩因子，无因次；$\rho_a$—吸附相密度，$kg/m^3$；$\rho_g$—游离相密度，$kg/m^3$；$\rho_s$—页岩密度，$kg/m^3$；$\phi$—孔隙度，%；$\phi_s$—吸附气所占的孔隙度，%。

参考文献

［1］邹才能，董大忠，王玉满，等．中国页岩气特征、挑战及前景（二）［J］．石油勘探与开发，2016，43（2）：166–178.

［2］邹才能，董大忠，王社教，等．中国页岩气形成机理、地质特征及资源潜力［J］．石油勘探与开发，2010，37（6）：641–653.

［3］左罗，熊伟，郭为，等．页岩气赋存力学机制［J］．新疆石油地质，2014，35（2）：32–36.

［4］赵文智，李建忠，杨涛，等．中国南方海相页岩气成藏差异性比较与意义［J］．石油勘探与开发，2016，43（4）：499–510.

［5］郭为，熊伟，高树生，等．温度对页岩等温吸附/解吸特征影响［J］. 石油勘探与开发，2013，40（4）：101–105.

［6］熊伟，郭为，刘洪林，等．页岩的储层特征以及等温吸附特征［J］．天然气工业，2012，32（1）：113–116.

［7］Yuan W，Pan Z，Li X，et al.Experimental study and modeling of methane adsorption and diffusion in shale［J］.Fuel，2014，117（117）：509–519.

［8］腾格尔，申宝剑，俞凌杰，等，四川盆地五峰组—龙马溪组页岩气形成与聚集机理［J］．石油勘探与开发，2017，44（1）：69–78.

［9］Gasparik M，Ghanizadeh A，Bertier P，et al.High–pressure methane sorption isotherms of black shales from the Netherlands［J］.Energy & Fuels，2012，26（8）：4995–5004.

［10］Gasparik M，Gensterblum Y，Ghenizadeh A，et al.High–pressure high–temperature methane sorption measurements on Carbonaceous shales by the manometric method experimental and data evaluation considerations for improved accuracy［J］.SPE Journal，2015，790–809.

［11］Merkel A，Fink R，Littke R.High pressure methane sorption characteristics of lacustrine shales from the Midland Valley Basin，Scotland［J］.Fuel，2016，182：361–372.

［12］Rexer TFT，Benham MJ，Aplin AC，et al.Methane adsorption on shale under simulated geological temperature and pressure conditions［J］.Energy & Fuels，2013，27（1）：3099–3109.

［13］赵天逸，宁正福，曾彦．页岩与煤岩等温吸附模型对比分析［J］．新疆石油地质，2014，35（3）：319–323.

［14］李相方，蒲云超，孙长宇，等．煤层气与页岩气吸附/解吸的理论再认识［J］．石油学报，2014，35（6）：1113–1129.

［15］Singh H，Javadpour F.Langmuir slip–Langmuir sorption permeability model of shale［J］.Fuel，2016，181：1096–1096.

［16］侯吉瑞，赵凤兰．界面化学及其在 EOR 中的应用［M］．北京：科学出版社，2014.

［17］Luo Z.A new method to calculate the absolute amount of high–pressure adsorption of supercritical fluid。［J］.Iranian Journal of Chemistry & Chemical Enginee.2015，34（2）：61–71.

［18］周尚文，王红岩，薛华庆，等．页岩过剩吸附量与绝对吸附量的差异及页岩气储量计算新方法［J］．天然气工业，2016，36（11）：12–20.

［19］Zuo L，Wang Y，Guo W，et al.Methane adsorption on shale insights from experiments and a simplified Local Density Model［J］.Adsorption Science & Technology，2014，32（7）：535–556.

［20］张庆玲．页岩容量法等温吸附实验中异常现象分析［J］．煤田地质与勘探，2015，43（5）：31-33.

［21］Chareonsuppanimit P，Mohammad SA，JR RLR，et al.High-pressure adsorption of gases on shales：Measurements and modeling［J］.International Journal of Coal Geology，2012，95（2）：34-46.

［22］Liu Y，Zhu Y，Li W，et al.Molecular simulation of methane adsorption in shale based on grand canonical Monte Carlo method and pore size distribution［J］.Journal of Natural Gas Science and Engineering，2016，30：119-126.

［23］Ambrose RJ，Hartman RC，Campos MD，et al.New pore-scale considerations for shale gas in place calculations［R］.SPE 131772.

［24］左罗，王玉普，熊伟，等．页岩含气量计算新方法［J］．石油学报，2015，36（4）：469-474.

［25］薛冰，张金川，杨超，等．页岩含气量理论图版［J］．石油与天然气地质，2015，36（2）：339-346.

Approximate semi–analytical modeling of transient behavior of horizontal well intercepted by multiple pressure–dependent conductivity fractures in pressure–sensitive reservoir

Junlei Wang, Ailin Jia, Yunsheng Wei, Yadong Qi

(Research Institute of Petroleum Exploration and Development, CNPC)

Abstract: Pressure sensitivity is a fundamental issue in the investigation of transient performance analysis in unconventional reservoirs.In this study, the focus is put on the transient behavior analysis for multi–fractured system with the effect of pressure sensitivity in fracture and reservoir. A generalized model is primarily formulated, where the details of fracture properties are taken into account. The corresponding semi–analytical solution is proposed by coupling fracture–reservoir flow model on the basis of linear superposition principle in the terms of Pedrosa's transform formulation [9].Furthermore, an accurate and reliable algorithm is developed to solve the resulting nonlinear mathematical problem.The validation of semi–analytical model is demonstrated based on the fact that the results from this study agree well with those reported in the previous literatures, numerical simulation, and field performance.

On the basis of the model, the transient flow behavior is detected and analyzed in detail. Results in this investigation show that the system exhibits six typical flow regimes : pseudo–bilinear flow, pseudo–linear flow, pseudo–radial flow, compound–linear flow, compound–radial flow and fracture closure effect. The characterization of transient behavior depends on the intense of pressure sensitivity in the fracture and reservoir, and fracture property parameters including fracture conductivity, number, spacing and asymmetry factor. Likewise, the transient behavior has four unique features : (1) the pressure sensitivity causes a gradual increasing in both pressure drop and derivative over time due to the partial closure in fracture, until to the maximum ; (2) the deviation of derivative from the non–pressure–sensitive case can be used to investigate the period when the partial fracture closure occurs, and the occurrence of maximal value on pressure derivative corresponds to the occasion that the fracture is completely closed and the conductivity approximately declines to zero ; (3) the influence of pressure–sensitive effect is more remarkable in the disadvantageous condition of fracture property, corresponding to considerable pressure depletion, which leads to great conductivity decay

rate ; (4) the existence of reservoir pressure-sensitive effect can significantly amplify the influence of fracture pressure-sensitive effect on the intermediate-and late-time flow regimes, while having a negligible influence on the early-time flow regime. This work provides a comprehensive knowledge and insight into the interpretation of fracturing evaluation and performance estimations of multi-fractured system in pressure-sensitive reservoirs.
Keywords: Multiple fractured horizontal well (MFHW) ; permeability modulus ; pressure sensitivity ; nonlinear equations ; transient pressure response ; fracture closure

With the large-scale field development of unconventional resource, the transient behavior analysis has been attracting more and more attention to predict well performance production and evaluate stimulation effectiveness. In the filed practice of some unconventional reservoirs (i.e.shale gas, naturally fractured and overpressured reservoirs), productivity loss is a remarkable feature of production dynamic ; for example, Haynesville shale displays typical productivity loss with extremely high initial production rates followed by steep declines [1]. This phenomenon is universally attributed to geomechanism factor, namely the pressure-sensitive nature of some formation and fracture [2-7]. However, transient analysis using conventional techniques to evaluate fractured well generally generates low-quality even incorrect estimates of the fracture-formation properties. For example, Clarkson et al. [5] noted that if permeability losses are not corrected in rate transient analysis, the resulting OGIP estimate would be pessimistic ; Qanbari et al. [8] demonstrated that the calculation error of initial formation permeability is approximately 32.1% if using conventional analysis method.

The exact mechanisms on stress sensitivity are commonly collapsed to the comprehensive result of proppant crushing and embedment in fractures and flow channel deformation in reservoir/matrix caused directly by pressure depletion during production process. Pressure depletion in porous media results in effective stress change which, in turn, changes fracture and formation permeability. The problem with the dependency of rock properties on stress was investigated by proposing the concept of permeability modulus (γ), which was first presented by Pedroza [9] to allow quantifying the permeability dependence on pressure and then widely extended to various fields of science and engineering [3-5, 10, 11].

The dependence of permeability on pore pressure makes the diffusion equation governing fluid flowing strongly nonlinear. In addition to numerical solution [12-16], analytical and semi-analytical approaches have been developed to deal with the nonlinear problem. These approaches are mostly proposed based on the well-known pseudo-function method [17] and perturbation technique [9], which serve the purpose of approximately or fully linearizing the diffusivity equation. The pseudo-function based approach succeeds in generating the required benchmark analytical solution. Ozkan et al. [18] incorporated stress-dependent permeability in fracture network into pseudo-pressure definition for modeling of fluid transfer from shale matrix to fracture network. In the work presented by Wang et al. [19], authors developed modified pseudo-pressure and pseudo-time which render the nonlinear equation amenable to analytical treatment, and provided appreciable numerical convolution method. Clarkson et al. [4] used modified pseudo-pressure and pseudo-time formulations to account for the stress-sensitive matrix permeability and applied time-dependent skin effect to consider fracture conductivity changes.Qanbari et al. [8] provided a correction factor to analyze transient linear flow in stress-sensitive tight oil reservoirs based on the modified pseudo-variables.This approach makes the diffusion equation subject to the classical linear equation,but the main difficulty of pseudo-variable method is to obtain the transient average pressure of dynamic drainage area, which is the requirement of calculating pseudo-time variable [20, 21] .By contrast,the perturbation technique universally or implicitly assumes that the fracture permeability modulus is identical to the matrix/reservoir ; for example, Wang [22] coupled the infinite conductivity fracture with matrix under the limitation of identical permeability modulus among multiple fractures ; in addition, identical permeability modulus between fracture and matrix is also applied in the literature presented by Chen et al. [23-25] .This particular assumption is convenient for directly coupling fracture and matrix after incorporating perturbation transformation, but it puts a constraint on the condition that the permeability modulus between fracture-fracture and fracture-matrix is notably different. In addition,Yao et al. [26] established a novel model to facilitate transient pressure analysis for hydraulically fractured well with stress-dependent conductivity,and solved the non-linear problem iteratively by discretizing source segments based on the work done by Zeng and Zhao's [27] work. However,Yao et al.'s [26] study focus only on the stress-dependent fracture conductivity without accounting for the effect of permeability changes in the matrix. In unconventional reservoirs,numerous experiments and field studies [4-5, 28-29] show that as the matrix is subjected to the stress-sensitive effect, the matrix is the parameter that is most strongly altered.

In addition,it is necessary in this study to emphasize the importance of fluid rheology in accurately interpreting pressure transient response. Power-law [30, 31] and Bingham fluids [32] are the most often used non-Newtonian fluids according to the rheological correlations. Flow behavior index n for power-law fluid model and pressure gradient G for Bingham fluid model are used to measure the level of difference between non-Newtonian and Newtonian fluids.In general,non-Newtonian fluid model is introduced to interpret pressure transient data in the field of polymer injection [33, 34],multi-phase displacement [35] and heavy oil flow [36] .On log-log curves of pressure and derivative,the type-curve characterization of non-Newtonian fluid is analogous to that caused by stress sensitivity system. Put another way, the transient response of non-Newtonian fluid also exhibits extra pressure depletion over Newtonian fluid. Hence,it may be more difficult to identify the characterization of transient pressure response caused by pressure sensitivity when using non-Newtonian fluid model.Moreover,the discussion of non-Newtonian fluid is also out of the scope of this paper,so the flow behavior of single-phase fluid is assumed to satisfy Newtonian flow behavior in the development of unconventional reservoirs in this work.

As analyzed from previous statement, there is still a lack of rigorous method to simulate the transient behavior of the system with multiple pressure-dependent conductivity fractures in the pressure-sensitive reservoirs. Therefore, the purpose of this study is to establish a semi-analytical model for capturing the transient behavior of such fractured system more rigorously and effectively. In the model, the fractures are presented with sufficient flexibility to consider the complex geometry including asymmetry factor and varying conductivity. The fundamental fracture flow is modeled on the scale of discrete segments and coupled with the analytical flow model in the reservoir. It is worth noting that fracture-fracture and fracture-reservoir flow are coupled on the basis of essential pressure-continuity condition, not presentative perturbation-continuity condition, which is different from the traditional treatments previously developed by Wang [22] and Chen et al. [23-25]. To verify the model in the multi-fractured system with pressure-sensitive effect, several numerical verifications are conducted with the help of commercial software (ECLIPSE and ECRIN). Case studies are generated to illustrate how the influence factors affect the performance of such system. Field case is used to justify the practical application of the approximate semi-analytical model.

1 Model description

Substantial amounts of experimental results show that the flowing path in fracture and reservoir is deformed and tends to close when the confining stress on rock increases as the pore pressure decreases (Fig.1). Thus, the permeability and porosity tend to behave a function of pore pressure. Based on the fact that the pressure dependence of porosity is less sensitive than it is for permeability [3, 10, 17, 37], the primary focus is on investigating how the pressure-dependent permeability influences dynamic characterization in the fracture-reservoir system.

Fig.2 shows the physical model of multiple fractured horizontal well used in this study, and the following assumptions are made:

- An isotropic, horizontal, slap reservoir is bounded by overlying and underlying impermeable strata;
- The reservoir possesses uniform thickness, porosity and initial permeability;
- Flow in reservoir matrix and fracture all satisfies Darcy's law, and the fluid follows the Newtonian flow behavior;
- The flux mass from the reservoir into horizontal wellbore is negligible, and the reservoir is produced only through a set of fully penetrating fractures;
- Fracture wings are asymmetrically spaced with regard to horizontal well, and all fractures are spaced perpendicular to horizontal wellbore;

- All fractures penetrate the formation completely, and convergent flow around the intersection of the wellbore and fracture could be accounted for by a skin factor [38]. Therefore the problem of deriving an analytical pressure transient solutions for MFHW is approximately two dimensional in essence ;
- The well keeps constant production rate. Wellbore storage effect, gravity effect, frictional resistance in wellbore and fracture face damage skin are not considered.

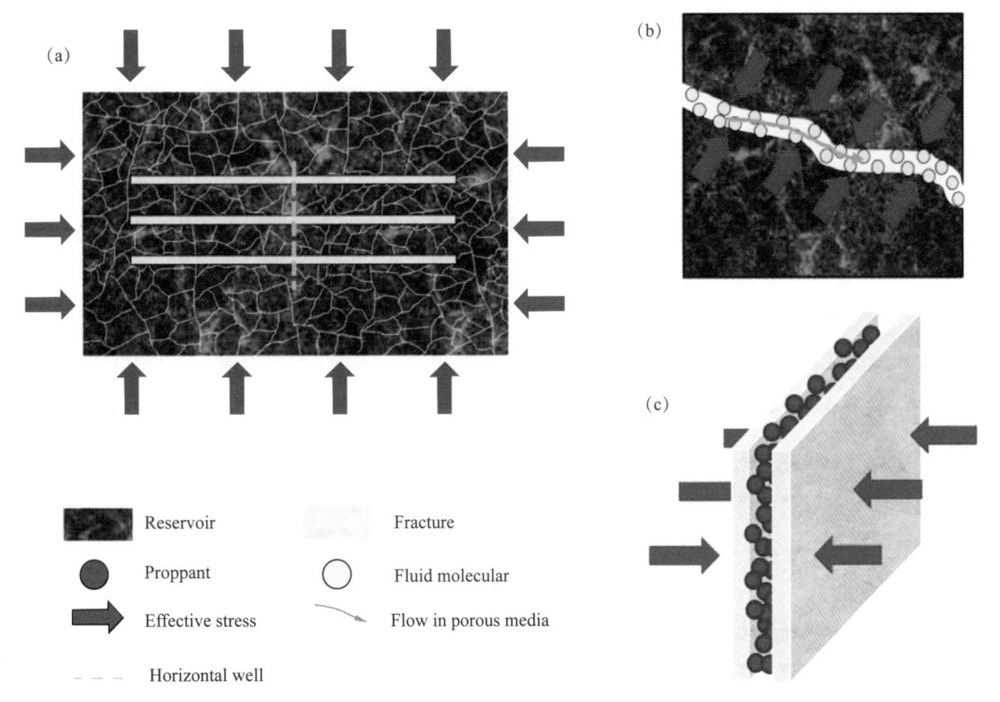

Fig.1 (a) Schematic of multiple fractured horizontal well ; (b) deformation of flowing path in reservoir ; (c) proppant crashing and embedment in fracture

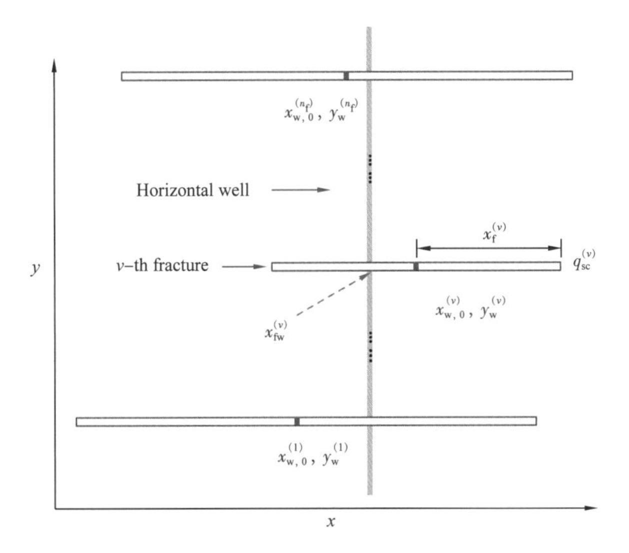

Fig.2 Location of fractures in infinite reservoir in 2D coordinate

2 Mathematical model

According to the feature of permeability-pressure relationship, the permeability in fracture and reservoir is expressed in the terms of permeability modulus. This definition assumes the following exponential relationship between permeability and pressure [Appendix A provides a fundamental discussion including experimental data and theoretical derivation to support the quotation of exponential dependence used in this paper], which is given as:

$$K_{\xi}=K_{\xi i}\exp\left[-\gamma_{\xi}\left(p_{i}-p_{\xi}\right)\right],\ \xi=\mathrm{f\ and\ m} \tag{1}$$

The analytical solution of Eq. (1) can capture general phenomenological trends of pressure sensitivity in terms of simple equation, which is widely applied in the field of pressure transient analysis by many experts[9, 14-16, 18]. The permeability modulus determines the intensity of the pressure-sensitive effect, which is in the range of 10^{-5} to 10^{-4} KPa^{-1}[18, 39]. It is noted that the fracture permeability modulus γ_f has a wider spectrum of behavior than the exponential tendency of formation permeability modulus γ_m, because fracture is particularly sensitive to pressure changes.

Here, in the scale of hydraulic fracture, the pressure dependence is directly related to the fracture conductivity rather than the fracture permeability. Therefore, the dimensionless conductivity for pressure-sensitive fracture is further given as:

$$C'_{\mathrm{fd}}\left(p_{\mathrm{f}}\right)=\frac{K_{\mathrm{fi}}w_{\mathrm{f}}}{K_{\mathrm{mi}}x_{\mathrm{f}}}\exp\left[-\gamma_{\mathrm{f}}\left(p_{\mathrm{i}}-p_{\mathrm{f}}\right)\right] \tag{2}$$

For simplicity, it needs to be stated that the physical variables in this study are all converted into dimensionless variables. As mentioned above, the Pedrosa's transform formulation[9] is introduced to weaken the nonlinearity caused by pressure-sensitive effect ($0<\gamma_{\xi D}\eta_{\xi D}<1$), where the relationship between pressure variable and Pedrosa's transformed variable η is satisfied as follows:

$$p_{\xi D}=-\frac{1}{\gamma_{\xi D}}\ln\left(1-\gamma_{\xi D}\eta_{\xi D}\right)\quad \xi=\mathrm{f\ and\ m} \tag{3}$$

Here, γ_{mD} is the dimensionless modulus of formation permeability, and γ_{fD} is of fracture permeability. Additionally, it is emphasized that the nonlinear Pedrosa's transformation of Eq. (3) is essentially a subset of a well-known procedure called the Cole-Hopf transformation[40-41]. The Cole-Hopf transformation could deal with the general nonlinear partial differential equation. In this work, the particular Cole-Hopf transformation corresponds to Eq. (3), and the nonlinear equations are given as Eq. (C2) and Eq. (D2).

2.1 Reservoir-fracture flow model

In this section, based on the assumption mentioned above, the diffusion equations governing fluid flowing in fracture and reservoir are respectively established. The flow in reservoir is described in the 2D coordinate(Fig.3a), and the flow inside fracture is described in the 1D coordinate(Fig.3b). For simplify, the mathematical models are written in the dimensionless form using dimensionless definitions in Appendix B.

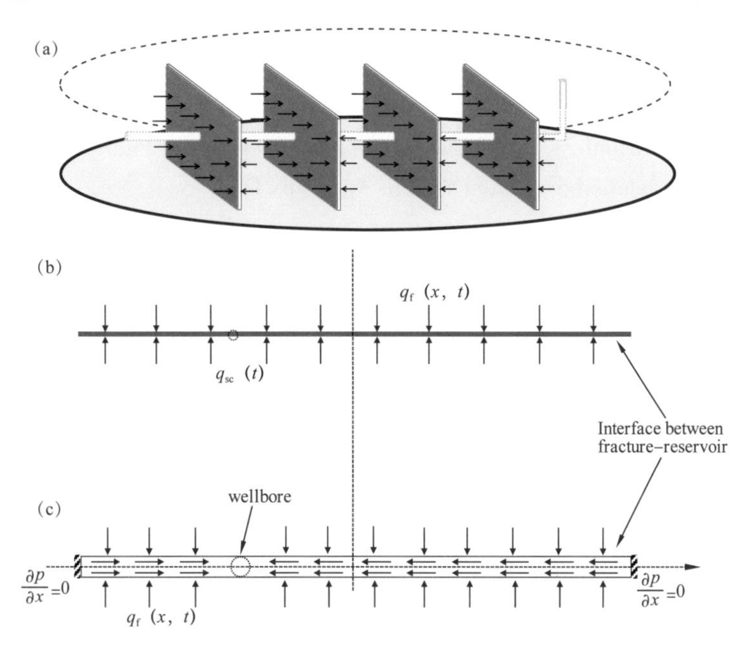

Fig.3 Schematic of (a) 3D reservoir–fracture flow model of MFHW ; (b) reservoir flow model ; (c) fracture flow model

In reservoir system, after incorporating Pedrosa's transform formulation, the diffusion equation with instantaneous source $q_{fD}\Delta x'_D$ is given as following dimensionless form,

$$\frac{\partial^2\eta_{mD}}{\partial x_D^2}+\frac{\partial^2\eta_{mD}}{\partial y_D^2}+\pi\Delta x'_D\sum_{v=1}^{n_f}q_{fD}^{(v)}\left(x'_D,t_D\right)\delta\left(x_D-x'_D,y_D-y_{wD}^{(v)}\right)=\frac{1}{1-\gamma_{mD}\eta_{mD}}\frac{\partial\eta_{mD}}{\partial t_D} \tag{4}$$

Eq. (4) is not fully linear because of the residual presence of the $1/(1-\gamma_{mD}\eta_{mD})$ term in the right hand side of diffusion equation in Eq. (4) .A perturbation technique is used to expand η_{mD} as a power series [$\eta_{mD}=\eta_{mD0}+\gamma_{mD}\eta_{mD1}+(\gamma_{mD}\eta_{mD2})^2+\cdots$] .Generally, the final solution for Eq. (4) is the sum of zero–and first–order perturbation series [9] .Further, zero–order term is sufficient for practical purposes to transient behavior in the stress–sensitive reservoir when $\gamma_{mD}<0.2$ [42], Wang et al. [43] and Chen et al. [25] compared a previous study presented by Pedrosa [9] and showed that a perturbation technique with a zeroth–order was able to produce results accurate enough for stress–sensitive system. By analyzing the magnitude of γ_{mD} based on the parameters in Tab.1, the magnitude of γ_{mD} is about in the range of 10^{-2}, generally smaller than 0.2.Therefore, Eq. (4) is fully linearized in terms of approximate zero–order series (seen Appendix C for the detailed derivation and result).

According to instantaneous sink/source function and linear superposition principle [44-45], the resulting pressure drop caused by n_f fractures in the forms of Pedrosa's transformed variable is satisfied as follows:

$$\eta_{mD}\left(x_D,y_D,t_D\right)=\pi\sum_{v=1}^{n_f}\int_0^{t_D}dt'_D\int_{x_{wD,0}^{(v)}-0.5\Delta x_D^{(v)}}^{x_{wD,0}^{(v)}+0.5\Delta x_D^{(v)}}\frac{q_{fD}^{(v)}\left(x'_D,t'_D\right)}{4\pi\left(t_D-t'_D\right)}\exp\left(-\frac{\left(x_D-x'_D\right)2+\left(y_D-y_{wD}^{(v)}\right)^2}{4\left(t_D-t'_D\right)}\right)dx'_D \quad (5)$$

In fracture system, the flowing equation is assumed to be incompressible [the unsteady-state flow in the fracture becomes insignificant in the case of less fracture volume relative to reservoir volume [46]]. The dimensionless governing equation is given in the following forms of Pedrosa's transformed variable (detailed derivation seen in Appendix D),

$$\frac{\partial^2\eta_{fD}^{(v)}}{\partial x_D^2}-\frac{\pi}{C_{fD}^{(v)}}q_{fD}^{(v)}\left(x_D,t_D\right)+\frac{2\pi}{C_{fD}^{(v)}}q_{scD}^{(v)}\left(t_D\right)\delta\left(x_D-\theta^{(v)}\right)=0,\ x_{wD,0}^{(v)}-x_{fD}^{(v)}\leqslant x_D\leqslant x_{wD,0}^{(v)}+x_{fD}^{(v)} \quad (6)$$

It is noted that Eq. (6) is fully linear because the right hand side is zero, which eliminates the residual presence of pressure-dependent term in the right hand side of Eq. (6). The accurate solution for Eq. (6) is obtained, not approximate solution like Eq. (5). Utilizing Green's function method (it is provided in Appendix C), the final expression within v-th fracture is given as

$$\eta_{fD}^{(v)}\left(x_D,t_D\right)=\eta_{wD}^{(v)}\left(t_D\right)+\frac{\pi}{C_{fD}^{(v)}}\int_{x_{wD,0}^{(v)}-x_{fD}^{(v)}}^{x_{wD,0}^{(v)}+x_{fD}^{(v)}}\left[N^{(v)}\left(x',x_D\right)-N^{(v)}\left(x',\theta^{(v)}\right)\right]q_{fD}^{(v)}\left(x',t_D\right)dx'$$

$$-\frac{\pi}{C_{fD}^{(v)}}2q_{scD}^{(v)}\left(t_D\right)\left[N^{(v)}\left(\theta^{(v)},x_D\right)-N^{(v)}\left(\theta^{(v)},\theta^{(v)}\right)\right] \quad (7)$$

2.2 Coupling reservoir and fracture model

In this study, the transient pressure solution must be semianalytically obtained by discretizing the fracture into several segments, as shown in Fig.4.

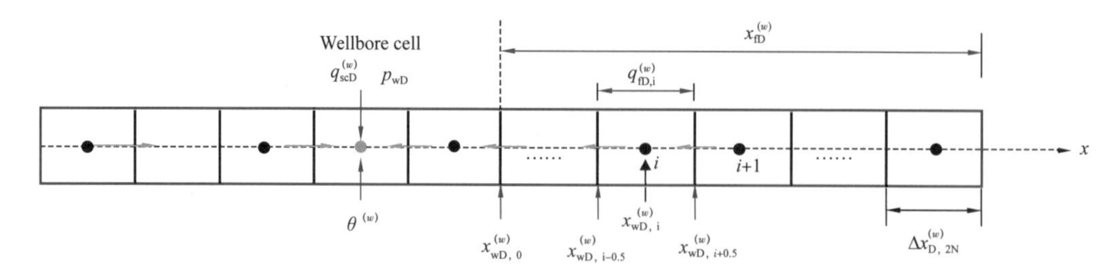

Fig.4 Discretization of fracture into N segments

On the m-th time step $t_D=t_{D,m}+0.5\Delta t_{D,m}$, on the interface between fracture and reservoir, the pressure drop of j-th segment on v-th fracture caused by other segments is expressed in the forms of Pedrosa's transformed variable. Therefore, Eq. (5) in the reservoir is discretized as,

$$\eta_{\mathrm{mD},j}^{(v,m)}=\sum_{w=1}^{n_{\mathrm{f}}}\sum_{i=1}^{2N_{\mathrm{w}}}q_{\mathrm{fD},i}^{(w,m)}\Delta\eta_{\mathrm{mD}}^{(m)}\left(x_{\mathrm{wD,j}}^{(y)},y_{\mathrm{wD}}^{(v)};x_{\mathrm{wD,i}}^{(w)},y_{\mathrm{wD}}^{(w)}\right)$$

$$-\sum_{w=1}^{n_{\mathrm{f}}}\sum_{i=1}^{2N_{\mathrm{w}}}\sum_{k=1}^{m-1}q_{\mathrm{fD},i}^{(w,k)}\left[\Delta\eta_{\mathrm{mD2}}^{(m)}\left(x_{\mathrm{wD},j}^{(v)},y_{\mathrm{wD}}^{(v)};x_{\mathrm{wD}}^{(w)},y_{\mathrm{wD}}^{(w)};k\right)-\Delta\eta_{\mathrm{mD1}}^{(m)}\left(x_{\mathrm{wD},j}^{(v)},y_{\mathrm{wD}}^{(v)};x_{\mathrm{wD},i}^{(w)},y_{\mathrm{wD}}^{(w)};k\right)\right] \tag{8}$$

Where, the elementary functions in Eq.(8)are provided in Appendix C, and the location of j–th segment is given as,

$$x_{\mathrm{wD,j}}^{(v)}=x_{\mathrm{wD,0}}^{(v)}-x_{\mathrm{fD}}^{(v)}+(j-0.5)\Delta x_{\mathrm{D}}^{(v)} \tag{9}$$

In fracture flow model, by discretizing Eq. (7), the pressure drop of j–th segment within v–th fracture is expressed as following series forms :

$$\eta_{\mathrm{fD},j}^{(v,m)}-\eta_{\mathrm{wD}}^{(v,m)}=\frac{\pi}{C_{\mathrm{fD}}^{(v)}}\sum_{i=1}^{2N_{v}}q_{\mathrm{fD},i}^{(v,m)}\int_{x_{\mathrm{wD},i}^{(v)}+0.5\Delta x_{\mathrm{D}}^{(v)}}^{x_{\mathrm{wD},i}^{(v)}+0.5\Delta x_{\mathrm{D}}^{(v)}}\left[N^{(v)}\left(x',x_{\mathrm{wD},j}^{(v)}\right)-N^{(v)}\left(x',\theta^{(v)}\right)\right]\mathrm{d}x'$$

$$-\frac{\pi}{C_{\mathrm{fD}}^{(v)}}\sum_{i=1}^{2N_{v}}q_{\mathrm{fD},i}^{(v,m)}\left[N^{(v)}\left(\theta^{(v)},x_{\mathrm{wD},j}^{(v)}\right)-N^{(v)}\left(\theta^{(v)},\theta^{(v)}\right)\right]\Delta x_{\mathrm{D}}^{(v)} \tag{10}$$

Where the total dimensionless influx rate of v–th fracture is satisfied as

$$q_{\mathrm{scD}}^{(v)}=\frac{1}{2}\sum_{i=1}^{2N_{v}}q_{\mathrm{fD},i}^{(v)}\Delta x_{\mathrm{D}}^{(v)} \tag{11}$$

On the interface between fracture and reservoir, the flux and pressure are continuous. It needs to be emphasized again that pressure variable p is continuous, not Pedrosa's transformed variable η. Here, there exist two kinds of distinct coupling conditions between fracture and reservoir, which is dependent on whether the reservoir is subject to stress–sensitive effect or not. The coupling condition is given as

$$\eta_{\mathrm{fD},j}^{(v,m)}=\begin{cases}1/\gamma_{\mathrm{fD}}^{(v)}-\left[1-\gamma_{\mathrm{mD}}\eta_{\mathrm{mD},j}^{(v,m)}\right]\gamma_{\mathrm{fD}}^{(v)}/\gamma_{\mathrm{mD}}/\gamma_{\mathrm{fD}}^{(v)}, & \gamma_{\mathrm{mD}}>0\\ 1/\gamma_{\mathrm{fD}}^{(v)}-\exp\left[-\gamma_{\mathrm{fD}}^{(v)}p_{\mathrm{mD},j}^{(v,m)}\right]/\gamma_{\mathrm{fD}}^{(v)}, & \gamma_{\mathrm{mD}}=0\end{cases} \tag{12a}$$

Besides, the pressure condition, not η, is continuous on the intersection between fracture and horizontal wellbore.

$$p_{\mathrm{wD}}^{(m)}=-\frac{1}{\gamma_{\mathrm{fD}}^{(1)}}\ln\left(1-\gamma_{\mathrm{fD}}^{(1)}\eta_{\mathrm{wD}}^{(1,m)}\right)\cdots=-\frac{1}{\gamma_{\mathrm{fD}}^{(n_{\mathrm{f}})}}\ln(1-\gamma_{\mathrm{fD}}^{(n_{\mathrm{f}})}\eta_{\mathrm{wD}}^{(n_{\mathrm{f}},m)}) \tag{12b}$$

The constraint condition of constant rate for multiple–fracture system yields the following expression,

$$0.5\sum_{v=1}^{n_{\mathrm{f}}}\sum_{i=1}^{2N_{v}}q_{\mathrm{fD},i}^{(v)}\Delta x_{\mathrm{D}}^{(v)}=1 \tag{13}$$

Additionally, the radial convergence of flow toward the wellbore within the transverse fracture can be taken into account following the method presented by Wang and Jia[47].

2.3 Semi-analytical solution

It is assumed that every fracture is discretized into a set of segments, $2N$.On the basis of continuity condition of Eq. (12), nonlinear equations can be established by substituting Eq. (8) into Eq. (10) on every segment (j=1,2, $\cdots$,$2N$) of every fracture (v=1,2, $\cdots$,n_f) .The corresponding solution is obtained numerically through incorporating Newton–Raphson method.This type of solution has been called semi analytical.

In each timestep, the unknown variables yield the following matrix form :

$$\boldsymbol{J}\delta\vec{\boldsymbol{X}}=-\vec{\boldsymbol{F}} \tag{14}$$

Where Jacobian matrix $\boldsymbol{J}$ is the matrix of partial derivatives, $\delta\vec{\boldsymbol{X}}$ is the correction of unknown variables, $\vec{\boldsymbol{F}}$ is the known vector, detailed information seen in Appendix F.To obtain the high accuracy results, the calculation procedure is given as follows :

● Step 1: Initial guess, including dimensionless influx distribution along fracture and wellbore pressure.The initial guess is critical for fast convergence.Here, the results of corresponding linear system are selected as initial guess,

$$\boldsymbol{A}\vec{\boldsymbol{X}}=\vec{\boldsymbol{B}} \tag{15a}$$

Eq. (15a) is solved by using Newton iteration method (Appendix E provides the details) .

● Step 2: Calculating Eq. (14) by using Newton–Raphson method, and obtaining the correction of unknown vector, which are included in the vector of $\vec{\boldsymbol{X}}$.

● Step 3: Updating unknowns in Eq. (14) with the correction vector, and then obtaining new unknowns according to Eq. (15b)

$$\vec{\boldsymbol{X}}_{\text{new}}=\vec{\boldsymbol{X}}_{\text{old}}+\delta\vec{\boldsymbol{X}} \tag{15b}$$

● Step 4: Terminate the iterative procedure if $\left|\delta\vec{\boldsymbol{X}}\right|<\varepsilon$ or $\left|\vec{\boldsymbol{F}}\right|<\varepsilon$; otherwise, back to Step 1 until the convergence is achieved.

The calculation shows that Eq. (14) can be solved within 5 iterations by use of the procedure mentioned above.It demonstrates the reliability of iterative method.

3 Results and discussion

In this section, the semi–analytical model is primarily verified and then the effect of main influencing factors on the characteristics of transient response is discussed, including pressure sensitivity in fracture and reservoir and fracture geometry property (i.e.fracture number, spacing, conductivity, and asymmetry factor) .Here, to make the transient response more obvious, the thickness is set a smaller value (2.9m) to approximately eliminate the convergence skin effect, and the equally–spaced fractures are set asymmetry with regard to wellbore.The parameters used for model validation is list in Tab. 1.

3.1 Model Verification

To verify the new solution, the transient pressure solutions obtained in this paper are compared with other theoretical results reported in the previous literature [37, 48] and commercial numerical simulator (ECLIPSE, ECRIN).

3.1.1 Verification of single fracture

The case of single asymmetry fracture without pressure-sensitive effect was verified by comparing with Berumen et al. [48].As seen in Fig.5, pressure solutions in this study are consistent with their results except the early-time response ($t_D<10^{-4}$) .This is caused by the steady-state assumption of fracture flow model in this study, differentiating from the unsteady state flow in numerical model provided by Berumen et al. [48].At the early-time period, flow regime is actually in transient state.The difference between unsteady- and steady-state becomes very small as long as the fracture reaches the pseudo-steady state. To further verify the model, the type curves with consideration of pressure-sensitive effect presented by Berumen [37] are compared with the semi-analytical model under different conditions of fracture conductivity.As shown in Fig.6, the solutions agree with these results throughout the full-life period, which also verify semi-analytical solutions in the paper.

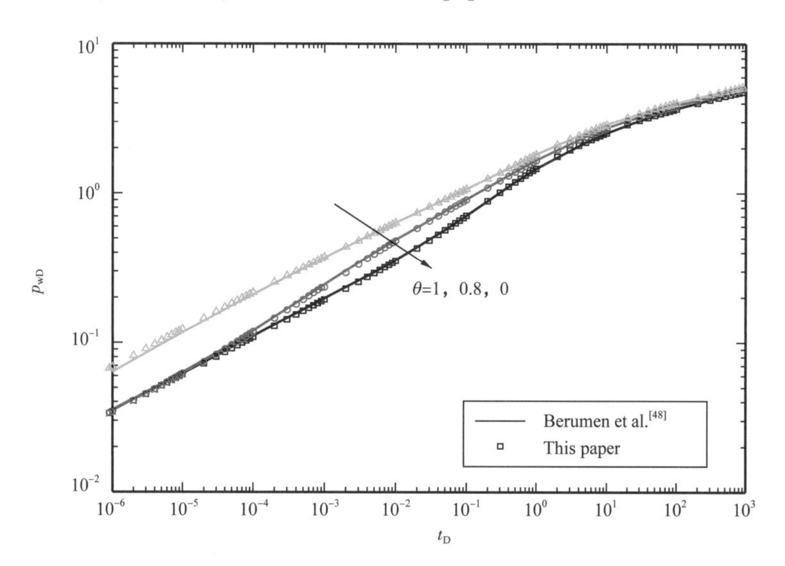

Fig.5 Comparison of results in this paper with numerical solutions presented by Berumen et al. [48] under different asymmetry factors (C_{fD}=5)

3.1.2 Verification of multiple fractures

The case of multiple asymmetry fractures without stress-sensitive effect was verified by comparing with commercial numerical simulator (ECLIPSE).To obtain accurate early-transient flow behavior, the local-grid-refinement (LGR) presented by Bennett et al. [49] is used to refine the block dimensions in the along-fracture direction and normal-to-fracture direction, and the top

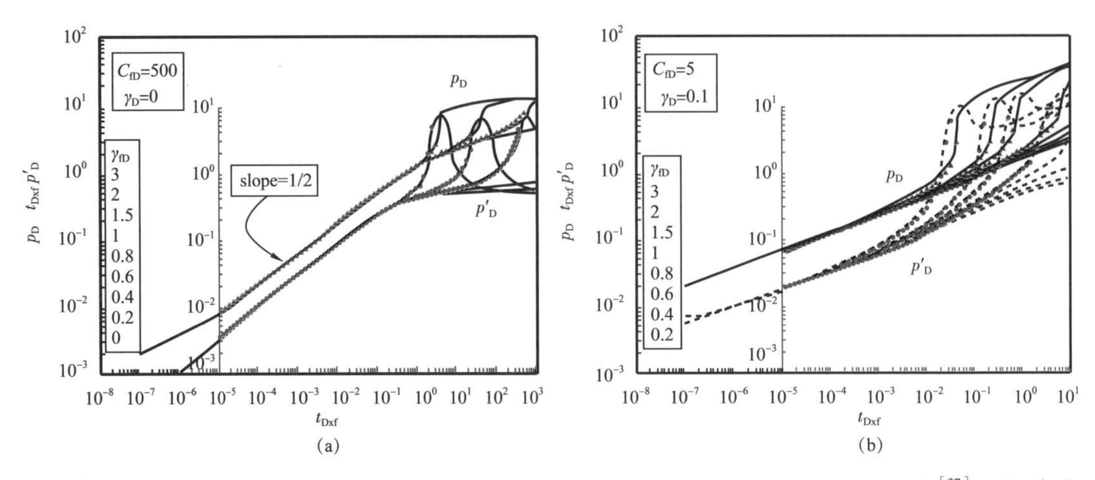

Fig.6 Comparison of the results in this study with numerical solutions presented by Berumen et al. [37] under (a) pressure-sensitivity fracture case (C_{fD}=500, θ=0, γ_{mD}=0, γ_{fD}=3, 2, 1.5) ; (b) pressue-sensitivity fracture & reservoir case (C_{fD}=5, θ=0, γ_{mD}=0.1, γ_{fD}=3, 2, 1.5, 1)

schematic is shown in Fig.7.The LGR configuration for global cells is set 339 × 341 × 7to generate refine cells with dimension of 0.1 × 0.1 × 0.1m to describe the hydraulic fracture.And all grid dimension are log distributed, ranging from 0.1 to 12.8m. Here, actual fracture width is 0.0127m, and fracture porosity is 0.35.In numerical simulator, equivalent fracture width is set to be smallest grid block, 0.1m.Therefore, the equivalent fracture porosity is calculated as 0.0445, according to the relationship of $\phi_e=\phi_f w_f/w_e$, and equivalent fracture permeability is calculated to be 120C_{fD} according to $k_f=C_{fD}$ ($k_m x_f/w_f$) .As seen in Fig.8a, semi-analytical results are consistent with numerical model in the whole time scope.The case of multi symmetry fractures with pressure-sensitive effect was verified by comparing with commercial well-testing software (ECRIN) .As seen from Fig.8b~Fig.8d, comparison of the solutions in this work with results from the numerical simulator shows a good agreement under different levels of pressure-dependent conductivity, no matter how intense the pressure sensitivity is in the reservoirs.

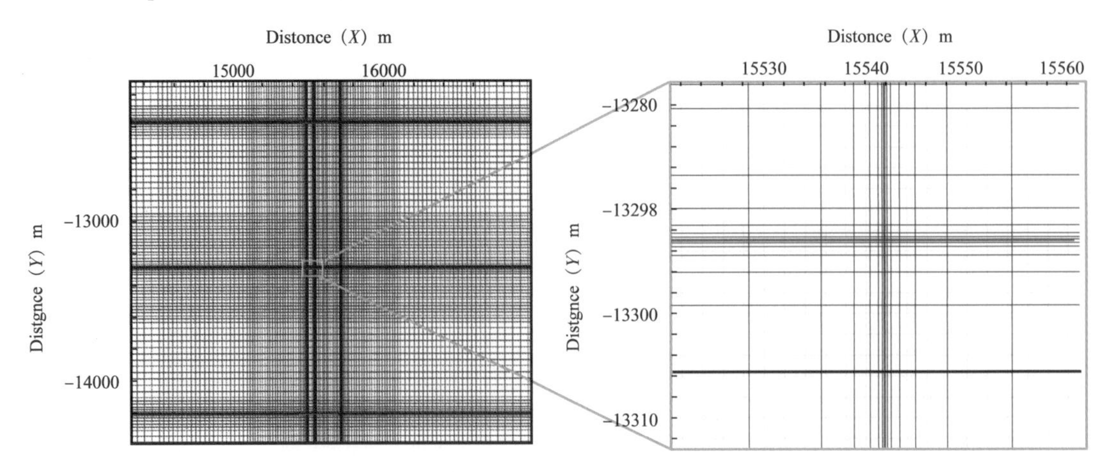

Fig.7 Schematic of LGR in numerical simulation with 3 asymmetrically-spaced fractures

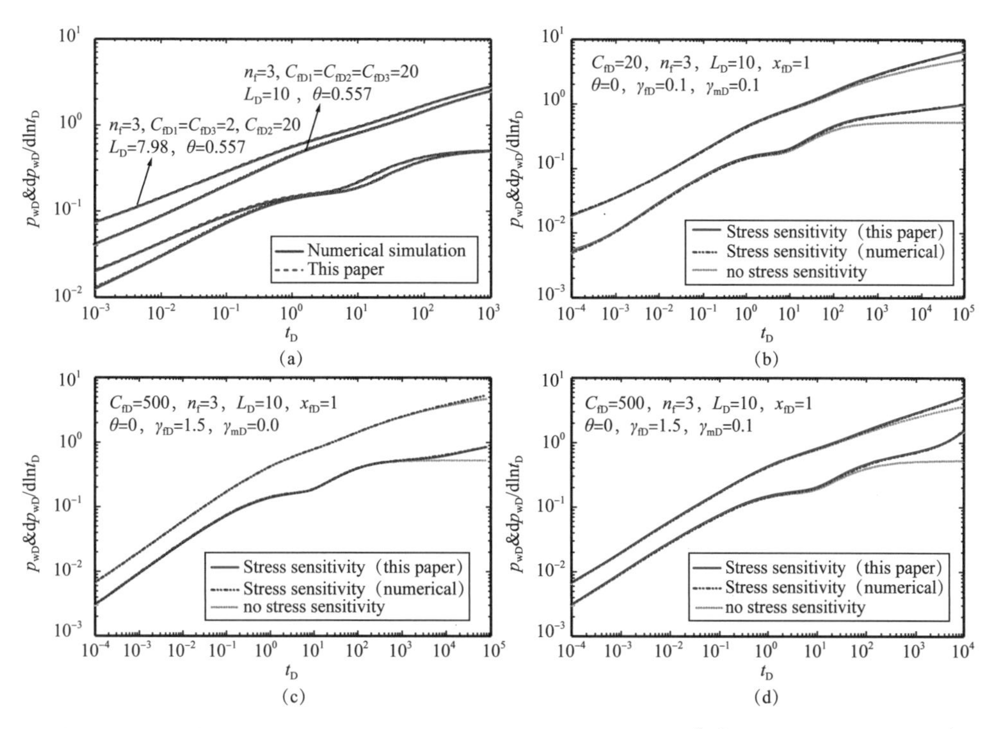

Fig.8 Comparison of the results in this paper with the numerical simulation (a) with asymmetry factor (n_f=3, θ=0.557, γ_{mD}=0, γ_{fD}=0); and under stress sensitivity case: (b) γ_{fD}=γ_{mD}, where this case is identical to the assumption in the work presented by Wang [22] and Chen et al. [23-24]; (c) $\gamma_{fD} \neq \gamma_{mD}$ and γ_{mD}=0, where the continuity condition is satisfied as the exponential expression in Eq. (12a); (d) $\gamma_{fD} \neq \gamma_{mD}$ and γ_{mD}>0, where the continuity condition is satisfied as the power-law expression in Eq. (12a)

3.2 Flow behavior characterization

Fig.9 depicts the transient pressure and derivative responses of a horizontal intercepted by three transverse symmetry fractures with pressure-sensitivity effect. As seen from Fig.9, the flow regime can be divided into six fundamental stages, which are described as follows:

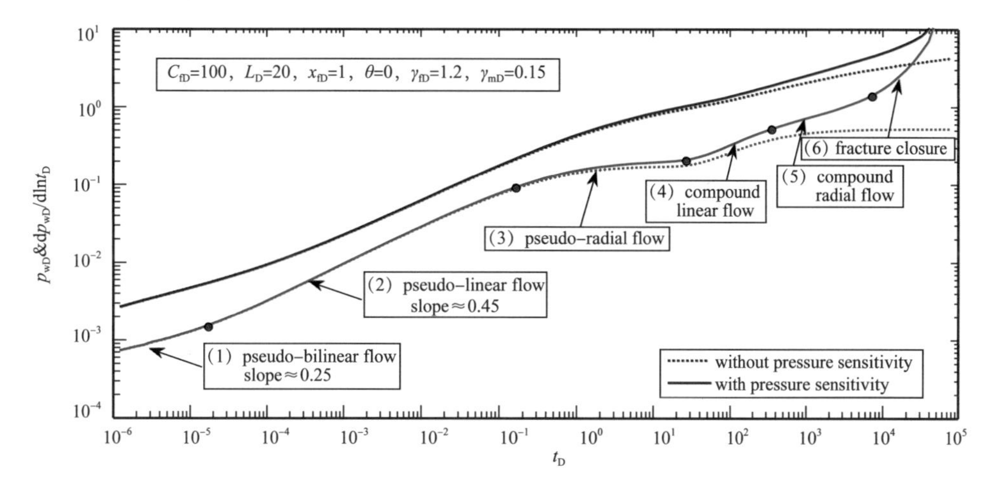

Fig.9 Transient pressure and derivative response during the whole production period

Stage 1 : *Pseudo-bilinear flow*.This flow behavior indicates simultaneous appearance of transient response in between fracture and reservoir, namely normal–to–fractureface linear flow in reservoir and linear flow in fracture.It is characterized by 0.25~0.35 slope straight line in log–log plot, not the constant 0.25 presented by Cinco–Ley et al. [50] .During this flow period, the pressure depletion along fracture is not considerable due to the shorter duration, so it is hard to detect the deviation of pressure and derivative (noted by solid line) from the classical case without pressure sensitivity (noted by dashed line), especially for high–conductivity case corresponding to shorter pseudo–bilinear duration. Based on the results of Breumen [37] and Zeazar et al [44], it is concluded that the slope of pseudo–bilinear flow is determined by fracture permeability modulus, fracture conductivity and fracture number, irrespective of the reservoir pressure–sensitive level.

Stage 2 : *Pseudo-linear flow*.During this flow period, flow occurs linearly from reservoir to individual fractures, and the pressure sensitivity would gradually begin to make difference on the transient response.The flow regime includes the effect of linear flow regime (for high conductivity, $C_{fD}>300$ defined by Gringarten et al. [51]) and the influence of finite–conductivity nature.Given that the magnitude of permeability modulus in fracture($\gamma_{fD}=1.2$)is larger than reservoir($\gamma_{mD}=0.15$), the extra pressure drop caused by fracture pressure–sensitive effect is more predominant than pressure drop caused by reservoir pressure–sensitive effect.In other word, the reservoir pressure–sensitive effect has a negligible influence on pseudo linear flow.Therefore, this flow regime is determined by pressure sensitivity in fracture and reservoir, fracture conductivity and fracture number.

Stage 3 : *Pseudo-radial flow*.In this region, radial flow pattern around fracture occurs directly from reservoir to fracture. Since this regime, in addition to be further influenced by fracture pressure–sensitive effect, the pressure sensitivity in reservoir begins to play a significant role in affecting the transient response.During pseudo–radial flow period, the pressure derivative shows the $1/n_f$ constant for the classical solution (the detailed description was presented by Luo et al. [52]) .Correspondingly, the slope of derivative with pressure sensitivity is variable, rather than a constant. As expected, if considering the pressure sensitivity, the curve of pressure drop distributes over the classical solution (noted by dashed line) .Put another way, as the pressure perturbation is more and more remarkably detected in the reservoir, the porous network would be further deformed and the fracture would tend to close, which causes a significant increase or deviation in pressure drop and derivative.

Stage 4 : *Compound-linear flow*.In this region, the interaction among fractures begins to take place.The transient response within the reservoir region surrounded by outermost fractures reaches the pseudosteady state. Meanwhile, pressure perturbation in reservoir further spreads towards the region beyond fracture tip.The flow pattern is predominately normal to the azimuth of the horizontal wellbore.The system behaves as a fractured vertical well, where the fractured length is equivalent to the horizontal wellbore and the fractured width is the actual fracture length.The characteristic slope of pressure derivative on log–log is larger than 0.36 which corresponds to the classical compound–linear flow, depending on the intense of pressure sensitivity in fracture and reservoir.

Stage 5 : *Compound-radial flow*.In this period, flow across the tip of horizontal well and fracture is simultaneously dominant.The flow pattern is anomalous to the long-term behavior of a fractured vertical well, which possesses the same geometry dimension as that mentioned in Stage 4. The corresponding slope of pressure derivative is not fixed at 0.5, rather than a variable with regard to fracture geometry and pressure-sensitivity level.

Stage 6 : *Fracture closure*. The pressure depletion leads to the fracture closing due to its deformation and consequent loss of conductivity along fracture.This phenomenon is reflected in the log-log pressure and derivative plot by the maximum value exhibited on these curves.The effect identified by the maximum value indicates that the fracture has lost most of conductivity, and the resulting conductivity is so tiny that fluid is hardly transported throughout fracture into wellbore.Put another way, the transient response terminates or finishes.

The characterization of flow regimes mentioned above would be more readily identified in the case of low-conductivity fracture with stronger intense of pressure sensitivity.Furthermore, the duration of some flow regime may elongate or shorten, even disappear.

3.3 Effect of pressure sensitivity

The transient pressure behavior will be discussed in detail by presenting type curves.These type curves show the influence of the pressure sensitivity, both in fracture and reservoir, on the transient pressure responses.

As explained above, the fracture closure results in an extra pressure drop, which is caused by the flow restriction along the fracture.Here the average conductivity ratio along fracture ($C_{\mathrm{fD,\ avg}}$) is introduced to describe this phenomenon, which is defined as the ratio of fracture conductivity C'_{fD} to the initial fracture conductivity C_{fD} as follows :

$$C_{\mathrm{fD,\ avg}}=\left(\frac{C'_{\mathrm{fD}}}{C_{\mathrm{fD}}}\right)=\frac{\int_{x_{\mathrm{wD0}}-x_{\mathrm{fD}}}^{x_{\mathrm{wD0}}+x_{\mathrm{fD}}}\exp\left[-\gamma_{\mathrm{fD}}p_{\mathrm{fD}}(x_{\mathrm{D}},t_{\mathrm{D}})\right]\mathrm{d}x_{\mathrm{D}}}{2x_{\mathrm{fD}}} \tag{16}$$

In addition, the fracture conductivity in the vicinity of wellbore ($C_{\mathrm{fD,\ vic}}$) is crucial to conduct the fluid throughout the fracture into wellbore.

3.3.1 Pressure sensitivity in fracture

Fig.10 depicts the pressure and its derivative with the consideration of pressure sensitivity in fracture.On these curves, the reservoir permeability is kept constant ($\gamma_{\mathrm{mD}}=0$) to distinctly detect the fracture pressure-sensitive effect.With time increasing, the pressure depletion leads to a gradual closing of fracture under the pressure-sensitive effect.On the log-log curve of pressure and derivative, it is reflected by the appearance of deviation from classical solution ($\gamma_{\mathrm{mD}}=\gamma_{\mathrm{fD}}=0$) .In general, the higher the fracture permeability modulus (or the more intense the pressure sensitivity in fracture), the earlier the effect shows up.With the fracture closing, the pressure derivative increases until reaches a maximum value.Thus, the pressure derivative could be used to detect the duration when the fracture closure take places.Likewise, a high value of γ_{fD} would contribute to a more intense effect of fracture closing than low value, and the average conductivity radio $C_{fD,\ avg}$ decreases rapidly until approximately reaches to 0, as shown in Fig.11.

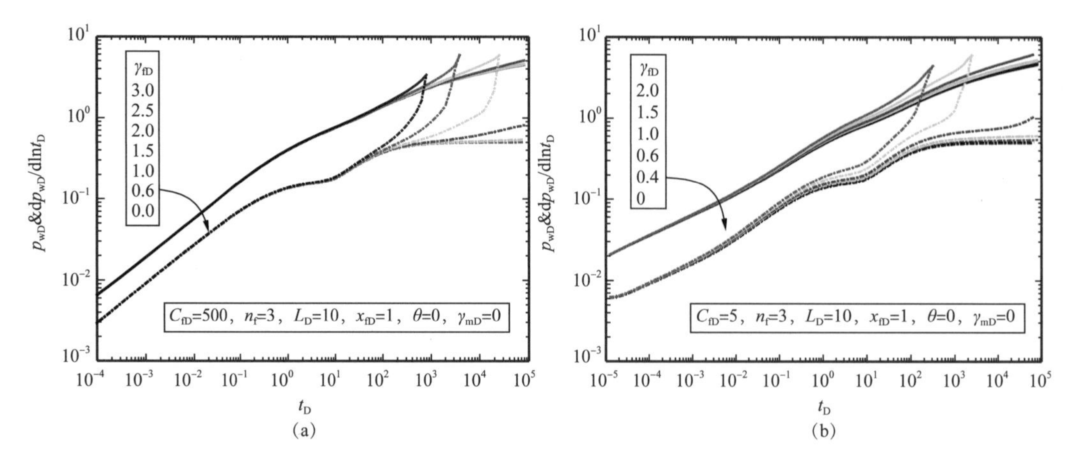

Fig.10 Type curve showing the effect of varying the fracture permeability modulus in the condition of (a) high fracture conductivity; (b) low fracture conductivity in non-pressure-sensitivity formation (γ_{mD}=0)

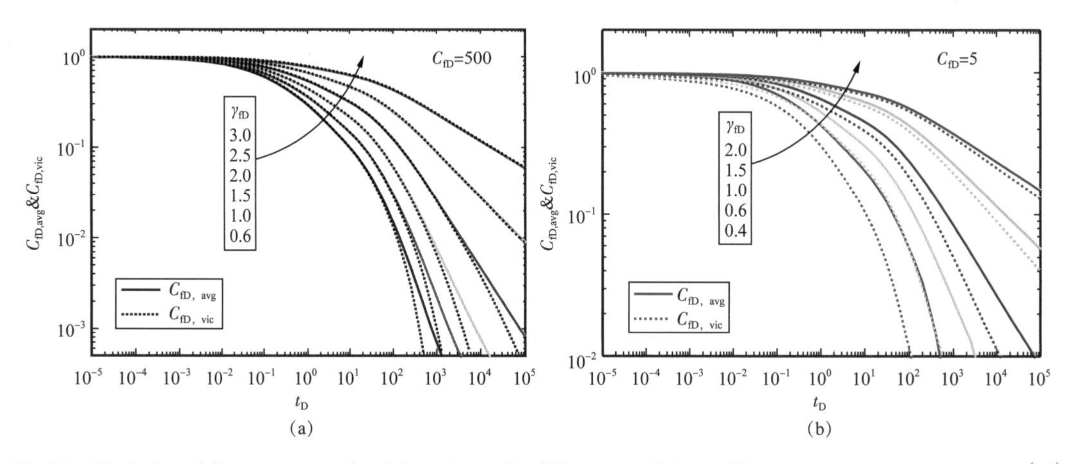

Fig.11 Variation of the average conductivity ratio under different conditions of fracture pressure sensitivity in (a) high-conductivity case; (b) low-conductivity case

Here it is needed to emphasis that the characterization of fracture closure in this study is different from the presentation of Berumen [37] because of the assumption that fracture permeability approximates to 0 when the pressure drop is extremely large in this study.

For high-conductivity case. It is readily seen by comparing Fig.10a~Fig.10b that the effect of pressure sensitivity is related to the degree of fracture conductivity. It is expected the deviation shows up later in higher conductivity fracture than lower conductivity fracture. Taking the case of γ_{fD}=2 for example, the occurring time of deviation is at t_D=20 under C_{fD}=500 in Fig.10a, while the deviation time is at $t_D=10^{-4}$under C_{fD}=5 in Fig.10b. Likewise, the transient response terminates later in the case of higher conductivity, i.e. fracture is closed completely [e.g., for γ_{fD}=2, the time when fracture completely closes is t_{Dend}=2 × 10^4under C_{fD}=500, while t_{Dend}=2 × 10^2under C_{fD}=5] .As seen in Fig.11a, the average conductivity radio ($C_{fD,\ avg}$) and vicinal fracture conductivity ($C_{fD,\ vic}$) is approximately overlapped during early–and intermediate–time period. It means that the distribution degree of pressure–dependent conductivity is relatively uniform along fracture. In addition, the higher γ_{fD} contributes to an earlier deviation of $C_{fD,\ avg}$ from $C_{fD,\ vic}$.The deviation time of $C_{fD,\ avg}$ from $C_{fD,\ vic}$ approximately corresponds to the time of deviation of pressure response with pressure sensitivity from without pressure sensitivity.

*For low–conductivity case.*Different from the characterization in high–conductivity case, Fig.10b shows that the fracture pressure sensitivity has a determinant influence on the flow regime in the whole period, even in small time scope.Fig.11b depicts that the average conductivity radio ($C_{fD,\ avg}$) overlays vicinal fracture conductivity ($C_{fD,\ vic}$) all the way, which means the distribution degree of pressure–dependent conductivity is nonuniform along fracture [this characterization can be interpreted based on the fact that the pressure drop along fracture is considerable in the low–conductivity case] .Even if in early time scope, the conductivity of the segment around wellbore is obviously smaller than other fracture segments, which indicates that the flow restriction would rapidly increases when throughout the segment around wellbore. This is the reason why the effect of pressure sensitivity makes significant difference during early–time period.

3.3.2 Pressure sensitivity in reservoir

To further detect the effect of pressure sensitivity in the reservoir, two cases are presented to perform pressure transient analysis under different conditions of fracture pressure–sensitive level with the consideration of reservoir pressure–sensitive effect seen in Fig.12, and under different conditions of reservoir pressure–sensitive level with the consideration of fracture pressure–sensitive effect seen in Fig.13.

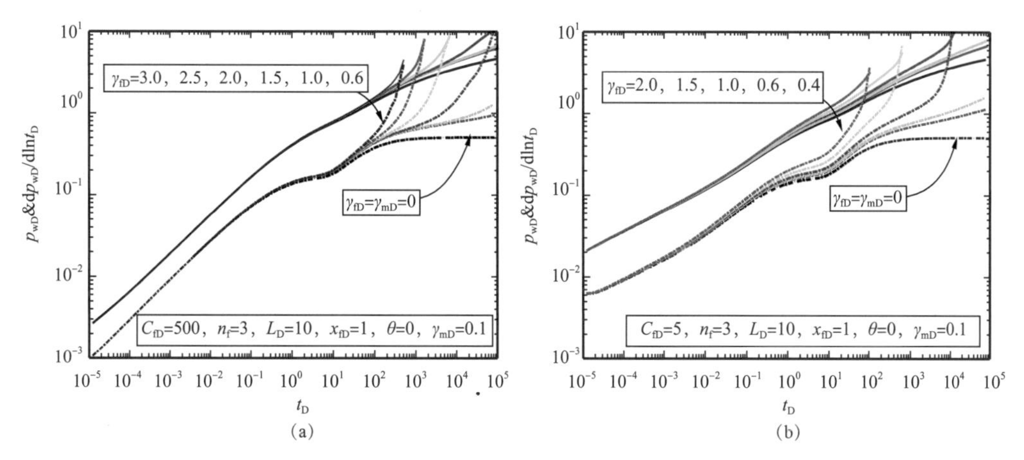

Fig.12 Type curve showing the effect of varying the fracture permeability modulus for (a) high fracture conductivity ; (b) low fracture conductivity in the pressure–sensitivity reservoir (γ_{mD}=0.1)

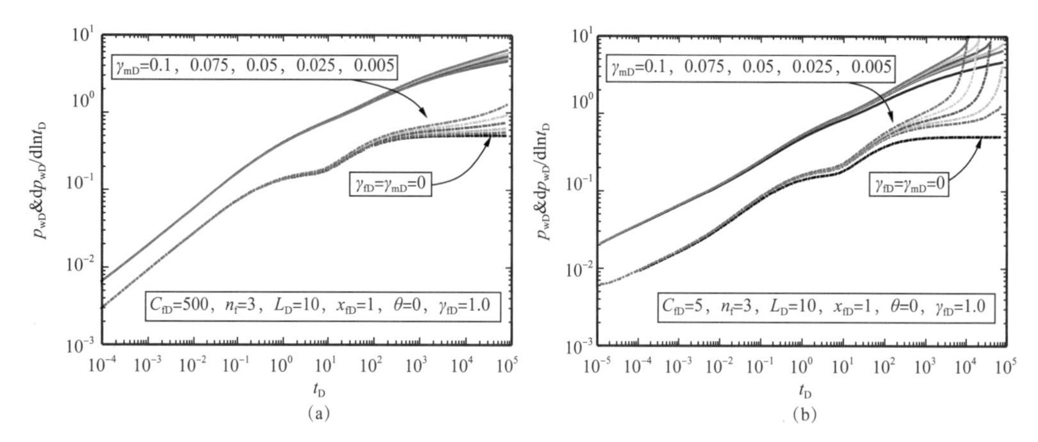

Fig.13 Type curve showing the effect of varying the reservoir permeability modulus for (a) high fracture conductivity; (b) low fracture conductivity in the pressure–dependent fracture (γ_{fD}=1)

In Fig.12, the characterization of transient pressure response is overall anomalous to that in Fig.10, especially for the time domain before pseudo–radial flow regime. In general, the pressure sensitivity in reservoir causes a greater pressure drop. Meanwhile, the degree of deviation subsequently becomes greater, and curves are more dispersedly distributed. Likewise, the terminal time appears earlier (e.g. for γ_{fD}=2, the time when fracture completely closes is $t_{Dend}=2\times10^4$ for the case without reservoir pressure–sensitivity effect in Fig.10a, while $t_{Dend}=1\times10^4$ in Fig.12a). The reason is that the extra pressure drop caused by reservoir deformation leads to a lower level of pressure profile along fracture; thus the fracture closure shows up in advance. For the cases of $\gamma_{fD}<1$ in Fig.12a, the pressure response with reservoir pressure–sensitive effect is more readily distinguished from the classical case ($\gamma_{mD}=\gamma_{fD}=0$) during compound–radial flow regime than the case in Fig.11a, which corresponds to more disperse distribution of pressure and derivative curves. These tendencies are also readily detected for low–conductivity case (i.e. C_{fD}=5 in Fig.12b).

Fig.13 reflects the effect of varying the reservoir pressure–sensitivity level under high–and low–conductivity cases. The larger the reservoir permeability modulus, the more severe is the pressure drop (i.e. fracture exhibits a greater conductivity decay rate). For high–conductivity fracture in Fig.13a, the pseudo–linear flow regimes are approximately independent from the reservoir permeability modulus, no matter how intense the pressure sensitivity is in fracture (Fig.10a also demonstrates a similar characterization), unlike low–conductivity case. Low conductivity has a detectable influence on the characterization of pseudo–linear flow regime, but not on pseudo–bilinear flow as seen in Fig.13b.

As previously analyzed, the pseudo–bilinear flow is determined by the fracture permeability modulus, irrespective of the reservoir pressure–sensitivity level. The lower the fracture conductivity, the more severe is the effect. Pseudo–linear flow is affected by the reservoir permeability modulus in addition to fracture permeability modulus and fracture geometry properties, which is magnified or remarkable in the case of low conductivity. The deformation in fracture and reservoir leads to the appearance of fracture closure, and the starting time of deviation appearance and terminal time of fracture completely closing are also determined by the pressure–sensitivity level, in addition to fracture geometry properties.

In summary, the fracture pressure-sensitivity level take effects on the whole flow regimes; the reservoir pressure-sensitivity level takes effect only on these beyond pseudo-bilinear flow. The existence of reservoir pressure sensitivity makes the effect of fracture pressure sensitivity more remarkable, especially for low conductivity cases. In addition, it is worth noting that the pressure sensitivity in fracture and reservoir has a negligible influence on the early-time period (the characteristic slope is approximately 1/2) for high-conductivity case (as demonstrated in the time scope of $t_D<0.5$ in Fig.10a, Fig.12a and Fig.13a).

3.4 Effect of fracture geometry

3.4.1 Fracture spacing

Fracture spacing between adjacent fractures determines the intense of fracture-production interaction. Before fracture interaction occurs, fracture behaves independently of the other fractures. The smaller fracture spacing, the earlier the interaction occurs. When the fracture spacing is smaller than fracture length, there would be a new flow regime: pseudo-pseudosteady state (the characteristic slope is nearly 1, which is caused by not-completely virtual closed boundary resulting from interaction between adjacent fractures [53]), as the case of L_D=0.5 and 1 shown in Fig.14.

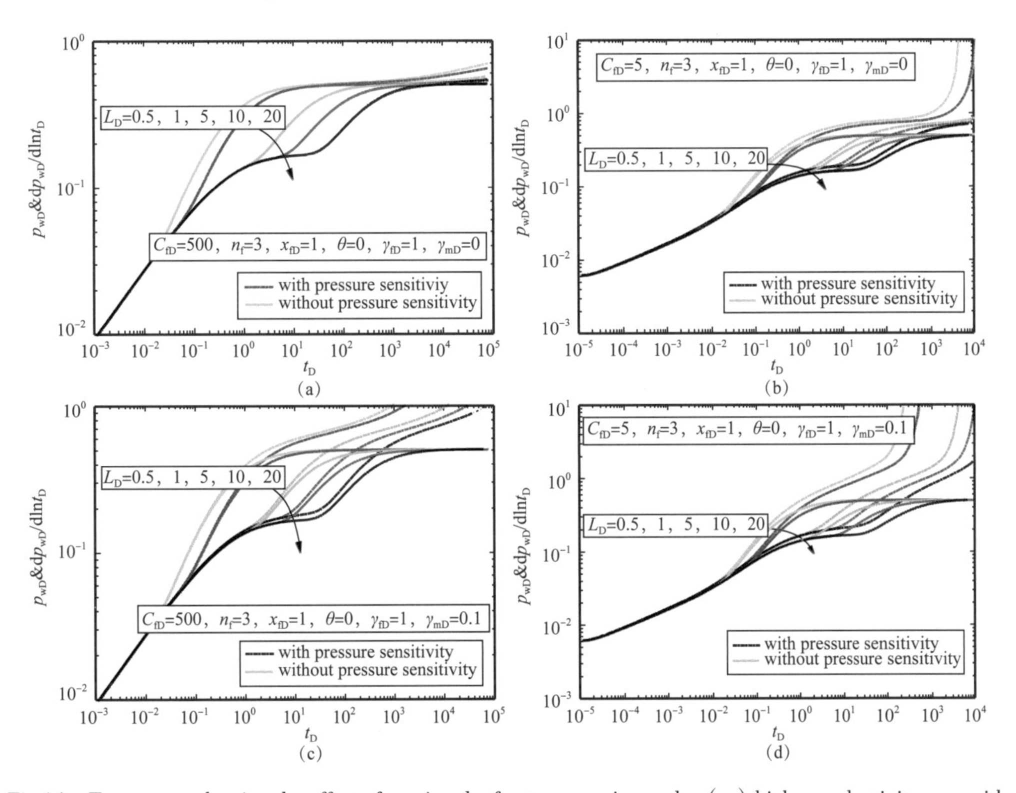

Fig.14 Type curve showing the effect of varying the fracture spacing under (a) high-conductivity case with pressure-sensitive effect in fracture (γ_{fD}=1, γ_{mD}=0, C_{fD}=500); (b) low-conductivity case with pressure-sensitive effect in fracture (γ_{fD}=1, γ_{mD}=0, C_{fD}=5); (c) high-conductivity case with pressure-sensitive effect in fracture and reservoir (γ_{fD}=1, γ_{mD}=0.1, C_{fD}=500); (d) low-conductivity case with pressure-sensitive effect in fracture and reservoir (γ_{fD}=1, γ_{mD}=0.1, C_{fD}=5)

On the curves from Fig.14a~Fig.14d, the pressure sensitivity in the condition of smaller fracture spacing is expected to have a great influence on flow behaviors and generate larger pressure drops to maintain constant production rate, especially for the case of low conductivity and the case with considering reservoir pressure–sensitivity effect. This is explained that small fracture spacing leads to great pressure depletion in the reservoir surrounded by fractures due to strong fracture interference ; as a consequence, fracture would reach a lower level of pressure profile and exhibit a greater conductivity decay rate.Therefore, the starting time of deviation appearance and the terminal time of fracture completely closure all shows up in advance.

By comparing Fig.14a and Fig.14c (also for Fig.14b and Fig.14d), the effect of pressure sensitivity in the reservoir is further detected under different conditions of fracture spacing.During the pseudo–pseudosteady flow period (for L_D=0.5, 1 in Fig.14c), the pressure in the reservoir is adequately depleted, so the pressure sensitivity in reservoir takes obvious effect on transient response [the smaller fracture spacing corresponds to a greater deviation degree of pressure derivative in Fig.14c] .Inversely, if the reservoir pressure–sensitivity effect is not taken into account, the curve of the pseudo–pseudosteady flow period is still overlapped with classical solution in the high–conductivity condition (for L_D=0.5 and 1 in Fig.14a) .It indicates that the degree of pressure depletion in the reservoir surrounded by fractures is not great enough to accelerate the conductivity decay rate on the premise of no reservoir deformation.With the increase of fracture spacing, which means weaker intense of fracture interaction, the magnitude of pressure depletion in the reservoir surrounded by fractures would rapidly reduce.Therefore, the starting time of deviation appearance and the terminal time of fracture completely closure shows up later.Moreover, the tendency is remarkable in the condition with the reservoir pressure–sensitive effect, which is reflected by dispersedly–distributed curves in Fig.14c.

By comparing Fig.14c and Fig.14d, the effect of fracture conductivity can be investigated with considering the pressure–sensitive effect. Increasing fracture spacing has a negligible influence on early–time flow regimes for both high–and low–conductivity cases.The effect of pressure sensitivity arrives earlier for low–conductivity case (e.g.the time of deviation is on t_D=0.03 for L_D=0.5 in high–conductivity case as seen in Fig.14c, but t_D=0.006 for L_D=0.5 in low–conductivity case as seen in Fig.14d) .The reason is stated that the low–conductivity case would cause greater pressure depletion than high conductivity case, which is characterized by a larger value of dimensionless pressure and derivative on log–log plot ; put another way, creating high conductivity is more advantageous than low conductivity to enhance productivity and elongate the production duration.

3.4.2 Fracture number

Here the distance between outermost fractures (L_{sD}) is set to be constant. Fracture number controls the connected area between fractures and reservoir in addition to the determination of interaction intense.Fig.15presents the effect of fracture number on transient response with the consideration of pressure sensitivity. The increase of fracture number means the increasing connected area of fracture & reservoir and the stronger intense of fracture interaction. In overall, the pressure drop has the tendency toward becoming decreasing as fracture number increases.

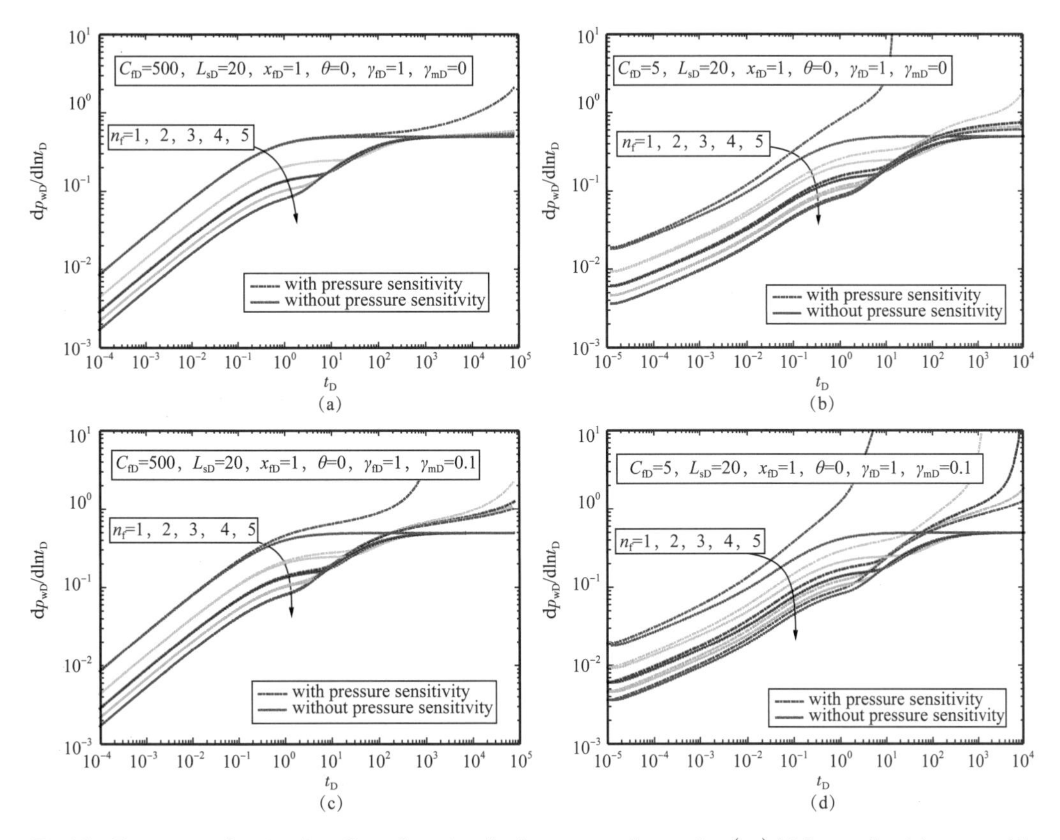

Fig.15 Type curve showing the effect of varying the fracture number under (a) high-conductivity case with pressure-sensitive effect in fracture (γ_{fD}=1, γ_{mD}=0, C_{fD}=500); (b) low-conductivity case with pressure-sensitive effect in fracture (γ_{fD}=1, γ_{mD}=0, C_{fD}=5); (c) high-conductivity case with pressure-sensitive effect in fracture and reservoir (γ_{fD}=1, γ_{mD}=0.1, C_{fD}=500); (d) low-conductivity case with pressure-sensitive effect in fracture and reservoir (γ_{fD}=1, γ_{mD}=0.1, C_{fD}=5)

By contrast with Fig.15a, Fig.15c indicates that the larger fracture number leads to a weaker deviation degree and a lower conductivity decay rate. Moreover, the duration of pseudo-linear flow elongates with the duration of pseudo-radial flow shortening. The effect of increasing fracture number is magnified with the consideration of reservoir pressure-sensitive effect. For example, the starting time of deviation changes from t_D=10 for n_f=1 to t_D=1000 for n_f=2 without considering pressure sensitivity in reservoir seen in Fig.15a, while the corresponding time is changes from t_D=0.1 for n_f=1 to t_D=0.3 for n_f=2 in Fig.15c. Likewise, the curves with considering reservoir pressure-sensitive effect in Fig.15c are distributed more dispersedly than Fig.15a. It indicates that the pressure drop can be more effectively reduced by increasing fracture number through relieving the deformation in the reservoir.

For low-conductivity case as shown in Fig.15b and Fig.15d, it is worth noting that changing fracture number has a more significant impact on transient behaviors in the consideration of reservoir pressure-sensitive effect, even in the early-time period. From the viewpoint of production distribution, the distribution coefficient of individual fracture becomes smaller as the fracture number increases. As the consequence, the pressure depletion in both fracture and reservoir would become weaker. Therefore, the pressure sensitivity does not take an important effect on non-interaction flow regimes (i.e. pseudo-bilinear, pseudo-linear and pseudo-radial flow regimes) in the condition of large fracture number [e.g. the curve of n_f=5 with pressure sensitivity are nearly overlapped with that of n_f=5 without pressure sensitivity in Fig.15b before fracture interaction takes place (t_D<0.5)], and the difference of deviation degree with and without pressure sensitivity is smaller in the case of n_f=5 than the case of n_f=1 in the late-time period.

3.4.3 Asymmetry factor

In Fig.16 the asymmetry factor allows the detection of the asymmetry degree of hydraulic fracture, ranging from θ=0, the symmetry fracture case with the well located at the half length of the fracture, to θ=1, the fully asymmetry case with the wellbore located at the tip of the fracture. As expected a bigger θ will lead to a bigger pressure drop during the full-life period. Besides, the asymmetry of fracture results in the occurrence of a new transition flow regime between pseudo-

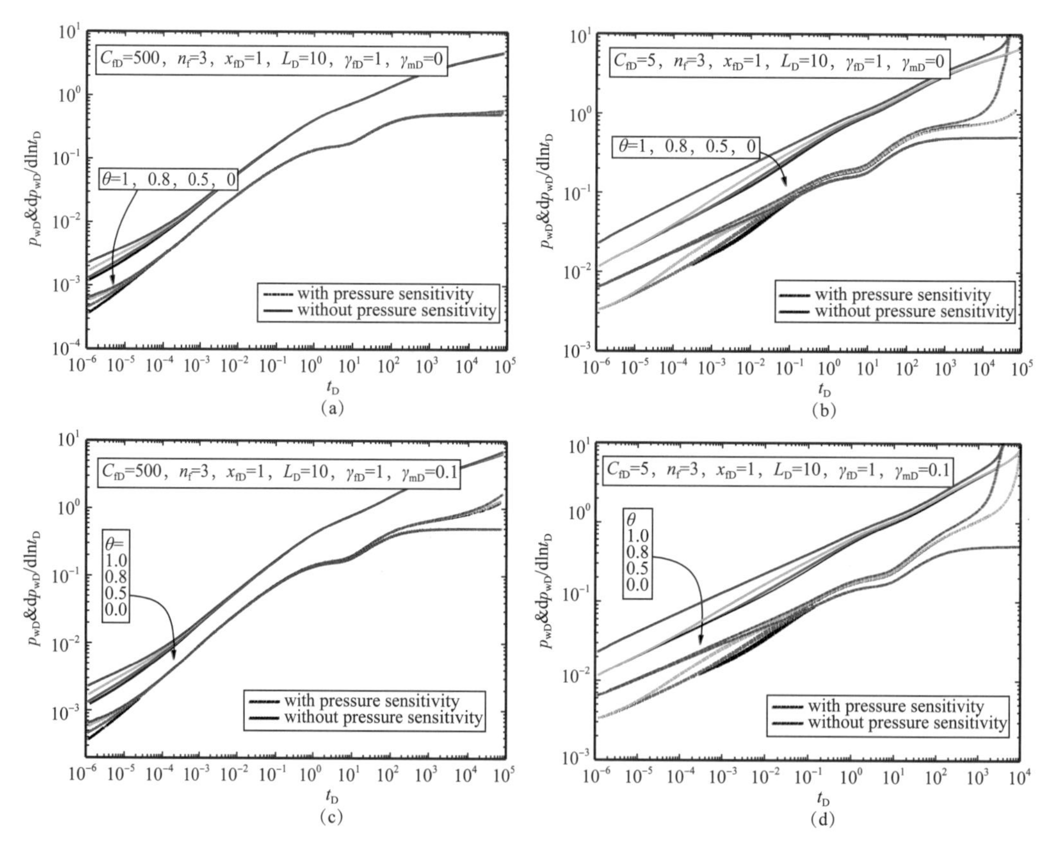

Fig.16 Type curve showing the effect of varying the fracture asymmetry factor under (a) high-conductivity case with pressure-sensitive effect in fracture (γ_{fD}=1, γ_{mD}=0, C_{fD}=500) ; (b) low-conductivity case with pressure-sensitive effect in fracture (γ_{fD}=1, γ_{mD}=0, C_{fD}=5) ; (c) high-conductivity case with pressure-sensitive effect in fracture and reservoir (γ_{fD}=1, γ_{mD}=0.1, C_{fD}=500) ; (d) low-conductivity case with pressure-sensitive effect in fracture and reservoir (γ_{fD}=1, γ_{mD}=0.1, C_{fD}=5)

bilinear and pseudo-radial flow regimes (see Resurreicao et al. [54] for a detailed description of "pseudo-linear flow" accounting for the effect of asymmetry) .As the conductivity of the fracture decreases, the effect of asymmetry on flow behavior becomes more accentuated.It is worth noting that the starting time and duration of flow regimes beyond "pseudo-linear flow" is independent of asymmetry of fracture.

On the analysis of Fig.16a, for high-conductivity fracture, pressure sensitivity in the fracture has a negligible influence on flow behavior under different conditions of asymmetry factor in addition to the early-time scope. Only in the late-time period (corresponding to compound-radial flow $t_D > 1000$), the effect of pressure sensitivity in fracture begins to take place remarkably.Taking the effect of pressure sensitivity in reservoir into account, the flow regimes of intermediate- and late-time period are significantly influenced by pressure sensitivity effect, as shown in Fig.16c ($t_D=1$ corresponding to the beginning of pseudo-radial flow) .Fig.16b shows that the deviation just arrives in early-time period, and a bigger asymmetry factor contributes to an earlier appearance of deviation from the classical case. For example, the starting time of deviation for $\theta=1$ is $t_D=10^{-4}$ while for $\theta=0$ is $t_D=10^{-3}$.The existence of reservoir pressure-sensitive effect will magnify the effect of asymmetry factor, which indicates an earlier appearance of deviation, where the starting time of deviation for $\theta=1$ is $t_D=10^{-5}$while for $\theta=0$ is $t_D=2\times10^{-4}$ in Fig.16d.

Here the following conclusions are obtained. For the high-conductivity case, the early-time flow behavior is independent of asymmetry factor in addition to the very smaller time scope, whether the pressure sensitivity in reservoir is considered or not ; the effect of asymmetry factor on late-time flow behavior is remarkable only when the reservoir pressure-sensitive effect is taken into account. For the low-conductivity case, the flow behavior during the full-life period is strongly dependent on the intense of pressure sensitivity, and the pressure sensitivity in reservoir strengthens the effect of pressure-dependent conductivity on transient response.

3.5 Model application to Field data

Example applications of approximate solution to filed practice are presented and demonstrated how to use it to conduct field performance analysis.The used parameters are list in Tab.2.For gas wells, pseudo-pressure and pseudo-time variables are introduced to account for gas property variations with pressure, given as :

$$m(p)=\frac{\mu_{gi}Z_{gi}}{p_i}\int_0^p\frac{\xi}{\mu_g(\xi)Z_g(\xi)}d\xi,t_a=\int_0^t\frac{\mu_{gi}c_{gi}}{\mu_g(\xi)c_g(\xi)}d\xi \tag{17}$$

Example 1-Transient pressure analysis.For shut-in condition, the "wellbore pressure" is used [55], and the calculation of pseudo-time is straightward because this pressure is measured at the well and is therefore known.In addition, the effect of wellbore storage and skin can be easily taken into account in Laplace domain [52], and therefore we could obtain the type curves according to Eq. (18), which is written as

$$m_{wD}(t_D)=L^{-1}\left(\frac{sL[p_{wD}(t_D)+S_k]}{s+C_D s^2\{sL[p_{wD}(t_D)]\}}\right) \tag{18}$$

where C is wellbore storage, m^3/Pa ; S_k is skin, dimensionless ; and dimensionless wellbore storage satisfies $C_D=C/(2\pi\phi h c_{ti} L^2)$ .L [ ] is the operator of Laplace transform, and L^{-1} [ ] is the operator of Laplace inverse transform. In this section, Roumboutsos and Stewart algorithm [56] is used to numerically transform the real time-domain result of p_{wD} (t_D) from Eq. (15) into Laplace domain ; Stehfest algorithm [57] is used to transform the Laplace-domain results from Eq. (18) into real time domain.

A pressure drawdown test of well A in Ordos Basin in China is selected to interpret transient pressure response.In this model, the horizontal well is intercepted by 3 fractures, where the dimensionless matched curve corresponds to the following parameters : γ_{mD}=0.2, γ_{fD}=1.9, L_{fD}=8.6, C_{fD}=10.Fig.17 shows the log-log diagnostic plot of actual field data and theoretical type curves.As seen from Fig.17, the matching effect is satisfactory.The results of relevant unknown parameters are explained as follows : $\gamma_m=2.02\times10^{-5}KPa^{-1}$, $\gamma_m=1.92\times10^{-5}KPa^{-1}$, K_{mi}=0.384mD, x_f=156.11m, L_e=1358m, F_c=599.4mD · m.It is confirmed that the latest performance is consistent with the forecast based on the model using above parameters.

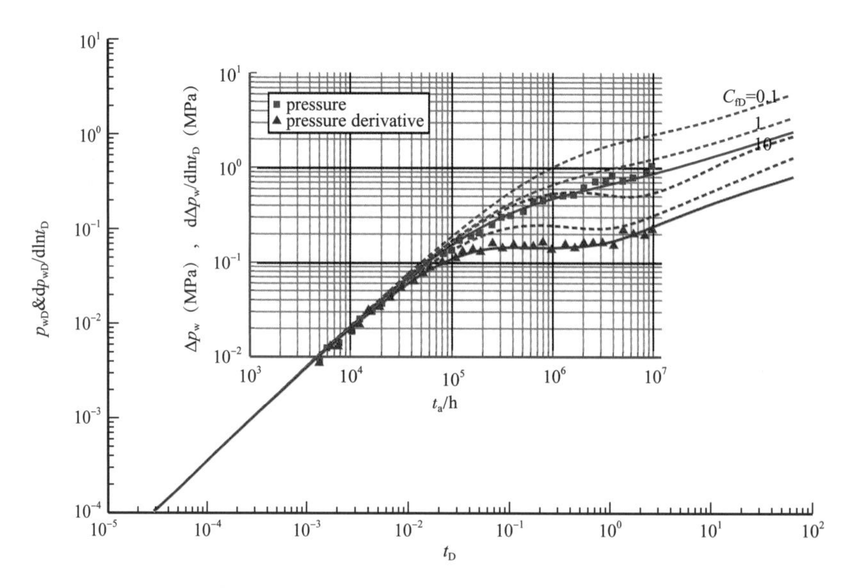

Fig.17 Transient pressure analysis of well A

Example 2-Transient rate analysis.Material balance pseudo-time and rate-normalized pseudo-pressure variables are used to account for the variable production rates and flowing pressure at the well due to operational changes, where material balance pseudo-time is given as,

$$t_{mba}=\frac{\mu_{gi}c_{gi}}{q_{sc}(t)}\int_0^t \frac{q_{sc}(\tau)}{\mu_{avg}(\tau)c_{avg}(\tau)}d\tau \tag{19}$$

Where μ_{avg} and c_{avg} respectively indicate the viscosity and compressibility at the average pressure in the region of influence during transient flow, and the detailed calculation could refer to the work presented by Clarkson et al. [4, 5] and Tabatabaie et al. [39].

Well B is a multi-fractured horizontal gas well with 51 stages for 534 days, located in Sichuan Basin in China. Fig.18 shows the result of performance analysis of well B. In the used model, $\gamma_m=\gamma_f=2.908\times10^{-5}KPa^{-1}$, effective horizontal well length L_e=1120m, dimensionless fracture conductivity $C_{fD}>300$, and formation permeability $K_{mi}=2.56\times10^{-4}$mD. As expected, the fracture half-length obtained by the history matching with considering pressure sensitivity was higher than the value without considering pressure sensitivity effect, namely $x_{f,\ with}$=114.95m and $x_{f,\ without}$=79.7m. Microseismic mapping indicates that the fracture half-length is approximately in the range of 100 to 200 meters.

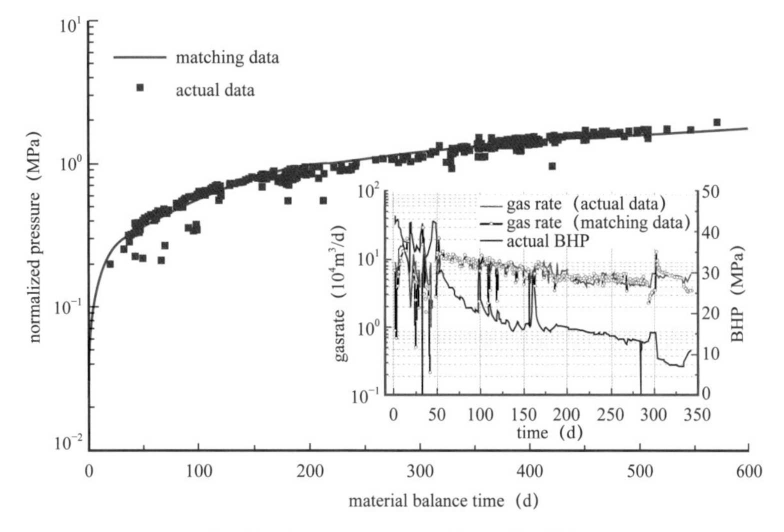

Fig.18 Rate transient analysis of well B

These two examples mentioned above justify the reliability of the model in this paper on the scale of practical field performance.

4 Conclusions

In this study, a general model has been developed to address the problem of pressure transient analysis for the multi-fractured system with pressure-sensitivity in the fracture and reservoir. The algorithm of the resulting nonlinear equations is presented to obtain semi-analytical solutions by using initial guess of corresponding linear equations. It could be fast converged to generate high-accuracy results with 5 iterations. Here, some important conclusions suggested by the results of this study are further emphasized below:

(1) In such fractured system, the transient pressure response exhibits six flow regimes: pseudo-bilinear flow, pseudo-linear flow, pseudo-radial flow, compound-linear flow, compound-radial flow and fracture closure. The characterization of transient response is determined by the fracture-reservoir pressure sensitivity, and fracture geometry properties including fracture

conductivity, number, spacing, asymmetry factor.

(2) The pressure sensitivity is the most essential parameter of interest in this study. The influence of pressure sensitivity can be magnified in the condition of serious pressure depletion. It indicates that the properties of fracture geometry are disadvantageous in terms of productivity evaluation, such as the case of smaller fracture conductivity, spacing, number, and stronger degree of asymmetry factor.

(3) The effect of pressure sensitivity becomes remarkable in the intermediate-and late-time period. For finite conductivity fracture, the early-time flow regime is influenced by fracture pressure-sensitive effect in addition to fracture geometry properties, irrespective of the reservoir pressure-sensitivity level. For high conductivity fracture, the characterization of early-time flow regime is approximately independent of the pressure-sensitive effect in fracture and reservoir, in spite of the measureable change in fracture conductivity and reservoir permeability.

(4) The partial fracture closing occurs during the entire process of pressure depletion, causing the loss of conductivity along fracture. It is detected by the deviation of pressure derivative from the non-pressure-sensitive case, and the value of pressure derivative increases over time until reaches the maximum when the fracture is completely closed and transient response is terminated.

Acknowledges

The authors are indebted to Dr. Xiaoda Liu in Texas A&M University for his great support. This article was supported by the National Major Research Program for Science and Technology of China (No.2016ZX05015).

Nomenclature

Field variables

c	Compressibility, Pa^{-1}
h	Formation height, m
K	Permeability, m^2
L	Fracture spacing, m
L_s	Distance between outermost fractures, m
n_f	Number of hydraulic fracture
p	Pressure, Pa
Q_{sc}	Production rate of MFHW, m^3/s
q_{sc}	Production rate of single fracture, m^3/s
q_f	Flux density along fracture face, m^2/s
t	Time, s
$x_{w,0}$	Fracture middle location in the x-coordinate, m
y_w	Fracture location in the y-coordinate, m
x_{fw}	Sink/source location within fracture, m

x_f	Half-hydraulic-fracture length, m
w_f	Hydraulic fracture width, m
γ_f	Permeability modulus in the fracture, Pa^{-1}
γ_m	Permeability modulus in the reservoir, Pa^{-1}
μ	Fluid viscosity, Pa · s
ϕ	porosity
ρ	Fluid density, g/m^3

Dimensionless variables

C_{fD}	Dimensionless fracture conductivity
$C_{fD,\ avg}$	Average conductivity ratio along fracture
$C_{fD,\ vic}$	Fracture conductivity around wellbore
t_D	Dimensionless time
p_D	Dimensionless pressure
θ	Fracture asymmetry factor
η_D	Pedrosa's transformed variable

Subscripts

f	Fracture property
i	Initial condition
m	Reservoir property
D	Dimensionless
sc	Standard condition
w	Wellbore property
ref	Reference variable

References

[1] Wilson K.Analysis of drawdown sensitivity in shale reservoirs using coupled-geomechanics models [C] //SPE Annual Technical Conference and Exhibition, 28-30 September 2015, Houston, Texas, USA.DOI : https : //doi.org/10.2118/175029-MS.

[2] Tao Q F, Ghassemi A, Ehlig-Economides C A.Pressure transient behavior for stress-dependent fracture permeability in naturallyfractured reservoirs [C] //International Oil and Gas Conference and Exhibition in China, 8-10 June 2010, Beijing, China. DOI : https : //doi.org/10.2118/131666-MS.

[3] Cho Y, Apaydin O G, Ozkan E.Pressure-dependent natural-fracture permeability in shale and its effect on shale-gas well production [C] //SPE Annual Technical Conference and Exhibition, 8-10 October 2012, San Antonio, Texas, USA.DOI : https : //doi.org/10.2118/159801-MS.

[4] Clarkson C R, Jensen J L, Chipperfield S. Unconventional gas reservoir evaluation : what do we have to consider? [J] Journal of Natural Gas Science and Engineering, 2012, 6: 9-33.

[5] Clarkson C R, Qanbari F, Nobakht M, et al.Incorporating geomechanical and dynamic hydraulic fracture property changes into rate-transient analysis : example from the Haynesville shale [C] //SPE Canadian

Unconventional Resources Conference, 30 October–1 November 2012, Calgary, Alberta, Canada.DOI : https : //doi.org/10.2118/162526–MS.

[6] Aybar U, Eshkalak M O, Sepehrnoori K, et al. Long term effect of natural fractures closure on gas production from unconventional reservoirs [C] //SPE Eastern Regional Meeting, 21–23 October 2014, Charleston, WV, USA. DOI : https : //doi.org/10.2118/171010–MS.

[7] Cao P, Liu J S, Leong Y K. A fully coupled multiscale shale deformation–gas transport model for the evaluation of shale gas extraction [J] .Fuel, 2016, 178: 103–117.

[8] Qanbari F, Clarkson C R. Analysis of transient linear flow in stress–sensitive formations [J] .SPE Reservoir Evaluation & Engineering, 2013, 17 (1) : 98–104.

[9] Pedrosa O A. Pressure transient response in stress–sensitive formations [C] // SPE California Regional Meeting, 2–4 April 1986, Oakland, California. DOI : https : //doi.org/10.2118/15115–MS.

[10] Ostensen R W.The effect of stress–dependent permeability on gas production and well testing [J] .SPE Formation Evaluation, 1986, 7: 227–235.

[11] Clarkson C R.Production data analysis of unconventional gas wells : Review of theory and best practices [J] . International Journal of Coal Geolgoy, 2013, 109–110: 101–146.

[12] Zhang M Y, Ambastha A K.New insights in pressure–transient analysis for stress–sensitive reservoirs [C] // SPE Annual Technical Conference and Exhibition, 25–28 September 1994, New Orleans, Louisiana. DOI : https : //doi.org/10.2118/28420–MS.

[13] Pedroso C A, Correa A C.A new model of a pressure–dependent–conductivity hydraulic fracture in a finite reservoir : constant rate production rate [C] //Latin American and Caribbean Petroleum Engineering Conference, 30 August–3 September 1997, Rio de Janeiro, Brazil.DOI : https : //doi.org/10.2118/38976–MS.

[14] Berumen S, Tiab D.Interpretation of stress damage on fracture conductivity [J] .Journal of Petroleum Science and Engineering, 1997, 17 (1–2) : 71–85.

[15] Raghavan R, Chin LY, 2002.Productivity changes in reservoirs with stress–dependent permeability [C] // SPE Annual Technical Conference and Exhibition, 29 September–2 October 2002, San Antonio, Texas, USA.DOI : https : //doi.org/10.2118/77535–MS.

[16] Zhang Z, He S L, Liu G F, et al.Pressure buildup behavior of vertically fractured wells with stress–sensitive conductivity [J] .Journal of Petroleum Science and Engineering, 2014, 122: 48–55.

[17] Raghavan R, Scorer J, Miller F G.An investigation by numerical methods of the effect of pressure–dependent rock and fluid properties on well flow tests [J] .Society of Petroleum Engineers Journal, 1972, 12 (3) : 267–275.

[18] Ozkan E, Raghavan R, Apaydin OG.Modeling of fluid transfer from shale matrix to fracture network [C] // SPE Annual Technical Conference and Exhibition, 19–22 September 2010, Florence, Italy.DOI : https : //doi.org/10.2118/134830–MS.

[19] Wang L, Wang X D. Modeling of pressure transient behavior for fractured gas wells under stress–sensitive and slippage effect [J] .International Journal of Oil, Gas and Coal Technology, 2014.

[20] Vardcharragosad P, Ayala L F.Production–data analysis of gas reservoirs with apparent–permeability and sorb–

phase effects : a density-based approach [C] //International Oil Conference and Exhibition in Mexico, 31 August-2 September 2006, Cancun, Mexico.DOI : https : //doi.org/10.2118/102197-MS.

[21] Zhang M, Ayala L F.Density-based production-data analysis of gas wells with significant rock-compressibility effects [J] .SPE Reservoir Evaluation & Engineering, 2015, 18 (2): 205-213.

[22] Wang H T.Performance of multiple fractured horizontal wells in shale gas reservoirs with consideration of multiple mechanisms [J] .Journal of Hydrology, 2014, 510: 299-312.

[23] Chen Z M, Liao X W, Zhao X L, et al.As new analytical method based on pressure transient analysis to estimate carbon storage capacity of depleted shales : a case study [J] .International Journal of Greenhouse Gas Control, 2015, 42: 46-58.

[24] Chen Z M, Liao X W, Zhao X L, et al.Performance of horizontal wells with fracture networks in shale gas formation [J] .Journal of Petroleum Science and Engineering, 2015, 133: 646-664.

[25] Chen Z M, Liao X W, Dou X J, et al.Development of a trilinear-flow model for carbon sequestration in depleted shale [J] .SPE Journal, 2016, 21 (4): 1-13.

[26] Yao S S, Zeng F H, Liu H, et al.A semi-analytical model for hydraulically fractured wells with stress-sensitive conductivities [C] //SPE Unconventional Resources Conference Canada, 5-7 November 2013, Calgary, Alberta, Canada. DOI : https : //doi.org/10.2118/167230-MS.

[27] Zeng F H, Zhao G. Semianalytical model for reservoirs with Forchheimer's non-Darcy flow [J] .SPE Reservoir Evaluation and Engineering, 2008, 11 (2): 280-291.

[28] Lorenz J C.Stress-sensitive reservoirs [J] .Journal of Petroleum Technique, 1999, 1: 61-62.

[29] Jin M, Somerville J, Smart B.Coupled reservoir simulation applied to the management of production induced stress-sensitivity [C] //International Oil and Gas Conference and Exhibition in China, 7-10 November 2000, Beijing, China.DOI : https : //doi.org/10.2118/64790-MS.

[30] Ikoku C U. Practical application of non-Newtonian transient flow analysis [C] //SPE Annual Technical Conference and Exhibition, 23-26 September 1979, Las Vegas, Nevada.DOI : https : //doi.org/10.2118/8351-MS.

[31] Omosebi A O, Igbokoyi A O.Boundary effect on pressure behavior of power-law non-Newtonian fluids in homogeneous reservoirs [J] .Journal of Petroleum Science and Engineering, 2016, 146: 838-855.

[32] Owayed J F, Tiab D.Transient pressure behavior of Bingham non-Newtonian fluids for horizontal wells [J] . Journal of Petroleum Science and Engineering, 2008, 61: 21-32.

[33] Hoek P J, Mahani H, Sorop T G, et al.Application of injection fall-off analysis in polymer flooding [C] // SPE Europec/EAGE Annual Conference, 4-7 June 2012, Copenhagen, Denmark.DOI : https : //doi.org/10.2118/154376-MS.

[34] Li Z T, Delshad, M.Development of an analytical injectivity model for non-Newtonian polymer solutions [C] //SPE Reservoir Simulation Symposium, 18-20 February 2013, The Woodlands, Texas, USA.DOI : https : //doi.org/10.2118/163672-MS.

[35] Wu Y S, Pruess K.A numerical method for simulating non-Newtonian fluid flow and displacement in porous media [J] .Advances in Water Resources, 1997, 21: 351-362.

[36] Wu Y S, Pruess K, Witherspoon P A.Flow and displacement of Bingham non-Newtonian fluids in porous media [J].SPE Reservoir Engineering, 1992, 8: 369-376.

[37] Berumen, S.Evaluation of fractured wells in pressure-sensitive formations. Norman : University of Oklahoma, 1995.

[38] Mukherjee H, Economides M.A parametric comparison of horizontal and vertical well performance [J]. SPE Formation Evaluation, 1991, 7: 209-216.

[39] Tabatabaie S H, Pooladi-Darvish M, Mattar L, et al. Analytical modeling of linear flow in pressure-sensitive formations [J].SPE Reservoir Evaluation & Engineering, 2017, 20 (1): 215-227.

[40] Cole J D. On a quasi-linear parabolic equation occurring in aerodynamics [J].Quarterly of Applied Mathematics, 1951, 3: 201-230.

[41] Hopf E. The partial differentialequation ut+uux=uxx [J].Communications on Pure and Applied Mathematics, 1950, 3: 201-230.

[42] Yeung

[43] Wang S H, Ding W Lin M L, et al.Approximate analytical-pressure studies on dual-porosity reservoirs with stress-sensitive permeability [J].SPE Reservoir Evaluation & Engineering, 2015, 12: 523-533.

[44] Zerzar A, Tiab D, Bettam Y.Interpretation of multiple hydraulically fractured horizontal wells [C] //Abu Dhabi International Conference and Exhibition, 10-13 October 2004, Abu Dhabi, United Arab Emirates.DOI : https : //doi.org/10.2118/88707-MS.

[45] Wang L, Wang X D, Li J Q, et al.Simulation of pressure transient behavior for asymmetrically finite-conductivity fractured wells in coals reservoirs [J].Transport in Porous Media, 2013, 97 (3): 353-372.

[46] Zeng F H.Modeling of non-Darcy flow in porous media and its application [D].Regina : University of Regina, 2008.

[47] Wang J L, Jia A L.A general productivity model for optimization of multiple fractures with heterogeneous properties [J]. Journal of Natural Gas Science and Engineering, 2014, 21: 608-624.

[48] Berumen S, Tiab D, Rodriguez F.Constant rate solutions for a fractured well with an asymmetric fracture [J]. Journal of Petroleum Science and Engineering, 2000, 25 (1-2): 49-58.

[49] Bennett C O, Reynold A C, Raghavan R, et al.Performance of finite conductivity, vertical fractured wells in single-layer reservoirs [J].SPE Formation Evaluation, 1986, 1 (4): 399-412.

[50] Cinco-Ley H, Samanigeo F, Dominguez N.Transient pressure behavior for a well with a finite-conductivity vertical fracture [J].SPE Journal, 1978, 18 (4): 253-264.

[51] Gringarten A C, Ramey H J, Raghavan R.Unsteady-state pressure distributions created by a well with a single infinite-conductivity vertical fracture [J].SPE Journal, 1974, 8: 347-360.

[52] Luo W J, Tang C F, Wang X D.Pressure transient analysis of a horizontal well intercepted by multiple non-planar vertical fractures [J].Journal of Petroleum Science and Engineering, 2014, 124: 232-242.

[53] Song B, Economides M J, Ehlig-Economides C. Design of multiple transverse fracture horizontal wells in shale gas reservoirs [C] //SPE Hydraulic Fracturing Technology Conference, 24-26 January 2011, The Woodlands,

Texas, USA. DOI : https : //doi.org/10.2118/140555–MS.

[54] Resurreicao C E, Fernando R.Transient rate behavior of finite–conductivity asymmetrically fractured wells producing at constant pressure [C] // SPE Annual Technical Conference and Exhibition, 6–9 October 1991, Dallas, Texas, USA. DOI : https : //doi.org/10.2118/22657–MS.

[55] Agarwal

[56] Roumboutsos A, Stewart G. A direct deconvolution of convolution algorithm for well test analysis [C] // SPE Annual Technical Conference and Exhibition, 2–5 October 1988, Houston, Texas, USA. DOI : https : //doi.org/10.2118/18157–MS.

[57] Stehfest H. Numerical inversion of Laplace transform [J] . Comm. ACM, 1970, 13 (1) : 47–49.

[58] Walsh J B. Effect of pore pressure and confining pressure on fracture permeability [J] . International Journal of Rock Mechanics and Mining Sciences & Geomechanics Abstracts, 1981, 18 (5) : 429–435.

[59] Gangi A F.Variation of whole and fractured porous rock permeability with confining pressure [J] . International Journal of Rock Mechanics and Mining Sciences & Geomechanics Abstracts, 1978, 15 (5) : 249–257.

[60] Ferrell H H, Felsenthal M. Effect of overburden pressure on flow capacity in a deep oil reservoir [J] . Journal of Petroleum Technology, 1962, 14 (9) : 962–966.

[61] Vairogs J, Rhoades V W. Pressure transient tests in formations having stress–sensitive permeability [J] . Journal of Petroleum Technology, 1973, 25 (8) : 965–970.

[62] Wall J D, Nur A, Bourbie T. Effects of pressure and partial water saturation on gas permeability in tight sands : experimental results [J] .Journal of Petroleum Technology, 1982, 34 (4) : 930–936.

[63] Bourbie T, Walls J. Pulse decay permeability : analytical solution and experimental test [J] .SPE Journal, 1982, 22 (5) : 719–721.

[64] Kikani J, Pedrosa O A. Perturbation analysis of stress–sensitive reservoirs [J] .SPE Formation Evaluation, 1991, 6 (3) : 379–388.

[65] Kwon O, Kronenberg A K, Gangi A F, et al. Permeability of Wilcox shale and its effective pressure law [J] . Journal of Geophysical Research : Solid Earth, 2001, 106 (B9) : 19339–19353.

[66] Kang Y L, Lin C, You L J, et al. Impact of oil–based drilling/completion invasion on the stress sensitivity of shale reservoirs [J] .Natural Gas Industry, 201535 (6) : 64–68.

[67] Zhang R, Ning Z F, Yang F, et al. Experimental study of stress sensitivity of shale reservoirs [J] . Chinese Journal of Rock Mechanics and Engineering, 2015, 34 (S1) : 2617–2622.

[68] Zhang R, Ning Z F, Yang F, et al. Shale stress sensitivity experiment and mechanism [J] .Acta Petrolei Sinica, 2015, 36 (2) : 224–232.

[69] Alramahi B, Sundberg M I, 2012. Proppant embedment and conductivity of hydraulic fractures in shales [C] //46th U.S. Rock Mechanics/Geomechanics Symposium, 24–27 June 2012, Chicago, Illinois, USA.

[70] Greenberg M D. Application of Green's functions in Science and Engineering [J] . Journal of the Franklin Institute, 1975, 300 (3) : 240–241.

Appendix A：Evaluation of exponential dependence of permeability on pressure

A1 Theoretical background

Generally, rock properties such as permeability and porosity are functions of effective stress or effective pressure. Walsh [58] and Gangi [59] presented that the cubic root of the permeability $k^{1/3}$ is linearly related to the logarithm of the effective pressure $\ln p_e$. Here, the effective pressure is usually defined as

$$p_{\text{eff}} = p_c - bp \tag{A1}$$

where p is the pore pressure, b is a parameter called as effective coefficient (its magnitude lies between 0 and 1), p_c is the confining pressure and p_{eff} is the effective pressure. It is noted that the confining pressure and pore pressure have an opposite effect on rock properties [37].

Experimental investigation on pressure dependence of rock properties has confirmed that the permeability behaves in an exponential manner with the pressure variation [58, 60-61]. In the work of Wall et al. [62], they detected that the change of permeability with regard to pore pressure was greater than confining pressure. Thus, it is reasonable to set that the pore pressure is used to characterize the rock deformation under constant confining pressure condition [62, 63]. After fully incorporating previous ideas and results, Pedrosa [9, 64] developed the important concept of permeability modulus：

$$\gamma = \frac{1}{K}\frac{dK}{dp} \tag{A2}$$

This parameter γ measures the dependence of permeability on pore pressure, which is obtained through laboratory measurements. From the view of practical purpose, γ can be assumed as a constant [4-5, 15]. Through integration with regard to Eq. (A2), the concept of permeability modulus implied an exponential expression for permeability -pressure as following：

$$K = K(p_i)\exp(-\gamma\Delta p) \tag{A3}$$

In 1997 work of Pedroso and Correa [13], Eq. (A3) was also presented in forms of confining stress σ_{EFF} to express the fracture permeability variation：

$$K = K(p_i)\exp(\gamma\Delta\sigma_{\text{EFF}}) \tag{A4}$$

Considering the propping agents package inside a fracture as a completely unconsolidated medium, for a propped fracture, the following relation between effective stress and confining stress is valid (effective coefficient b was taken to be unity in the works of Ostensen [10] and Kwon [65])：

$$\sigma_{\text{EFF}} = \sigma_c - p \tag{A5}$$

Combing Eq. (A4) and Eq. (A5), the relation is rewritten as:

$$K = K(p_i)\exp\{\gamma[(\sigma_c - p)_t - (\sigma_c - p)]\} \tag{A6}$$

The confining stress would vary only if tectonic modifications occur; put another way, the confining stress could be assumed to be constant under most practical condition. Therefore Eq. (A6) is consistent with Eq. (A2) in differential notation.

A2 Experimental verification

Pressure dependence of formation permeability. The linear relationship according to Walsh [58] and Gangi [59] is directly expressed as

$$K^{1/3}=A\ln p_{eff} + B \tag{A7}$$

Based on the laboratory measurement data from Longmaxi shale samples in China, Fig.A1 shows the linear relationship between the cubic root of the permeability $(K/K_i)^{1/3}$ and the logarithm of the effective pressure $\ln p_{eff}/p_{eff,i}$. After taking the initial condition into account, Eq. (A7) could rewritten as the following dimensionless form:

$$(K/K_i)^{1/3} = A'\ln(p_{eff}/p_{eff,i}) + 1 \tag{A8}$$

where the coefficient A' is satisfied as $A'=A/K_i^{1/3}$. It is seen from Fig.A1 that the value of A' is in the range of 0.3~0.4 for the Longmaxi Shale. Many works have confirmed that A' varies from 0.2 to 0.5 in most field cases [66–68]. Fig.A2 shows that Eq. (A8) could be characterized in the forms of alternative exponential expression, analogous to Eq. (1).

Pressure dependence of fracture conductivity. Fracture conductivity is strongly dependent on the pore pressure due to proppant embedment. Based on the laboratory measurement data provided by Alramahi et al. [69], Fig.A3 shows the relationship between normalized fracture conductivity and confining pressure for different shale samples from stiff shale to soft shale. As shown in Fig.A3, the soft shale is more strongly dependent on the pressure change than stiff shale, which means the proppant trends to embed into the fracture face more likely.

According to the experimental data in Fig.A3, for propped fracture, the general expression of fracture conductivity is given as

$$\lg(F_c/F_{c,i})=Ap_{eff}+B \tag{A9}$$

Here, the coefficient b is assume to be unity, and then combining Eq. (A1) and Eq. (A7) could lead to the following form:

$$\lg(F_c/F_{c,i}) = A(p_{eff} - p_{eff,i}) = A[(p_c - p) - (p_c - p_i)] = A(p_i - p) \tag{A10}$$

Put another way, Eq. (A10) is further written as follows:

$$F_c = F_{c,i}\exp[-\gamma(p_i - p)] \tag{A11}$$

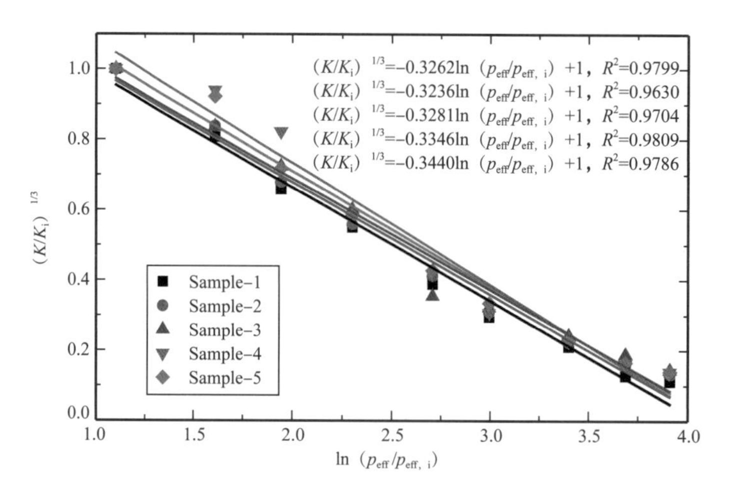

Fig.A1 Linear relation between $(K/K_i)^{1/3}$ and p_{eff} from actual rock samples

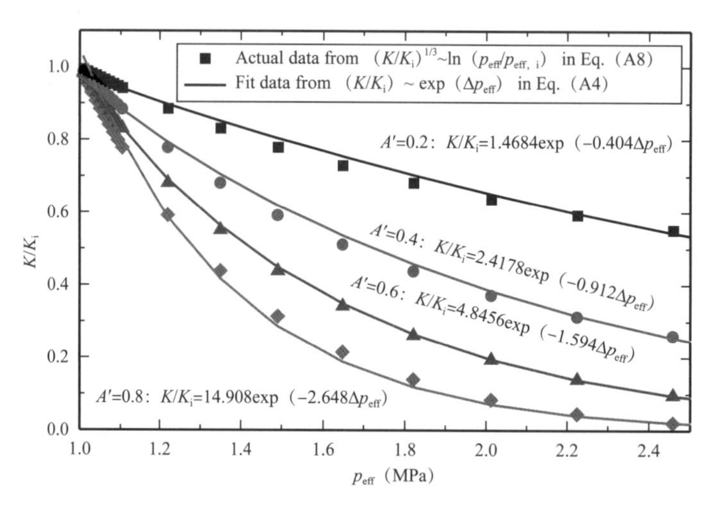

Fig.A2 Fitting relationship between exponential Eq. (A4) and linear Eq. (A8)

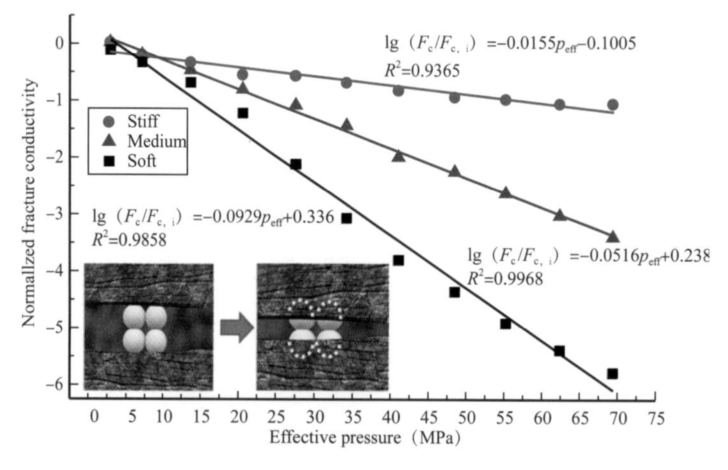

Fig.A3 The relation between normalized fracture conductivity and effective pressure for different shale samples

Where $\gamma=|A|\cdot\ln 10$. As shown in Fig.A3, the magnitude of A is about $10^{-5}KPa^{-1}$, and this phenomenon is consistent with the range of 10^{-5} to 10^{-4} KPa^{-1} provided by Ozkan et al. [18]. These illustrations mentioned above can support the reliability of using exponential expression of Eq. (1).

Appendix B: Dimensionless Variables

Dimensionless pressure

$$P_{\xi D}=\frac{2\pi K_{mi}h(p_i-p_\xi)}{Q_{sc}\mu} \tag{B1}$$

Dimensionless time

$$t_D=\frac{K_{mi}t}{\phi_m\mu c_{tm}L^2_{ref}} \tag{B2}$$

Dimensionless flux distribution along fracture

$$q_{fD}=\frac{2L_{ref}q_f}{Q_{sc}} \tag{B3}$$

Dimensionless production rate of individual fracture

$$q_{scD}=\frac{q_{sc}}{Q_{sc}} \tag{B4}$$

Dimensionless length variables

$$\xi_D=\frac{\xi}{L_{ref}}\xi=x,y,x_w,y_w,x_f \tag{B5}$$

Dimensionless fracture conductivity

$$C_{fD}=\frac{K_{fi}w_f}{K_{mi}L_{ref}} \tag{B6}$$

Dimensionless fracture/reservoir permeability modulus

$$\gamma_{\xi D}=\frac{Q_{sc}\mu}{2\pi K_{mi}h}\gamma_\xi,\xi=f\text{ and }m \tag{B7}$$

Appendix C: Reservoir Solution

Based on the assumption mentioned in Section 2, a two-dimension reservoir flow model with n_f point sources of elementary segment is established,

$$\frac{\partial}{\partial x}\left(\rho\frac{k_m}{\mu}\frac{\partial p_m}{\partial x}\right)+\frac{\partial}{\partial y}\left(\rho\frac{k_m}{\mu}\frac{\partial p_m}{\partial y}\right)+\sum_{v=1}^{n_f}\frac{\rho q_f^{(v)}(x',t)\Delta x'}{h}\delta(x-x',y-y_w^{(v)})=\frac{\partial(\phi_m\rho)}{\partial t} \tag{C1}$$

Through using permeability variation in Eq. (1), Eq. (C1) is recast as the following dimensionless form:

$$\gamma_{mD}\exp(-\gamma_{mD}p_{mD})\left[\left(\frac{\partial p_{mD}}{\partial x_D}\right)^2+\left(\frac{\partial p_{mD}}{\partial y_D}\right)^2\right]+\exp(-\gamma_{mD}p_{mD})\nabla^2 p_{mD}$$

$$+\pi\Delta x'_D\sum q_{fD}^{(v)}(x'_D,t_D)\delta(x_D-x'_D,y_D-y_{wD}^{(v)})=\frac{\partial p_{mD}}{\partial t_D} \quad \text{(C2)}$$

By substituting Eq. (3) into Eq. (C2), linear equation of Eq. (C2) is obtained as Eq. (4). To obtain an approximate analytical solution for Eq. (4), perturbation technique is used to rewrite η_{mD} in the form of power series

$$\eta_{mD}=\eta_{mD0}+\gamma_{mD}\eta_{mD1}+\gamma_m^2\eta_{mD2}+\cdots \quad \text{(C3)}$$

and the first term in the right hand side of Eq. (4) is also expanded as power series,

$$\frac{1}{1-\gamma_{mD}\eta_{mD}}=1+\gamma_m\eta_{mD}+(\gamma_m\eta_{mD})^2+(\gamma_m\eta_{mD})^3+\cdots \quad \text{(C4)}$$

Substituting Eq. (C3) ~ (C4) into Eq. (4), the following equation is given as:

$$\left(\frac{\partial^2\eta_{mD0}}{\partial x_D^2}+\frac{\partial^2\eta_{mD0}}{\partial y_D^2}+\pi\Delta x'_D\sum_{v=1}^{n_f} q_{fD}^{(v)}(x'_D,t_D)\delta(x_D-x'_D,y_D-y_{wD}^{(v)})-\frac{\partial\eta_{mD0}}{\partial t_D}\right)$$

$$+\gamma_{mD}\left(\frac{\partial^2\eta_{mD1}}{\partial x_D^2}+\frac{\partial^2\eta_{mD1}}{\partial y_D^2}-\frac{\partial\eta_{mD1}}{\partial t_D}-\eta_{mD0}\frac{\partial\eta_{mD0}}{\partial t_D}\right)$$

$$+\gamma^2{}_{mD}\left(\frac{\partial^2\eta_{mD2}}{\partial x_D^2}+\frac{\partial^2\eta_{mD2}}{\partial y_D^2}-\frac{\partial\eta_{mD2}}{\partial t_D}-\frac{\partial(\eta_{mD0}\eta_{mD1})}{\partial t_D}-\eta_{mD0}\frac{\partial\eta_{mD0}}{\partial t_D}\right)+\cdots=0 \quad \text{(C5)}$$

According to the conclusion of Pedrosa [9], the first term on the left right hand side of Eq. (C5) is regarded as the approximate expression when $\gamma_{mD}<0.2$. Thus, Eq. (4) is approximately anomalous to classical diffusion equation, and the corresponding solution is given by Eq. (5). Accordingly, the perturbation solution for elementary segment under constant rate condition could be directly obtained as follows:

$$\Delta\eta_{mD}^{(m)}(x_{wD,j}^{(v)},y_{wD}^{(v)};x_{wD,i}^{(w)},y_{wD}^{(w)})=\int_{x_{wD,i}^{(w)}-0.5\Delta x_D^{(w)}}^{x_{wD,i}^{(w)}+0.5\Delta x_D^{(w)}}\left[-\frac{1}{4}\mathrm{Ei}\left(-\frac{\left(x_{wD,j}^{(v)}-x'_D\right)^2+(y_{wD}^{(v)}-y_{wD}^{(w)})^2}{4\Delta t_{D,m}}\right)\right]dx'_D \quad \text{(C6)}$$

$$\Delta\eta_{mD}^{(m)}\left(x_{wD,j}^{(v)},y_{wD}^{(v)};x_{wD,i}^{(w)},y_{wD}^{(w)};k\right)=\int_{x_{wD,i}^{(w)}+0.5\Delta x_D^{(w)}}^{x_{wD,i}^{(w)}+0.5\Delta x_D^{(w)}}\left[-\frac{1}{4}\mathrm{Ei}\left(-\frac{\left(x_{wD,j}^{(v)}-x'_D\right)^2+(y_{wD}^{(v)}-y_{wD}^{(w)})^2}{4\left[t_{D,m}-t_{D,k}+0.5(\Delta t_{D,m}-\Delta t_{D,k})\right]}\right)\right]dx'_D \quad \text{(C7)}$$

$$\Delta\eta_{\mathrm{mD2}}^{(m)}(x_{\mathrm{wD,j}}^{(v)},y_{\mathrm{wD}}^{(v)};x_{\mathrm{wD,j}}^{(w)},y_{\mathrm{wD}}^{(w)};k)=\int_{x_{\mathrm{wD,i}}^{(w)}-0.5\Delta x_{\mathrm{D}}^{(w)}}^{x_{\mathrm{wD,i}}^{(w)}+0.5\Delta x_{\mathrm{D}}^{(w)}}\left[-\frac{1}{4}\mathrm{Ei}\left(-\frac{(x_{\mathrm{wD,j}}^{(v)}-x_{\mathrm{D}}')^2+(y_{\mathrm{wD}}^{(v)}-y_{\mathrm{wD}}^{(w)})^2}{4\left[t_{\mathrm{D,m}}-t_{\mathrm{D,k}}+0.5(\Delta t_{\mathrm{D,m}}+\Delta t_{\mathrm{D,k}})\right]}\right)\right]\mathrm{d}x_{\mathrm{D}}' \quad (C8)$$

Appendix D : Fracture Flow Model

Based on the principle of mass conservation, the differential equation describing the flow in fracture is stated as the following equation with the sink/source terms,

$$-\frac{\partial}{\partial x}\left(-\rho\frac{K_{\mathrm{f}}}{\mu}\frac{\partial p_{\mathrm{f}}}{\partial x}\right)+\frac{\rho q_{\mathrm{f}}(x,t)}{w_{\mathrm{f}}h}-\frac{\rho_{\mathrm{sc}}q_{\mathrm{sc}}(t)}{w_{\mathrm{f}}h}\delta(x-x_{\mathrm{fw}})=0 \quad (D1)$$

Where the source term $q_{\mathrm{f}}(x)$ represents the flux density along fracture flowing from reservoir to fracture ; the sink term $q_{\mathrm{sc}}(t)$ represents the fluid mass leaving the system.

Through using permeability variation in Eq. (1), Eq. (D1) is expanded as the following dimensionless form :

$$-\gamma_{\mathrm{fD}}^{(v)}\exp(-\gamma_{\mathrm{fD}}^{(v)}p_{\mathrm{fD}}^{(v)})\left(\frac{\partial p_{\mathrm{fD}}^{(v)}}{\partial x_{\mathrm{D}}}\right)^2+\exp(-\gamma_{\mathrm{fD}}^{(v)}p_{\mathrm{fD}}^{(v)})\frac{\partial^2 p_{\mathrm{fD}}^{(v)}}{\partial x_{\mathrm{D}}^2}-\frac{\pi}{C_{\mathrm{fD}}^{(v)}}q_{\mathrm{fD}}^{(v)}(x_{\mathrm{D}})+\frac{2\pi}{C_{\mathrm{fD}}^{(v)}}q_{\mathrm{scD}}^{(v)}\delta(x_{\mathrm{D}}-\theta^{(v)})=0 \quad (D2)$$

Using Eq. (3) makes Eq. (D2) linearized in the following forms of Pedrosa substitution

$$\frac{\partial^2\eta_{\mathrm{fD}}^{(v)}}{\partial x_{\mathrm{D}}^2}-\frac{\pi}{C_{\mathrm{fD}}^{(v)}}q_{\mathrm{fD}}^{(v)}(x_{\mathrm{D}},t_{\mathrm{D}})+\frac{2\pi}{C_{\mathrm{fD}}^{(v)}}q_{\mathrm{scD}}^{(v)}(t_{\mathrm{D}})\delta(x_{\mathrm{D}}-\theta^{(v)})=0 \quad (D3)$$

Here Green's function $N(x',x_{\mathrm{D}})$ is used to solve Eq. (D3) .Now the key is to obtain appropriate Green's function.According to the theory of Green's function, $N(x',x_{\mathrm{D}})$ satisfies the following math model :

$$\begin{cases}N''(x',x_{\mathrm{D}})=\delta(x'-x_{\mathrm{D}})+F\\ N'(x_{\mathrm{wD,0}}^{(v)}-x_{\mathrm{fD}}^{(v)},x_{\mathrm{D}})=0\\ N'(x_{\mathrm{wD,0}}^{(v)}+x_{\mathrm{fD}}^{(v)},x_{\mathrm{D}})=0\end{cases} \quad (D4)$$

Where constant F is the correction factor, and the general solution is given as [70]

$$N(x',x_{\mathrm{D}})=Ax'+\frac{F}{2}x'^2+(x'-x_{\mathrm{D}})H(x'-x_{\mathrm{D}}) \quad (D5)$$

Where $H(x'-x_{\mathrm{D}})$ is Heaviside step function.On the basis of boundary conditions and continuity condition on $x'=x_{\mathrm{D}}$, the particular solution for Eq. (D5) is given as

$$N^{(v)}(x',x_{\mathrm{D}})=\begin{cases}-\dfrac{1}{4x_{\mathrm{fD}}^{(v)}}\left[x'-(x_{\mathrm{wD},0}^{(v)}-x_{\mathrm{fD}}^{(v)})\right]^2-\dfrac{1}{4x_{\mathrm{fD}}^{(v)}}\left[x_{\mathrm{D}}-(x_{\mathrm{wD},0}^{(v)}+x_{\mathrm{fD}}^{(v)})\right]2+\dfrac{\left(x_{\mathrm{wD},0}^{(v)}-x_{\mathrm{fD}}^{(v)}\right)^2}{3x_{\mathrm{fD}}^{(v)}},\\ x_{\mathrm{wD},0}^{(v)}-x_{\mathrm{fD}}^{(v)}\leqslant x'<x_{\mathrm{D}}\\ \dfrac{1}{4x_{\mathrm{fD}}^{(v)}}\left[x'-(x_{\mathrm{wD},0}^{(v)}+x_{\mathrm{fD}}^{(v)})\right]2-\dfrac{1}{4x_{\mathrm{fD}}^{(v)}}\left[xD-(x_{\mathrm{wD},0}^{(v)}-x_{\mathrm{fD}}^{(v)})\right]2+\dfrac{\left(x_{\mathrm{wD},0}^{(v)}-x_{\mathrm{fD}}^{(v)}\right)^2}{3x_{\mathrm{fD}}^{(v)}},\\ x_{\mathrm{D}}<x'\leqslant x_{\mathrm{wD},0}^{(v)}+x_{\mathrm{fD}}^{(v)}\end{cases} \quad \text{(D6)}$$

Now, Green's function of Eq. (D6) is used to integrate Eq. (D3) from $x_{\mathrm{wD},0}-x_{\mathrm{fD}}$ to $x_{\mathrm{wD},0}+x_{\mathrm{fD}}$. After finishing the integration, the final solution for Eq. (D3) is obtained, as shown in Eq. (7). It is noted that Eq. (D6) is the general solution of Green's function, suitable for the case with arbitrary values of fracture length and location. This is an important extension presented by Resurreicao et al. [54] and Wang et al. [45].

Appendix E: Coupled solution for linear system

This method is proposed to solve the linear equations caused by multiple fracture interaction. The corresponding solution is regarded as the initial guess of nonlinear equations. On the m-th time step, the system of linear equations is given in the form of matrix,

$$\boldsymbol{A}\vec{\boldsymbol{X}}=\vec{\boldsymbol{B}} \quad \text{(E1)}$$

Where matrix $\boldsymbol{A}$ is a coefficient matrix of dimension $(2N\cdot n_{\mathrm{f}}+1)\times(2N\cdot n_{\mathrm{f}}+1)$, $\vec{\boldsymbol{X}}$ is the vector of unknowns, and $\vec{\boldsymbol{B}}$ is the known vector which represents the time cumulative effect from 0 to $(m-1)$-th time step. These matrix and vectors are further expanded as following submatrix and subvectors:

$$\boldsymbol{A}=\begin{bmatrix}\boldsymbol{A}_{1,1} & \cdots & \boldsymbol{A}_{1,w} & \cdots & \boldsymbol{A}_{1,nf} & \boldsymbol{A}_{1,nf+1}\\ \vdots & & \vdots & & \vdots & \vdots\\ \boldsymbol{A}_{v,1} & \cdots & \boldsymbol{A}_{v,w} & \cdots & \boldsymbol{A}_{v,nf} & \boldsymbol{A}_{v,nf+1}\\ \vdots & & \vdots & & \vdots & \vdots\\ \boldsymbol{A}_{nf,1} & \cdots & \boldsymbol{A}_{nf,w} & \cdots & \boldsymbol{A}_{nf,nf} & \boldsymbol{A}_{nf,nf+1}\\ \boldsymbol{A}_{nf+1,1} & \cdots & \boldsymbol{A}_{nf+1,w} & \cdots & \boldsymbol{A}_{nf+1,nf} & \boldsymbol{A}_{nf,nf+1}\end{bmatrix} \quad \text{(E2)}$$

$$\vec{\boldsymbol{X}}=(\boldsymbol{X}_1\cdots\boldsymbol{X}_v\cdots\boldsymbol{X}_{nf}\boldsymbol{X}_{nf+1})^{\mathrm{T}} \quad \text{(E3)}$$

$$\vec{\boldsymbol{B}}=(\boldsymbol{B}_1\cdots\boldsymbol{B}_v\cdots\boldsymbol{B}_{nf}\boldsymbol{B}_{nf+1})^{\mathrm{T}} \quad \text{(E4)}$$

Submatrix $\boldsymbol{A}_{v,w}$ represents the vector of pressure drop on v-th fracture caused by w-th fracture, so $\boldsymbol{A}_{v,w}$ is further expressed as the following form:

$$
\boldsymbol{A}_{v,w}=\begin{bmatrix} \alpha_{1+2(v-1)N,1+2(w-1)N} & \cdots & \alpha_{1+2(v-1)N,i+2(w-1)N} & \cdots & \alpha_{1+2(v-1)N,2N+2(w-1)N} \\ \vdots & & \vdots & & \vdots \\ \alpha_{j+2(v-1)N,1+2(w-1)N} & \cdots & \alpha_{j+2(v-1)N,i+2(w-1)N} & \cdots & \alpha_{j+2(v-1)N,2N+2(w-1)N} \\ \vdots & & \vdots & & \vdots \\ \alpha_{2N+2(v-1)N,1+2(w-1)N} & \cdots & \alpha_{2N+2(v-1)N,i+2(w-1)N} & \cdots & \alpha_{2N+2(v-1)N,2N+2(w-1)N} \end{bmatrix} \tag{E5}
$$

and

$$
\boldsymbol{A}_{v,w}=\begin{cases} (0.5\Delta x_{\mathrm{D}}^{(w)}\cdots 0.5\Delta x_{\mathrm{D}}^{(w)}\cdots 0.5\Delta x_{\mathrm{D}}^{(w)})\, v=n_{\mathrm{f}}+1 \text{ and } v\leqslant n_{\mathrm{f}} \\ (1\cdots 1\cdots 1)^{T}\, w=n_{\mathrm{f}}+1 \text{ and } v\leqslant n_{\mathrm{f}} \\ 0\ w=n_{\mathrm{f}}+1 \text{ and } v=n_{\mathrm{f}}+1 \end{cases} \tag{E6}
$$

The element $\alpha_{j+2(v-1)N,\ i+2(w-1)N}$ in sub matrix $\boldsymbol{A}_{v,\ w}$ is given as following:

$$
\alpha_{j+2(v-1)N,i+2(w-1)N}=\Delta\eta_{\mathrm{mD}}^{(m)}(x_{\mathrm{wD,j}}^{(v)},y_{\mathrm{wD}}^{(v)};x_{\mathrm{wD,i}}^{(w)},y_{\mathrm{wD}}^{(w)})
$$

$$
+\begin{cases} \dfrac{\pi}{C_{\mathrm{fD}}^{(v)}}\left(\begin{aligned}&\int_{x_{\mathrm{wD,i}}^{(v)}-0.5\Delta x_{\mathrm{D}}^{(v)}}^{x_{\mathrm{wD,i}}^{(v)}+0.5\Delta x_{\mathrm{wD}}^{(v)}}\left[N^{(v)}(x';x_{\mathrm{wD,j}}^{(v)})-N^{(v)}(x';\theta^{(v)})\right]\mathrm{d}x' \\ &+\left[N^{(v)}(\theta^{(v)};x_{\mathrm{wD,j}}^{(v)})-N^{(v)}(\theta^{(v)};\theta^{(v)})\right]\Delta x_{\mathrm{D}}^{(v)}\end{aligned}\right) & v=w \\ 0 \quad v\neq w \end{cases} \tag{E7}
$$

The unknown subsector of $\boldsymbol{X}_{\mathrm{v}}$ represents the flux distribution along v-th fracture, which is given as

$$
\boldsymbol{X}_{v}=\begin{cases} \left(q_{\mathrm{fD,1}}^{(v,m)}\cdots q_{\mathrm{fD,j}}^{(v,m)}\cdots q_{\mathrm{fD,2N}}^{(v,m)}\right)^{\mathrm{T}} & v\leqslant n_{\mathrm{f}} \\ \eta_{\mathrm{wD}}^{(m)} & v=n_{\mathrm{f}}+1 \end{cases} \tag{E8}
$$

The known subsector of $\boldsymbol{B}_{\mathrm{v}}$ represents the cumulative pressure drop effect caused by n_{f} fracture from 1-th to (m-1) -th time steps, which is expressed as following:

$$
\boldsymbol{B}_{v}=\left(\beta_{1+2(v-1)N}\cdots\beta_{j+2(v-1)N}\cdots\beta_{2N+2(v-1)N}\right)^{\mathrm{T}} \tag{E9}
$$

The element $\beta_{j+2(v-1)N}$ represents the cumulative effect of j-th segment on v-th fracture caused by all segments, which is given as follows:

$$
\beta_{j+2(v-1)N}=\sum_{w=1}^{n_f}\sum_{i=1}^{2N_w}\sum_{k=1}^{m-1}q_{\mathrm{fD,i}}^{(w,k)}\left[\Delta\eta_{\mathrm{mD2}}^{(m)}(x_{\mathrm{wD,j}}^{(v)},y_{\mathrm{wD}}^{(v)};x_{\mathrm{wD,i}}^{(w)},y_{\mathrm{wD}}^{(w)};k)-\Delta\eta_{\mathrm{mD1}}^{(m)}(x_{\mathrm{wD,j}}^{(v)},y_{\mathrm{wD}}^{(v)};x_{\mathrm{wD,i}}^{(w)},y_{\mathrm{wD}}^{(w)};k)\right] \tag{E10}
$$

Appendix F: Coupled solution for nonlinear system

According to the theory of Newton-Raphson method, a set of linear equations for the unknown corrections $\delta\vec{X}$ are presented, namely

$$J\delta\vec{X}=-\vec{F} \tag{F1}$$

Where J is the Jacobian matrix with dimension of $(2N \cdot n_\theta+1) \times (2N \cdot n_\theta+1)$, sector $\vec{F}$ denotes the residue form of Eq. (8) ~Eq. (13), as follows:

$$\vec{F}=(F_1 \cdots F_v \cdots F_{nf} \cdots F_{nf}+1)^{\mathrm{T}} \tag{F2}$$

The subsector F_v represents the residue form of v-th fracture

$$F_v=(F_1^{(v)} \cdots F_j^{(v)} \cdots F_{2N}^{(v)})^{\mathrm{T}} \tag{F3}$$

and the Jacobian matrix J is further expressed as

$$J=\begin{bmatrix} \frac{\partial F_1}{\partial X_1} & \cdots & \frac{\partial F_1}{\partial X_w} & \cdots & \frac{\partial F_1}{\partial X_{nf}} & \frac{\partial F_1}{\partial X_{nf+1}} \\ \vdots & & \vdots & & \vdots & \vdots \\ \frac{\partial F_v}{\partial X_1} & \cdots & \frac{\partial F_v}{\partial X_w} & \cdots & \frac{\partial F_v}{\partial X_{nf}} & \frac{\partial F_v}{\partial X_{nf+1}} \\ \vdots & & \vdots & & \vdots & \vdots \\ \frac{\partial F_{nf+1}}{\partial X_1} & \cdots & \frac{\partial F_{nf+1}}{\partial X_w} & \cdots & \frac{\partial F_{nf+1}}{\partial X_{nf}} & \frac{\partial F_{nf+1}}{\partial X_{nf+1}} \end{bmatrix} \tag{F4}$$

The sub matrix $\partial F_v / \partial X_w$ denotes the derivative of the functional residue on the v-th fracture with regard to w-th unknown variables

$$\frac{\partial F_v}{\partial X_w}=\begin{bmatrix} \frac{\partial F_1^{(v)}}{\partial x_1^{(w)}} & \cdots & \frac{\partial F_1^{(v)}}{\partial x_i^{(w)}} & \cdots & \frac{\partial F_1^{(v)}}{\partial x_{2N}^{(w)}} \\ \vdots & & \vdots & & \vdots \\ \frac{\partial F_j^{(v)}}{\partial x_1^{(w)}} & \cdots & \frac{\partial F_j^{(v)}}{\partial x_i^{(w)}} & \cdots & \frac{\partial F_j^{(v)}}{\partial x_{2N}^{(w)}} \\ \vdots & & \vdots & & \vdots \\ \frac{\partial F_{2N}^{(v)}}{\partial x_1^{(w)}} & \cdots & \frac{\partial F_{2N}^{(v)}}{\partial x_i^{(w)}} & \cdots & \frac{\partial F_{2N}^{(v)}}{\partial x_{2N}^{(w)}} \end{bmatrix} \tag{F5}$$

Where the element of $F_j^{(v)}$ indicates the function expression of the j-th segment on v-th fracture with regard to the unknown variables.Likewise, $\partial F_j^{(v)} / \partial x_i^{(w)}$ indicates the derivative of $F_j^{(v)}$ with regard to the unknown variable $x_i^{(w)}$, where the unknown corresponds to the variable of i-th segment on w-th fracture.

页岩气储层压裂水平井气水两相渗流模型

郭小哲[1]，王　晶[1]，刘学锋[2]

（1. 中国石油大学石油工程学院；2. 中海油能源发展工程技术分公司）

摘　要：针对定量计算页岩气储层水平井压裂后含水对产气量的影响程度，应用渗流力学、油藏数值模拟及油层物理学等基础理论与方法，建立了考虑解吸—吸附、扩散、滑脱及应力敏感的双重介质气水两相渗流模型，并进行了差分离散及方程组系数线性化处理，建立了数值模型。通过新建模型的实例应用，进行了气水产量的预测，对比分析了不同含水饱和度、压裂缝网参数、扩散、滑脱及应力敏感等对气井生产的影响。结果表明：原始含水饱和度决定产量回升幅度和稳产时间，原始含水饱和度越小，压裂液返排率对产量影响越大；当裂缝导流能力和缝网区域半长较大时，再提高他们的值则增产幅度会减小；扩散作用弱于滑脱作用，应力敏感作用要明显大于扩散作用和滑脱作用，增产稳产目标要重点考虑应力敏感。

关键词：页岩气藏；气水两相渗流；数值模拟；压裂；扩散；滑脱；应力敏感

随着对页岩气储层开发认识的不断深入，用于描述各种复杂渗流机理的数学模型[1-7]及压裂产能计算模型[8-14]已经建立了很多。但是这些模型中重点考虑了天然气单相渗流，忽略了压裂后残留水或地层水对气相渗流的影响。但现场实际情况是多数页岩气井在实际生产中一直伴随着产水，并且生产水气比一直较高。以威 201-H1 井的生产情况为例，初始产水量高达 $70m^3/d$，在 400 天之后的产水保持在 $2\sim5m^3/d$，此时水气比为 $5m^3/10^4m^3$，属于高水气比生产井，气水两相渗流影响明显。

关于页岩气储层气水两相渗流的研究也有相应的文献，尹虎等[15]建立了页岩气藏气水两相渗流数学模型，未考虑压裂液返排后储层饱和度的分布；王怀龙[16]建立了较为综合的数学模型，没有考虑滑脱效应和应力敏感等。在以上研究的基础上，建立了考虑解吸—吸附、扩散、滑脱、应力敏感的页岩气储层压裂水平井双重介质气水两相渗流的数值模型，结合压裂后储层含水变化，研究了气水产出规律及关键影响因素，可以实现产能预测及压裂参数优化功能。

1　气水两相渗流数学模型的建立

页岩气储层压裂水平井地质模型假设为裂缝—孔隙双重介质；在基质系统中多存在纳米级孔隙，根据气水相渗关系，其缚束水饱和度一般较大，甚至普遍存在高于原始含水饱和度的现象，在压裂过程中，压裂液中的水会渗吸进入基质，使基质与裂缝接触面很小范围内的含水饱和度变大，由于基质渗透率很低并且束缚水饱和度较大，受渗吸影响区域的渗入水不会改变单相渗流的状态，水的吸入仅仅是把含水饱和度提高到缚水饱和度，气相渗透率仍为“1”，故基质中假设只存在天然气单相渗流，并且渗透率不改变；从多尺度渗流机理考虑，基质中存在吸附—解吸、扩散渗流和滑脱渗流，作用较为明显的应力敏感存在于裂缝系统中。基于此，分别设计基质和裂缝的渗流模型。

1.1 基质系统渗流模型

基岩中气体总量包含吸附气量和游离气量两部分。根据质量守恒定律，建立其连续性方程为

$$\nabla\left[\rho_{m}v_{m}\right]-q_{mfg}=\frac{\partial}{\partial t}\left(\phi_{m}\rho_{m}+\rho_{gsc}\frac{V_{L}p_{m}}{p_{L}+p_{m}}\right) \tag{1}$$

其中，

$$v_{m}=-\frac{K_{m}}{\mu_{m}}\nabla p_{m}$$

$$q_{mfg}=-\frac{\sigma K_{m}\rho_{m}}{\mu_{m}}(p_{m}-p_{f})$$

整理得到基质渗流基本微分方程：

$$\nabla\left[\alpha\frac{K_{m}\rho_{m}}{\mu_{m}}\nabla p_{m}\right]-\alpha\frac{\sigma K_{m}\rho_{m}}{\mu_{m}}(p_{m}-p_{f})=\alpha\frac{\partial}{\partial t}\left(\phi_{m}\rho_{m}+\rho_{gsc}\frac{V_{L}p_{m}}{p_{L}+p_{m}}\right) \tag{2}$$

基质综合渗透率受扩散和滑脱双重作用影响。应用 Knudsen 公式[17-19]，扩散作用的基质渗透率等效值 K_{m1} 为

$$K_{m1}=K_{m0}\left[1+\frac{128}{15\pi^{2}}K_{n}\arctan(4K_{n}^{0.4})\right]\left[1+\frac{4K_{n}}{1+K_{n}}\right] \tag{3}$$

应用 KlinKenberg 公式[20]，滑脱作用的基质渗透率等效值 K_{m2} 为

$$K_{m2}=K_{m0}\left(1+\frac{b}{p_{m}}\right) \tag{4}$$

则基质综合渗透率值等于原始渗透率与扩散和滑脱带来的渗透率增大值之和，即为

$$K_{m}=K_{m0}+(K_{m1}-K_{m0})+(K_{m2}-K_{m0})=K_{m1}+K_{m2}-K_{m0} \tag{5}$$

1.2 裂缝系统渗流模型

1.2.1 气相渗流方程

根据质量守恒定律可以得到：

$$\nabla\left[\rho_{fg}v_{fg}\right]+q_{g}+q_{mfg}=\frac{\partial}{\partial t}(\phi_{f}\rho_{fg}S_{g}) \tag{6}$$

其中，

$$v_{fg}=-\frac{K_f K_{rg}}{\mu_{fg}}\nabla p_{fg}$$

整理得到裂缝系统气相渗流基本微分方程为

$$\nabla\left[\alpha\frac{K_f K_{fg}\rho_{fg}}{\mu_{fg}}\nabla p_{fg}\right]+\alpha\frac{\sigma K_m\rho_m}{\mu_m}(p_m-p_{fg})+q_g=\alpha\frac{\partial}{\partial t}(\phi_f\rho_{fg}S_g) \tag{7}$$

1.2.2 水相渗流方程

同理得到裂缝中水相的渗流基本微分方程为

$$\nabla\left[\alpha\frac{K_f K_{rw}\rho_{fw}}{\mu_{fw}}\nabla p_{fw}\right]+q_{vw}=\alpha\frac{\partial}{\partial t}(\phi_f\rho_{fw}S_w) \tag{8}$$

1.2.3 裂缝渗透率处理

应力敏感作用于裂缝系统，裂缝渗透率 K_f 应用指数拟合公式[21]为

$$K_f=K_{f0}e^{-c(p_i-p_{wf})} \tag{9}$$

1.2.4 裂缝含水饱和度处理

假设储层中有基质原始含水饱和度 S_{mwo}，压裂时，由于渗吸作用使与人工裂缝接触的基质面较小深度范围内的含水达到基岩的束缚水饱和度 S_{mwc}，当井生产时，此饱和度不会降低（一般页岩基质的 $S_{mwo}<S_{mwc}$），由此得到裂缝系统的含水饱和度 S_w 的计算式为

$$S_w=S_{mwo}+\frac{V_{in}(1-R_r)-V_{ms}}{LBh\phi_f} \tag{10}$$

其中，

$$\phi_f=\frac{(V_{in}-V_{ms})\beta}{LBh};V_{ms}=2L_{tF}hd(S_{mwc}-S_{mwo})$$

$$d=\sqrt{\frac{K_{m0}t_m}{\pi\phi_m\mu_w C_t}};L_{tF}=\frac{LB\phi_f}{w_f}$$

2 气水两相渗流数学模型的数值解

对渗流基本微分方程进行差分离散并整理，得到差分方程为

基质气相渗流：

$$b_m\delta p_{m_{i-1,j}}+a_m\delta p_{m_{i-1,j}}+d_m\delta p_{m_{i,j-1}}+c_m\delta p_{m_{i,j-1}}+e_m\delta p_{m_{i,j}}+e_{mg}\cdot\delta p_{f_{i,j}}=g_m \tag{11}$$

裂缝气相渗流：

$$b_{\mathrm{g}}\delta p_{\mathrm{f}_{i+1,j}}+a_{\mathrm{g}}\delta p_{\mathrm{f}_{i-1,j}}+d_{\mathrm{g}}\delta p_{\mathrm{f}_{i,j+1}}+c_{\mathrm{g}}\delta p_{\mathrm{f}_{i,j-1}}+e_{\mathrm{g}}\delta p_{\mathrm{f}_{i,j}}+e_{\mathrm{gm}}\cdot\delta p_{\mathrm{m}_{i,j}}+b_{\mathrm{sg}}\cdot\delta s_{\mathrm{w}_{i-1,j}}+a_{\mathrm{sg}}\cdot\delta s_{\mathrm{w}_{i-1,j}}$$

$$+d_{\mathrm{sg}}\cdot\delta s_{\mathrm{w}_{i,j-1}}+c_{\mathrm{sg}}\cdot\delta s_{\mathrm{w}_{i,-1}}+e_{\mathrm{sg}}\cdot\delta s_{\mathrm{w}_{i,j}}=g_{\mathrm{g}} \tag{12}$$

裂缝水相渗流：

$$b_{\mathrm{w}}\cdot\delta p_{\mathrm{f}_{i+1,j}}+a_{\mathrm{w}}\cdot\delta p_{\mathrm{f}_{i-1,j}}+d_{\mathrm{w}}\cdot\delta p_{\mathrm{f}_{i,j+1}}+c_{\mathrm{w}}\cdot\delta p_{\mathrm{f}_{i,j-1}}+e_{\mathrm{w}}\cdot\delta p_{\mathrm{f}_{i,j}}+b_{\mathrm{sw}}\cdot\delta s_{\mathrm{w}_{i-1,j}}+a_{\mathrm{sw}}\cdot\delta s_{\mathrm{w}_{i-1,j}}$$

$$+d_{\mathrm{sw}}\cdot\delta s_{\mathrm{w}_{i,j-1}}+c_{\mathrm{sw}}\cdot\delta s_{\mathrm{w}_{i,j-1}}+e_{\mathrm{sw}}\cdot\delta s_{\mathrm{w}_{i,j}}=g_{\mathrm{w}} \tag{13}$$

以上差分方程$\delta p_{\mathrm{m}},\delta p_{\mathrm{f}},\delta s_{\mathrm{w}}$为未知数，因此构成3个未知数和3个方程，联立可以进行求解，差分方程受基质气、裂缝气和裂缝水多因素约束，求解方法很复杂，应用IMPES方法会导致物质平衡误差，对模型计算结果影响较大，但仅对单井渗流区域范围内使用IMPES方法，设定规则化的网格序列和边界，则与其他方法相比较误差基本可忽略，因此，为了简化运算，模型对系数按显式处理，求解方法应用IMPES方法，对各节点按块中心网格顺序排列，联立渗流方程，利用Gauss-Seidel迭代法顺序求解n+1时刻的基质压力、裂缝压力和裂缝含水饱和度分布。

3　应用实例

以某页岩气储层压裂水平井为例，应用所建立模型及计算方法，分别计算对比含水饱和度、压裂参数及综合渗流对产能的影响程度。

该井主要参数见表1。概念模型大小为1400m×1000m×45m，设计dX=dY=33.3m的均匀网格系统，纵向上1个小层dZ=45m，网格数为42×30×1=1260个。

表1　压裂水平井关键参数

参数	数值	参数	数值
D（m）	2100	L（m）	1200
h（m）	45	S_{mw0}	0.1
ϕ_{m}	0.04	B（m）	300
p_{i}（MPa）	23	K_{m0}（μD）	0.1

3.1　含水饱和度影响

受渗吸影响，基质渗吸区域内含水饱和度为束缚水饱和度，裂缝含水饱和度则与原始含水饱和度、压裂液返排率和渗吸深度有关。对于渗吸深度影响裂缝含水饱和度的情况见表2。闷井1小时和闷井20小时相比，含水饱和度的变化相对于原始含水饱和度和压裂液返排率对裂缝含水饱和度的影响要小很多，况且压裂后闷井时间多为1天左右，因此，渗吸深度对裂缝含水饱和度的分析可暂不考虑。

表 2　渗吸深度对裂缝含水饱和度的影响

序号	t（h）	d（m）	S_w
1	1	0.0000892	0.59
2	10	0.00028	0.54
3	20	0.000399	0.51
4	40	0.000564	0.47

设压裂后闷井 24 小时，则基质渗吸深度为 0.437mm，由此分别研究不同原始含水饱和度和不同压裂液返排率的影响。

3.1.1　不同原始含水饱和度

保持压裂液返排率 0.5 不变，根据式（10）计算得到的裂缝含水饱和度随着原始含水饱和度的变大而呈线性增大趋势，变化范围由 S_{mw0}=0 对应的 S_w=0.36 到 S_{mw0}=0.35 对应的 S_w=0.86。

对不同的原始含水饱和度进行模拟，得到气水产量如图 1 所示。

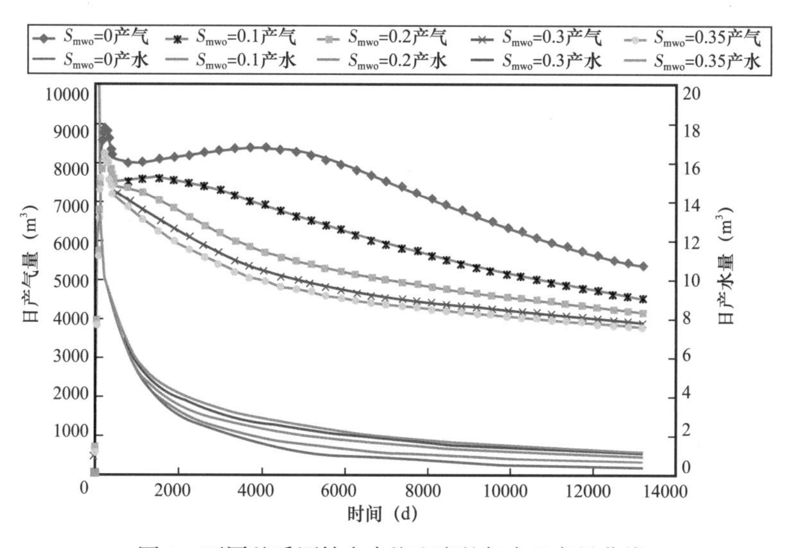

图 1　不同基质原始含水饱和度的气水日产量曲线

在返排率一定的条件下，基质原始含水饱和度越大，日产气量越小，渗吸进入基质的压裂液越少，裂缝中含水饱和度越高，日产水量也越大；从产气量曲线上看，当基质含水饱和度较小时，日产气量在大量排水后有一个回升的态势，含水饱和度越小其回升的幅度越大，维持时间越长，后期日产气量越大，累计产气量同样明显较大；同理，当含水饱和度较大时，产气量回升幅度减弱或者消失，产量递减较快，后期日产气量明显变小，累计产气量也将较小。

3.1.2　不同压裂液返排率

保持基质初始含水饱和度 0.1 不变，计算得到裂缝含水饱和度随着压裂液返排率的增大呈线性增减趋势，变化范围由 R_r=0.1 对应的 S_w=0.91 到 R_r=0.9 对应的 S_w=0.11。

对不同的压裂液返排率进行模拟，得到气水日产量如图 2 和图 3 所示。

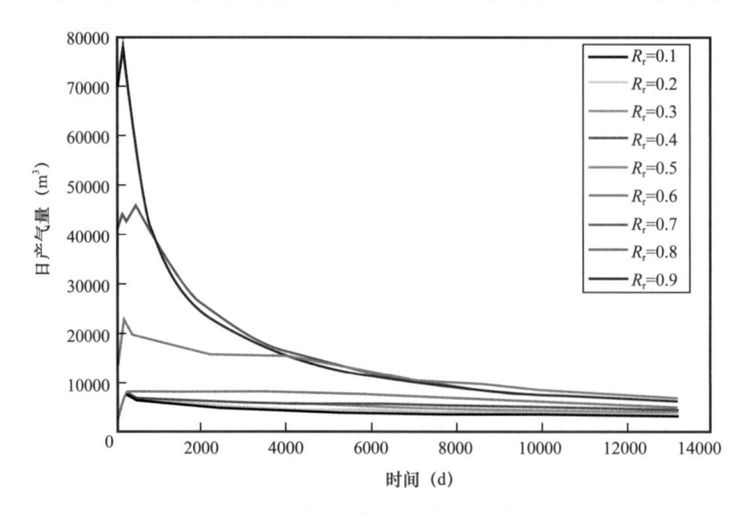

图 2　不同返排率的日产气量曲线

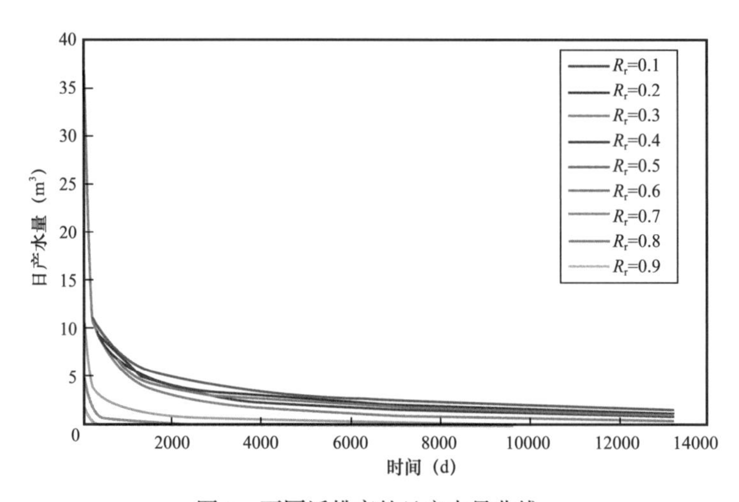

图 3　不同返排率的日产水量曲线

当返排率低于 0.5 时，前期日产气量较低，各个时间段的产气量差异不明显，日产水量较大，差异亦不明显；当返排率处于 0.5～0.8 时，前期日产气量随着返排率增加而大幅度增加，但递减也较快，中后期生产差别不明显，仍比返排率小于 0.5 的情况好，此区间产水量也较明显的随返排率增高而降低；当返排率大于 0.8 时，日产气量和产水量变化不大。这说明，压裂液返排率在 0.5～0.8 利于提高开发效果。

3.2　压裂参数影响

保持其他参数不变的情况下，研究裂缝系统导流能力、缝网半长对生产的影响。

3.2.1　不同裂缝导流能力

对不同裂缝导流能力进行模拟，如图 4 和图 5 所示。

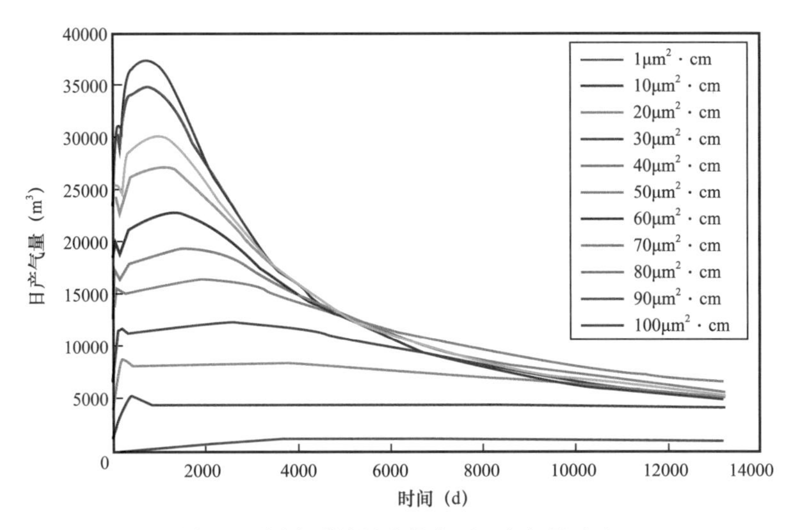

图 4　不同主裂缝导流能力下日产气量对比

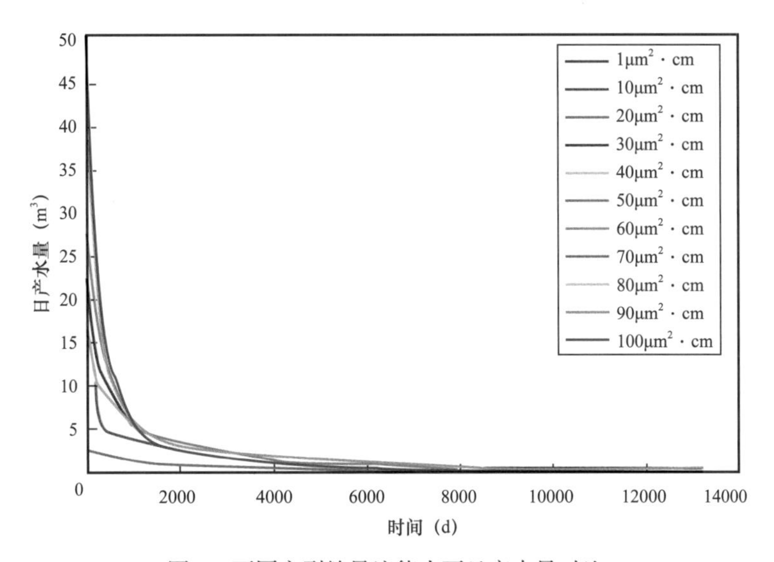

图 5　不同主裂缝导流能力下日产水量对比

裂缝导流能力低于 $20\mu m^2\cdot cm$ 时，该井日产气量不足 $1\times10^4m^3$，产水量也较低并且维持时间长，不利于排水采气；当裂缝导流能力大于 $20\mu m^2\cdot cm$ 时，前期的日产气量随着导流能力的增大明显升高，初期产水量也明显增大，利于生产；当裂缝导流能力大于 $40\mu m^2\cdot cm$ 时，前期产量确实有较大差异，但中后期基本无差异，产水量亦然。这说明，压裂时导流能力在 $40\mu m^2\cdot cm$ 即可达到利于生产的目的，若追求前期较大产量则需要进一步提高导流能力。

3.2.2　不同缝网半长

对不同缝网半长进行模拟，结果如图 6 所示。

当缝网半长小于 200m 时，前期日产气量差异不大，后期分化明显，缝网区域越大，日产气越高，日产水也较高；当缝网半长大于 200m 时，日产气量在整个生产过程中都有明显增高，但增加幅度不大，产水亦然。这说明，缝网半长要大于200m，越大越有利于生产。

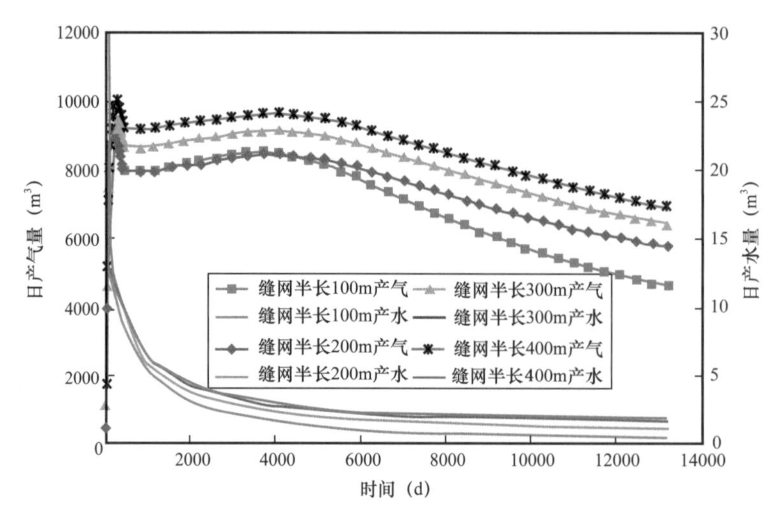

图6　不同缝网半长下气水日产量对比

3.3　综合渗流影响

3.3.1　扩散影响

对不同扩散系数进行模拟，如图7所示。

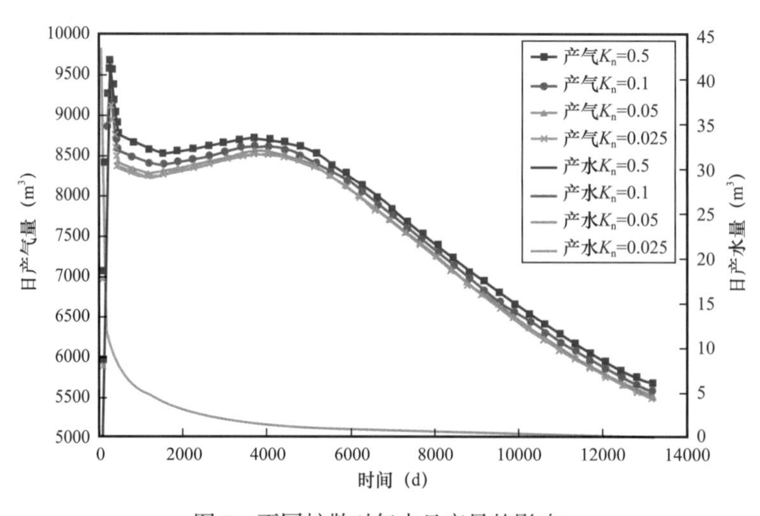

图7　不同扩散对气水日产量的影响

当K_n在0.05～0.5范围内时，克努森系数引起的气体产量增加量可达到250m³/d，当K_n小于0.05时，克努森系数对产量的影响并不显著，说明随着孔隙直径的减小，扩散作用引起的气体产量增加量也越来越大，所以对于具有纳米级较小孔隙的页岩气藏而言，扩散作用的影响是不容忽视的；由于基质中的水很难采出，故克努森系数对产水量影响很小，表现为不同克努森系数的产水量曲线重合。

3.3.2 滑脱影响

对不同滑脱系数进行模拟，如图 8 所示。

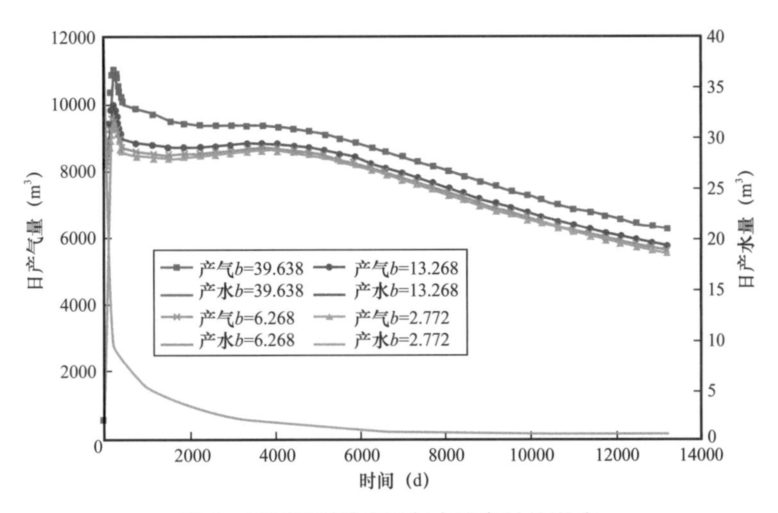

图 8　不同滑脱效应对气水日产量的影响

当滑脱系数大于 6.268 时，滑脱引起的气体产量增加量可达到 1400m³/d，其影响程度高于扩散作用；当其小于 6.268 时，滑脱系数引起的气体产量增加量很小。由此，说明随着孔隙直径的减小，滑脱作用越来越显著，所以在页岩气藏数值模拟时必须要考虑滑脱效应的影响。与扩散作用类似，滑脱作用对产水量的影响非常小。

3.3.3 应力敏感影响

对不同应力敏感系数进行模拟，如图 9 所示。

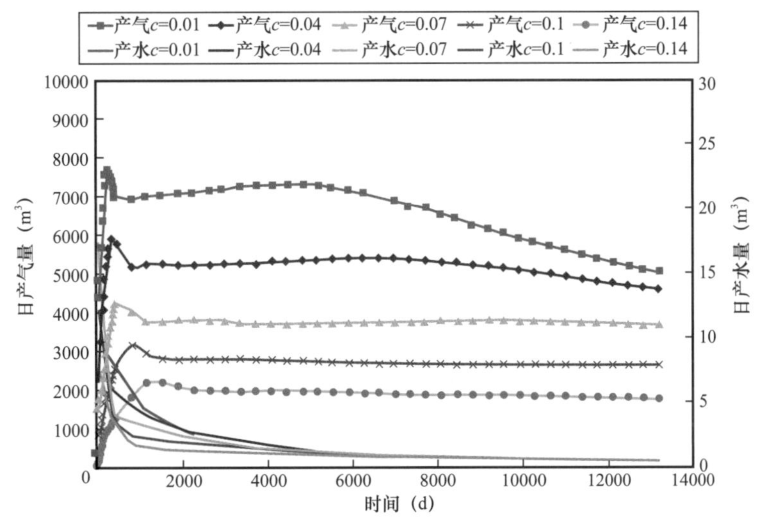

图 9　不同应力敏感对气水日产量的影响

随着地层压力的降低，压力敏感作用使裂缝闭合，降低导流能力，进而使气井产量下降。因此，由图 9 可知，应力敏感对产气量和产水量影响都很大，应力敏感系数每增加

0.03 时，产气量减少值高达 2000m^3/d，最少也能达到 900m^3/d，而且应力敏感系数与日产气量降低基本成正比，与扩散作用和滑脱效应相比较，应力敏感对页岩气井的气水产量影响更为显著。因此，提高压后裂缝效果的保持相当关键。

4 结论

（1）建立的考虑页岩气藏解吸—吸附、扩散、滑脱、应力敏感的页岩气压裂水平井双重介质气水两相渗流的数学模型和数值模型可应用于多因素分析，模型得到的气水产出规律与现场实际基本一致，适用于现场压裂水平井的产量预测，但需要参数较多，参数取值可能对结果产生较大的偏差。

（2）数值模拟结果表明：基质原始含水饱和度越小，压裂液返排后产水对生产的影响越大；压裂液返排率在 0.5～0.8、裂缝导流能力大于 40$\mu m^2 \cdot cm$，缝网半长大于 200m 均能达到较好的气井开发效果；扩散作用弱于滑脱作用，应力敏感作用要明显大于前两者，生产时弱化应力敏感作用是关键。

符号注释：

下标 m，f 分别代表基质系统和裂缝系统；下标 g，w 分别代表气相和水相；下标 i,j 分别为平面网格坐标；ρ—天然气密度，kg/m^3；p—压力，MPa；ϕ—孔隙度，小数；V_L—Langmuir 体积，m^3/m^3；p_L—Langmuir 压力，MPa；ρ_{gsc}—标准状态天然气密度，kg/m^3；σ—形状因子，无因次；μ—气体黏度，mPa·s；K—渗透率，mD；t—时间，s；α—维数系数，无因次；K_{m0}—基质原始渗透率，mD；K_n—克努森系数，小数；K_{m1}—基质扩散渗透率等效值，mD；b—滑脱因子，MPa；K_{m2}—基质滑脱渗透率等效值，mD；K_r—相对渗透率，小数；q_g—气体的质量流量，kg/s；S_w—裂缝中含气饱和度，小数；q_{vw}—水的体积流量，m^3/s；c—应力敏感系数，小数；K_{f0}—裂缝初始渗透率，mD；p_i—原始地层压力，MPa；p_{wf}—井底压力，MPa；V_{in}—压裂液注入体积，m^3；S_{mwo}—基质系统的原始含水饱和度，小数；R_r—压裂液返排率，小数；L—水平井长度，m；h—储层厚度，m；B—缝网区域宽度，m；β—裂缝效率系数，小数；V_{ms}—渗吸入基质体积，m^3；d—渗吸深度，m；L_{tf}—裂缝总长度，m；S_{mwc}—基质系统的束缚水饱和度，小数；t_m—压裂后闷井时间，h；C_t—基质储层和液体水的综合压缩系数，1/MPa；w_f—裂缝平均宽度，m；δp—压力增量，MPa；dX、dY、dZ—x、y、z 方向的网格步长，m；D—井深，m；差分方程中带下角标 m、g、w 的 a、b、c、d、e、e_{gm}、e_{mg}、a_s、b_s、c_s、d_s、e_s 分别为基质气相、裂缝气相、裂缝水相的未知量系数，无量纲；g_m、g_f、g_w—基质气相、裂缝气相、裂缝水相的常数项，无量纲。

参考文献

［1］杜殿发，王妍妍，付金刚，等．页岩气藏渗流机理及压力动态分析［J］．计算物理，2015，32（1）：51-57.

［2］夏阳，金衍，陈勉．页岩气渗流过程中的多场耦合机理［J］．中国科学：物理学 力学 天文学，2015，45（9）：094703.

［3］樊冬艳，姚军，孙海，等．页岩气藏分段压裂水平井不稳定渗流模型［J］．中国石油大学学报：自

然科学版，2014，38（5）：116–123.
[4] 段永刚，李建秋．页岩气无限导流压裂井压力动态分析[J]．天然气工业，2010，30（10）：26–29.
[5] Wu Yushu, Wang Cong, Ding D.Transient pressure analysis of gas wells in unconventional reservoirs [R]. SPE 160889，2012.
[6] Chen C C, Rajagopal R.A multiply-fractured horizontal well in a rectangular drainage region [R]. SPE 37072, 1997.
[7] 高杰，张烈辉，刘启国，等．页岩气藏压裂水平井三线性流试井模型研究[J]．水动力学研究与进展A辑，2014，29（1）：108–113.
[8] 温庆志，翟学宁，罗明良，等．页岩气藏压裂返排参数优化设计[J]．特种油气藏，2013，20（5）：137–140.
[9] 程远方，李友志，时贤，等．页岩气体积压裂缝网模型分析及应用[J]．天然气工业，2013，33（9）：53–59.
[10] 时贤，程远方，蒋恕，等．页岩储层裂缝网络延伸模型及其应用[J]．石油学报，2014，35（6）：1130–1137.
[11] 胡嘉，姚猛．页岩气水平井多段压裂产能影响因素数值模拟研究[J]．石油化工应用，2013，32（5）：34–39.
[12] 梁利平．页岩气藏体积压裂评价及产能模拟研究[D]．西安：西北大学，2014.
[13] Wu K L, Li X F, Yu W, et al. A model for surface diffusion of adsorbed gas in nanopores of shale gas reservoirs [R]. SPE 25662, 2015.
[14] Sun H, Chawathe A, Hoteit H, et al.Understanding shale gas flow behavior using numerical simulation [R]. SPE 167753, 2015.
[15] 尹虎，王新海，张芳，等．吸附气对气水两相流页岩气井井底压力的影响[J]．断块油气田，2013，20（1）：74–76.
[16] 王怀龙．页岩气藏渗流理论及单井数值模拟研究[D]．成都：西南石油大学：2015.
[17] 郭小哲，周长沙．基于扩散的页岩气藏压裂水平井渗流模型研究[J]．西南石油大学学报：自然科学版，2015，37（3）：38–44.
[18] 盛茂，李根生，黄中伟，等．考虑表面扩散作用的页岩气瞬态流动模型[J]．石油学报，2014，35（2）：347–352.
[19] 吴克柳，李相方，陈掌星．页岩气纳米孔气体传输模型[J]．石油学报，2015，36（7）：837–848.
[20] 郭小哲，周长沙．基于滑脱的页岩气藏压裂水平井渗流模型及产能预测[J]．石油钻采工艺，2015，37（3）：61–65.
[21] 郭小哲，周长沙．考虑应力敏感的页岩气藏压裂水平井渗流模型建立与分析[J]．长江大学学报：自科科学版，2015，12（5）：44–50.

变压力扩展指数模型法在页岩气井递减分析中的应用

陈　满[1,2]，谢维扬[1,2]，张小涛[3]，蒋　鑫[1,2]，蒋　睿[1,2]，张　舟[4]

（1. 西南油气田公司勘探开发研究院；2. 页岩气评价与开采四川省重点实验室；
3. 四川页岩气勘探开发有限责任公司；4. 四川长宁天然气开发有限责任公司）

摘　要：近年来四川盆地页岩气的勘探开发取得了突破性进展，水平井与体积压裂主体工艺技术被广泛应用于川南地区长宁区块的页岩气开发中。页岩气分段压裂水平井的产量递减规律与常规气井差异大：在初期的压裂液返排阶段，气井产量递减较快，与人造裂缝性低渗气藏气井的早期生产动态相近；后期由于页岩气的解吸附作用增强，气井长期处于低压低产阶段，产量递减慢。北美出现了多种针对页岩气井的产量递减分析方法，但这些方法局限于定压生产的假设，与页岩气井实际生产中受邻井压裂、地面管网建设等因素影响而出现的变压力变产量生产的情况不相符。通过引入压力重整产量，对扩展指数法进行改进，形成变压力扩展指数模型法（简称 VP-YM-SEPD 法，即 Variable-Pressure drop YM-SEPD 法）来解决页岩气井变压力生产数据分析的难题。应用 VP-YM-SEPD 法分析生产数据，需要气井生产史达半年以上，可以分析得到可靠的递减参数。该方法成功用于分析川南地区长宁区块某页岩气井的生产数据，解决了页岩气井变压力生产的递减分析难题。

关键词：页岩气；变压力生产；产量递减分析；扩展指数法；四川盆地；长宁区块

页岩气是一种主要以吸附态和游离态赋存于富含有机质泥页岩及其夹层中的一种自生自储的非常规天然气。页岩气藏目前采用水平井与体积压裂主体工艺技术进行开发[1]，其流动过程涉及解吸附、克努森扩散、达西渗流、非达西渗流等。页岩气在储层中的特殊流动规律及复杂产出机理导致页岩气井的产量递减规律与常规气井有所不同。基于现场百余口页岩气井实际生产数据的统计与分析发现，根据产出流体的产量变化规律可将页岩气井排液采气的生产过程划分为 5 个阶段（图 1）。

阶段①：井筒及大裂缝早期线性流动阶段（液量快速上升，见微气或不见气）；

阶段②：微裂缝早期线性流动阶段（液量上升至峰值，气量快速上升）；

阶段③：微裂缝晚期线性流动阶段（液量快速下降，气量上升至峰值）；

阶段④：以游离气为主的基质补充流动阶段（低液量，气量快速递减，水气比通常小于 $5m^3/10^4m^3$）；

阶段⑤：基质解吸气与游离气共同补充流动阶段（低压低产）。

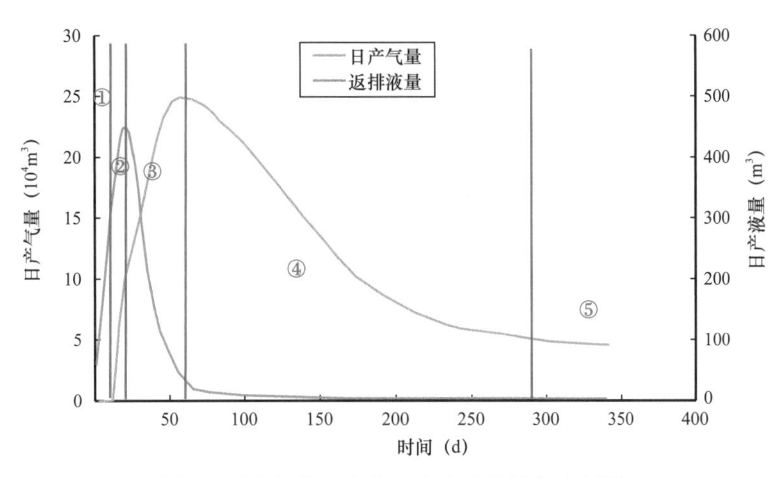

图 1　页岩气井 5 个典型流动阶段划分示意图

近年来，根据页岩气产量初期递减快，后期递减慢的特点，北美出现了多种针对页岩气井产量递减分析的经验公式，包括改进的双曲递减分析方法[2, 3]、Duong 法[4, 5]、扩展指数法[6–9]、幂率指数损失率法[10–12]、LGM 法[13–15]。笔者在应用这些方法对国内页岩气井进行数据分析时发现，由于受限于定压生产的假设，未考虑现场实际生产中压力的变化，北美经验公式仅能分析产量递减时的生产数据，而在第①～③阶段（图 1），产量还处于上升期，地层能量的衰减主要表现为井口压力的快速递减；第④阶段一旦出现关井的情况，上述的经验公式均不适用。而国内页岩气开发目前正处于规模上产阶段，在生产过程中，由于受邻井压裂和地面管线建设的影响，投产气井会限产或频繁关井（图 2）。因此亟须研发一种针对变压力变产量生产气井的产量递减分析方法，解决复杂生产制度下页岩气井生产动态分析的难题。

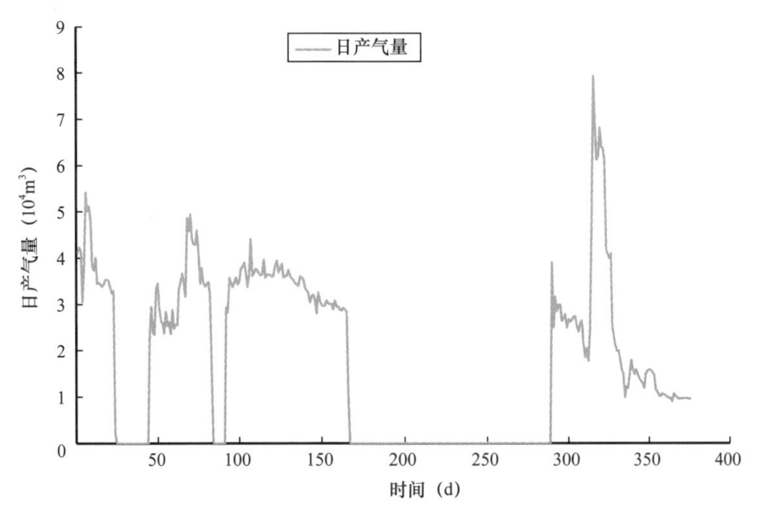

图 2　长宁区块某页岩气井投产初期采气曲线

为了解决北美经验方法的局限性，构建了改进的扩展指数法，预测页岩气井在变流压生产条件下的产量递减趋势：

回顾了北美近年来主要的页岩气井产量递减分析方法，对比了它们的优缺点。

基于现代产量递减分析中提出的压力重整产量的概念，改进了扩展指数法，并用该方法分析了数值模拟产生的单井生产史，预测结果与历史数据吻合度高，随着生产史越长，预测精度越高，证实了新方法的有效性。

应用该法分析了国内长宁区块某页岩气水平井的生产数据，并与行业内普遍认可的 Blasingame 法、解析模型法所得的结果进行了比对，进一步说明了该方法的可靠性。

1 页岩气井产量递减分析方法

页岩气藏是人工压裂改造后方可获得工业气流的人造气藏，气井初期产气量较高，且产量、压力很快将同时下降。后期解吸气作为一种能量补充，导致气井产量递减逐渐变缓，呈现长期小产量下低压稳产的特点。针对页岩气井典型的曲棍球式的产量递减特点，北美学者研究了多种适用于页岩气井产量递减分析的方法。

常规的 Arps 双曲递减模型[2]，如式（1），应用到页岩气井时，会得到较大的 b 值，通常大于 1。而改进的双曲递减分析方法[3]，通过定义最小递减率下限 $D_{\min}$，避免了 b 值过大引起的后期产量预测过高。

$$q=q_{\mathrm{i}}\frac{1}{(1+D_{\mathrm{i}}bt)^{1/b}} \tag{1}$$

Duong 法[4]，如式（2），则是假设了整个页岩气井生产周期都处于线性流阶段。随着生产时间的增加，压裂形成的缝网中的压力逐渐降低，从而激活了周围微裂缝中的游离气。该法预测生产早期的页岩气井产量数据效果较好，而对生产后期的页岩气井产量数据进行预测时，结果通常偏大。

$$\begin{cases} q=\hat{q}_{\mathrm{i}}t^{-m}\exp\left[\dfrac{a}{1-m}(t^{1-m}-1)\right]+q_{\infty} \\ \dfrac{q}{G_{\mathrm{p}}}=at^{-m} \end{cases} \tag{2}$$

扩展指数法[6]，如式（3），与幂率指数损失率法[11]，如式（4），均有依据页岩气井递减率随时间增加而逐渐降低的特点，在常规指数递减法基础上，将递减率改写成了时间的函数，来表征页岩气井产量初期递减快，后期递减慢的特点。Yu 等[7]提出了一种简便高效的直线分析方法（YM-SEPD 法）来确定扩展指数法的模型参数，见式（5），不仅可以提高该方法的分析效率，而且有效地提高了预测精度。

$$q=\hat{q}_{\mathrm{i}}\exp\left[-D_{\infty}t-\hat{D}_{\mathrm{i}}t^{n}\right] \tag{3}$$

$$q=\hat{q}_{\mathrm{i}}\exp\left[-D_{\infty}t-\hat{D}_{\mathrm{i}}t^{n}\right] \tag{4}$$

$$\ln\left[\frac{\hat{q}_{\mathrm{i}}}{q}\right]=\tau^{-n}t^{n} \tag{5}$$

LGM 法[13]，如式（6），是根据逻辑增长曲线来拟合累计产气量的方法对页岩气井的产量进行分析预测。该法可以根据体积法或其他方法计算得到估算的最终可采储量（EUR）来约束模型参数，预测结果受 EUR 预估值影响较大。

$$\begin{cases} G_{\mathrm{p}}=\dfrac{Kt^{n}}{c+t^{n}} \\ q=\dfrac{\mathrm{d}G_{\mathrm{p}}}{\mathrm{d}t}=\dfrac{Knct^{n-1}}{(c+t^{n})^{2}} \end{cases} \tag{6}$$

式（1）～式（6）中，q 为日产气量，$10^4\mathrm{m}^3$；q_{i} 为初始日产气量，$10^4\mathrm{m}^3$；D_{i} 为初始递减率，百分数；b 为双曲递减指数，无量纲；$\hat{q}_{\mathrm{i}}$ 为日产气量截距，$10^4\mathrm{m}^3/\mathrm{d}$；$a$，$m$ 为 Duong 模型拟合参数，无量纲；q_{∞} 为 Duong 模型初始产量，$10^4\mathrm{m}^3/\mathrm{d}$；$\tau$ 为弛豫时间，d；n 为时间指数，无量纲；D_{∞}，$\hat{D}_{\mathrm{i}}$ 为幂率指数损失模型拟合参数，无量纲；G_{p} 为累产气量，$10^4\mathrm{m}^3$；c 为 LGM 模型拟合参数，无量纲；K 为 LGM 模型单井最大累计产气量，$10^4\mathrm{m}^3$。

上述递减方法只能分析较为平滑的生产数据，对于生产制度较为复杂的气井，应用上述方法得到的结果多解性强，不确定性高。因此，在川南长宁区块页岩气水平井生产动态分析中，需要构建考虑压力变化的递减分析方法。本文研究成果对于分析复杂生产制度下的页岩气井递减规律和进行产能预测具有重要意义。

2 改进的扩展指数法

为了考虑压力变化对产量递减分析的影响，行业内主要有两种解决方法：一种是引入压力重整产量$\dfrac{q}{\Delta p}$[16, 17]，例如 Blasingame 现代产量递减分析方法中的拟压力重整化产量；另一种是利用卷积的方法，例如试井解释中变产量生产的一口气井，其压力响应为恒定单位产量生产条件下的单井压力响应解与该井不同时刻产量的卷积。

通过引入压力重整产量$\dfrac{q}{\Delta p}$，将压力对产量变化的影响与改进的扩展指数模型 YM-SEPD[7] 相结合，得到变压力扩展指数模型（以下称变压力扩展指数模型法，即 VP-YM-SEPD 法），如式（7）。

$$\begin{cases} \dfrac{q}{\Delta p}=\dfrac{\hat{q}}{\Delta p_{\mathrm{i}}}\exp\left[-(t/\tau)^{n}\right] \\ \ln\left[\dfrac{\dfrac{\hat{q}}{\Delta p_{\mathrm{i}}}}{\dfrac{q}{\Delta p}}\right]=\tau^{-n}t^{n} \end{cases} \tag{7}$$

式中：$\Delta p=P_{\mathrm{i}}-p_{\mathrm{wf}}(t)$，其中 P_{i} 为初始地层压力，MPa；$p_{\mathrm{wf}}(t)$ 为 t 时刻的井底流压，MPa；$\dfrac{\hat{q}}{\Delta p_{\mathrm{i}}}=\mathrm{maximum}\left\{\dfrac{q}{\Delta p}\right\}$，即最大压力重整产量。利用式（7）的第二式作双对数诊断图（图 6），确定模型拟合参数 n、τ，即可对任意时刻的压力重整产量作出预测；同时，根据压力随时间的递减规律，进行井底流压的外推，即可算出任意时刻的产量。

另一种方法是通过卷积的方式改进扩展指数法（以下简称卷积法）来考虑压力的影响，如式（8）。

$$q(t)=\sum_{k=1}^{n}(\Delta p_{\mathrm{wf}_k}-\Delta p_{\mathrm{wf}_{k-1}})\frac{\hat{q}}{\Delta p_{\mathrm{i}}}\exp\left[-(t-t_{k-1})/\tau\right]^{n} \tag{8}$$

其中，Δp_{wf}为初始地层压力与井底流压的差值，MPa。

利用数值模拟软件模拟了一口页岩气井的生产数据[18]，模拟工区长1860m、宽797m、高100m，模拟井水平段长1100m，共压裂10段，设计裂缝半长均为100m，主裂缝宽度采用最小步长，即假设主裂缝宽度为0.03m。模型束缚水饱和度0.3，地层温度97.8℃，孔隙度2.5%～3.0%，兰格缪尔吸附气量0.7～2.0$\mathrm{m^3/t}$，兰格缪尔吸附压力6.8MPa，岩石弹性压缩系数为$1.0\times10^{-4}\,\mathrm{MPa^{-1}}$（图3）。该井前期采用定产生产的方式生产，之后进入递减阶段。以整个生产史数据为基础，通过作双对数诊断图，得到模型拟合参数τ、n，并对压力进行外推，代入式（7）、式（8）后，得到两种方法下该井的产量预测结果（图4）。

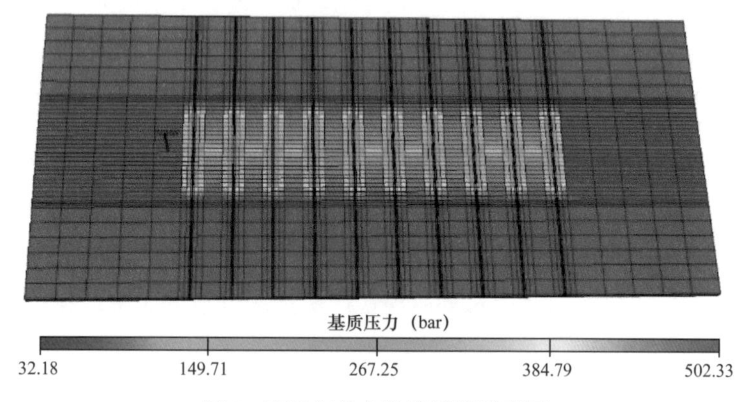

图3　页岩气单井数值模拟模型图

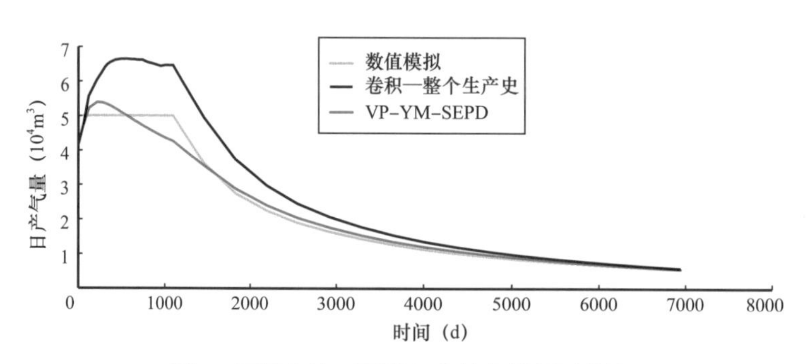

图4　不同方法下页岩气井的产量预测曲线

图4表明，利用整个生产史的数据进行分析，变压力扩展指数模型法的预测精度明显高于卷积法，因此将采用式（7）分析和预测页岩气井单井产量。为了进一步明确变压力扩展指数模型法的适用条件，有必要对不同时长生产史所预测的结果进行对比分析。

为此，利用页岩气单井数值模拟模型产生了一组变产量变压力的气井生产数据（图5）。该井初期依次以$10\times10^4\mathrm{m^3/d}$、$5\times10^4\mathrm{m^3/d}$、$3\times10^4\mathrm{m^3/d}$的产量生产，之后定压生产，井底流压控制在5MPa，其复杂的生产制度更接近于页岩气井的实际生产情况，预测20年，累

计产气量为 $13215.36\times10^4m^3$。图 5 表明，采用 VP-YM-SEPD 法，利用 20 年（7300 天）的生产数据预测的产量，与数值模拟数据基本一致，证实了新方法的可靠性；利用 180 天的生产数据预测的产量，也与数值模拟数据较接近，进一步证实了新方法在较短生产时长下预测的准确性。

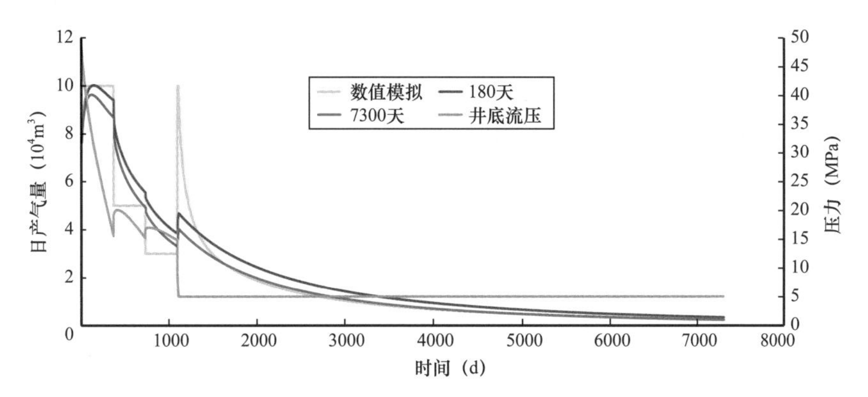

图 5　不同时长生产史 VP-YM-SEPD 法预测结果对比图

使用新方法进一步选取不同时长的生产史数据进行气井产量的预测，具体结果见表 1，其中 EUR 为气井生产 20 年的累计产气量。由表 1 可以看出：

表 1　不同时长生产史 VP-YM-SEPD 法 EUR 预测结果对比表

不同时长的模型	n	τ	EUR（10^4m^3）	误差
基础模型			13215.36	
180d	0.3414	46.8081	15454.91	16.9%
1a	0.3848	57.3013	12594.03	-4.7%
2a	0.4138	57.7261	9870.89	-25.3%
3a	0.4135	57.3745	9823.23	-25.7%
5a	0.3603	47.7239	12988.27	-1.7%
20a	0.3516	44.3999	13135.57	-0.6%

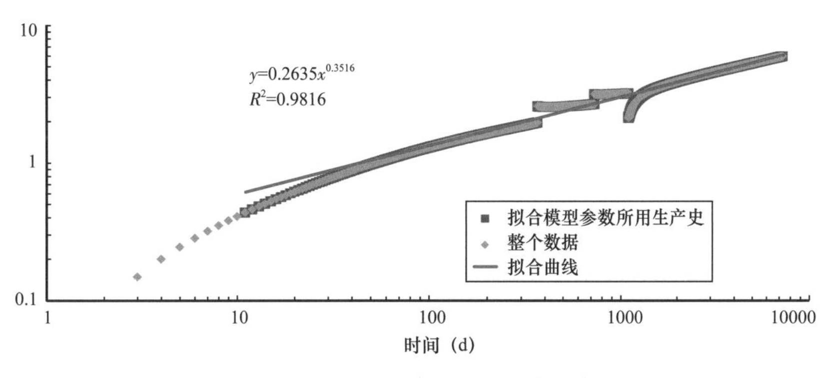

图 6　变产量变压力气井的双对数诊断图

使用 180 天的生产史进行预测，EUR 误差为 16.9%，表明气井在生产半年的情况下，运用该方法可快速评价气井 EUR；

使用 20 年的生产史进行预测，EUR 的误差仅为 -0.6%，且预测的产量与模拟数据基本一致，表明在足够长的生产史条件下，该方法预测精度高；

与 YM-SEPD 法类似[7]，在求取模型参数时，为了提高拟合效果，需要在去掉初期返排阶段的数据后，进行双对数诊断图拟合，拟合得到的页岩气井扩展指数 n 约 0.34～0.42；

在定产生产制度下，初期返排液量得到控制，压力递减较慢，产量稳定，导致拟合出的扩展指数 n 偏低，预测 EUR 偏高。第 2、第 3 年生产制度的改变使特征曲线出现波动，降低了预测精度，EUR 的误差约 -25%，第 5 年之后特征曲线逐渐平整，预测精度大大提高，EUR 误差仅为 -1.7%。

3 实例应用

改进的扩展指数法——变压力扩展指数模型法最大的优势在于引入了考虑压力变化的重整产量项，能够对复杂生产制度下的单井数据进行分析。A 井是川南地区长宁区块龙马溪组页岩气层的一口开发井。该井采用分段水力压裂改造，埋深约 2500m，水平段长 1680m，压裂 23 段，套管完井生产，半年累计产气超过 $3500\times10^4m^3$。运用变压力扩展指数模型法分别对 A 井 60 天、170 天的生产史进行了分析。

由图 6 可以看出，利用 60 天生产史的预测结果在初期阶段跟实际生产较为吻合，但后期的预测曲线明显低于实际数据，而利用 170 天生产史的预测结果，在初期返排阶段偏低，后期跟实际生产吻合较好。进一步将后期低压阶段的套压外推，得到该井的 EUR。图 7 表明，利用 60 天生产史预测的 EUR 明显偏低，利用 170 天生产史预测的 EUR 为 $9600\times10^4m^3$。实例井应用表明，在放压生产制度下，初期返排液量大，压力迅速递减，导致拟合出的扩展指数 n 偏高，预测 EUR 偏低，应用新方法时，气井需要有 180 天以上的生产史以得到可靠的递减参数。由表 2 可知，结合 Blasingame 法和商业软件中分段压裂水平井解析模型法的预测结果，综合评价该井的 EUR 为（9600～10286）$\times10^4m^3$。在实际应用中，应根据气井生产制度的改变，跟踪评价该方法预测结果，并结合其他产量递减分析方法，综合评价气井未来的产量递减趋势。

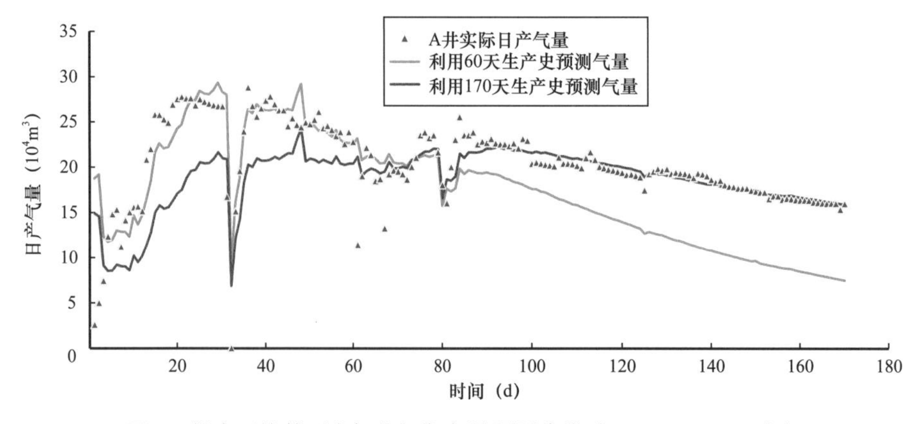

图 7　长宁区块某页岩气井初期产量预测曲线（VP-YM-SEPD 法）

表 2 不同方法预测的 EUR 统计表

模型	n	τ	裂缝半长（m）	渗透率（mD）	EUR（10^4m^3）
VP-YM-SEPD	0.3881	23.5185			9600
Blasingame			71.4	0.0020～0.1035	10286
分段压裂水平井解析模型			88.3	0.0004～0.0600	9982

4 结论

（1）目前北美主流的页岩气井产量递减分析方法，局限于定压生产的假设，对变压力生产的页岩气井并不适用，且要求生产数据平滑，对于实际开发中生产制度较为复杂的气井应用效果差。

（2）针对页岩气井变压力变产量生产的递减分析方法——变压力扩展指数模型法（VP-YM-SEPD 法），在一定时长条件下（半年以上），可用于气井变压力变产量生产时的产量递减预测。

（3）在实际应用中，应根据气井生产制度的改变，跟踪评价预测结果，并结合其他产量递减分析方法的应用，综合评价气井未来的产量递减趋势。

参 考 文 献

［1］杨洪志，张小涛，陈满，等. 四川盆地长宁区块页岩气水平井地质目标关键参数优化［J］. 天然气工业，2016，36（8）：60-65.

［2］Arps J J.Analysis of decline curves［J］. Transactions of the AIME，1945，160（1）：228-247.

［3］Harrell D，Hodgin J，Wagenhofer T.Oil and Gas Reserves Estimates：Recurring Mistakes and Errors［C］// SPE Annual Technical Conference and Exhibition，26-29 September 2004，Houston，Texas，USA.DOI：https：//doi.org/10.2118/91069-MS.

［4］Duong A N.An Unconventional Rate Decline Approach for Tight and Fracture-Dominated Gas Wells［C］// Canadian Unconventional Resources and International Petroleum Conference，19-21 October 2010，Calgary，Alberta，Canada. DOI：https：//doi.org/10.2118/137748-MS .

［5］Joshi K，Lee W J. Comparison of Various Deterministic Forecasting Techniques in Shale Gas Reservoirs［C］// SPE Hydraulic Fracturing Technology Conference，4-6 February 2013，The Woodlands，Texas，USA. DOI：https：//doi.org/10.2118/163870-MS.

［6］Valko P P，Lee W J.A Better Way To Forecast Production From Unconventional Gas Wells［C］// SPE Annual Technical Conference and Exhibition，19-22 September，Florence，Italy.DOI：https：//doi.org/10.2118/134231-MS.

［7］Yu S Y，Miocevic D J. An Improved Method to Obtain Reliable Production and EUR Prediction for Wells with Short Production History in Tight/Shale Reservoirs［C］// SPE/AAPG/SEG Unconventional

Resources Technology Conference, 12–14 August 2013, Denver, Colorado, USA. DOI: https://doi.org/10.15530/URTEC-1563140-MS.

[8] Statton J C. Application of the Stretched Exponential Production Decline Model to Forecast Production in Shale Gas Reservoirs [D]. Texas: Texas A&M University, 2012.

[9] Can B, Kabir C S. Probabilistic Performance Forecasting for Unconventional Reservoirs with Stretched-Exponential Model [C] // North American Unconventional Gas Conference and Exhibition, 14–16 June 2011, The Woodlands, Texas, USA. DOI: https://doi.org/10.2118/143666-MS.

[10] Ilk D, Rushing J A, Perego A D, et al. Exponential vs. Hyperbolic Decline in Tight Gas Sands: Understanding the Origin and Implications for Reserve Estimates Using Arps' Decline Curves [C] // SPE Annual Technical Conference and Exhibition, 21–24 September 2008, Denver, Colorado, USA. DOI: https://doi.org/10.2118/116731-MS.

[11] Ilk D, Blasingame T A. Decline Curve Analysis for Unconventional Reservoir Systems —Variable Pressure Drop Case [C] // SPE Unconventional Resources Conference Canada, 5–7 November 2013, Calgary, Alberta, Canada. DOI: https://doi.org/10.2118/167253-MS.

[12] IIk D, Rushing J A, Blasingame TA. Decline-Curve Analysis for HP/HT Gas Wells: Theory and Applications [C] // SPE Annual Technical Conference and Exhibition, 4–7 October 2009, New Orleans, Louisiana, USA. DOI: https://doi.org/10.2118/125031-MS.

[13] Clark A J, Lake L W, Patzek T W. Production Forecasting with Logistic Growth Models [C] // SPE Annual Technical Conference and Exhibition, 30 October–2 November 2011, Denver, Colorado, USA. DOI: https://doi.org/10.2118/144790-MS.

[14] Clark A J. Decline curve analysis in unconventional resource plays using logistic growth models [D]. Austin: The University of Texas at Austin, 2011.

[15] Clark A J. Determination of Recovery Factor in the Bakken Formation, Mountrail County, ND [C] // SPE Annual Technical Conference and Exhibition, 4–7 October 2009, New Orleans, Louisiana, USA. DOI: https://doi.org/10.2118/133719-STU.

[16] Zhang X T, Jiang X, Yang J Y, et al. A newly developed rate analysis method for a single shale gas well [J]. Energy Exploration & Exploitation, 2015, 33 (3): 309-316.

[17] 沈金川，刘尧文，葛兰，等. 四川盆地焦石坝区块页岩气井产量递减典型曲线建立 [J]. 天然气勘探与开发，2016，39 (2): 36-40.

[18] 张小涛，吴建发，冯曦，等. 页岩气藏水平井分段压裂渗流特征数值模拟 [J]. 天然气工业，2013，33 (3): 47-52.

四、生产应用篇

页岩气水平井产量影响因素分析

贾成业，贾爱林，何东博，位云生，齐亚东，王军磊

（中国石油勘探开发研究院）

摘　要：中国是全球第3个商业化开发页岩气的国家，到2030年页岩气规划产量为（800～1000）×10^8m^3，展现出良好的发展前景。长宁—威远和昭通示范区开发效果显示，目前该区页岩气水平井平均测试产量达到19×$10^4m^3/d$，但测试产量却参差不齐。为此，从地质和工程两方面分析了上述示范区页岩气水平井产量影响因素，提出了不同区块水平井提高单井产量的技术方向；根据含气量差异将优质页岩段进一步细分为4类储层，评价水平井Ⅰ类储层钻遇率；依据天然裂缝发育程度、主应力非均质性、脆性指数等工程参数来评价储层改造条件；建立压裂加液量、加砂量、施工排量和返排率与测试产量之间的相关关系；评价压裂形成裂缝复杂程度。研究结果表明：（1）长宁区块Ⅰ类储层钻遇率高于威远和昭通区块，且当水平井Ⅰ类储层钻遇率大于50%，可保障气井测试页岩气产量高于15×$10^4m^3/d$、预计单井最终可采储量（EUR）高于8000×10^4m^3；（2）昭通区块储层改造条件和压裂形成的裂缝复杂程度优于长宁和威远区块，但加砂量和施工排量等压裂施工参数偏低，制约了前者水平井的测试产量。结论认为，昭通和威远区块进一步提高Ⅰ类储层钻遇率、昭通区块进一步优化压裂施工参数是提高上述示范区页岩气水平井单井产量的主要技术方向。

关键词：页岩气；水平井；测试产量；EUR；储层分类；水平井钻遇率；影响因素；压裂参数

1　概述

页岩气是非常规天然气的主要类型之一，是指赋存于富有机质泥页岩及其夹层中，以吸附或游离状态为主要存在方式的非常规天然气，成分以甲烷为主，是一种清洁、高效的能源资源[1]。美国是最早商业化开发页岩气的国家，并推动了页岩气革命，基本实现了能源供需的自给自足，成为世界第一大天然气生产国。据美国能源信息管理局（EIA）公开数据，2015年美国页岩气产量达到15.21TCF（折合4307.83×10^8m^3），超过致密气成为开发规模最大的非常规天然气[2]。

中国是继美国和加拿大之后第3个实现页岩气商业化开发的国家。2009年12月18日，中国第一口页岩气评价井——威201井开钻并成功获得商业气流；2013年1月，国家发改委和国家能源局批准设立首批页岩气国家级示范区有效推动了页岩气勘探开发进程；到2015年12月，四川长宁—威远、颠黔北昭通和重庆涪陵3个国家级页岩气示范区已建成页岩气产能77×$10^8m^3/a$，实现年产量44.71×10^8m^3；2016年9月，国家能源局发布《页岩气发展规

划（2016—2020 年）》，提出 2030 年中国实现页岩气产量（800～1000）$\times 10^8 m^3$ 的目标，页岩气展现出良好的发展前景。目前，虽然长宁—威远和昭通区块页岩气水平井平均测试产量 $19 \times 10^4 m^3/d$，但是页岩气井测试产量分布范围在（3.4～35）$\times 10^4 m^3/d$，水平井测试产量与 EUR 间呈正相关关系（图 1、图 2）。进一步分析页岩气水平井产量影响因素，探索提高页岩气井产量的技术方法，是提高水平井 EUR、实现页岩气效益开发的技术关键。

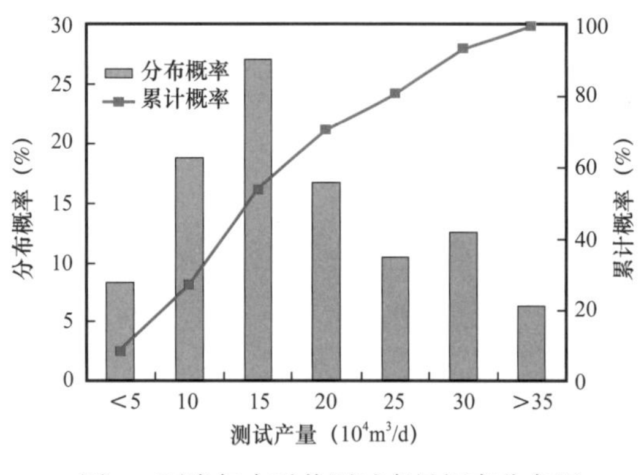

图 1　页岩气水平井测试产量概率分布图

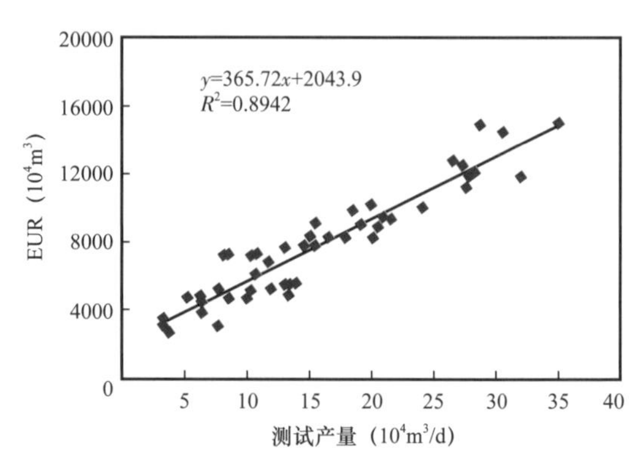

图 2　水平井测试产量与 EUR 关系图

地质因素和工程因素是非常规储层开发评价的关键技术指标[3]。水平井钻遇储层品质和体积压裂改造效果是制约页岩气开发效果的主要因素。基于四川盆地页岩气示范区实钻水平井钻遇储层品质、储层改造条件、压裂技术参数和压裂改造效果，从页岩储层地质参数和工程参数两个方面综合分析制约水平井产量的影响因素，提出针对不同区块提高气井产量的技术措施，明确页岩气水平井开发技术发展方向。

2　水平井钻遇储层分析

四川盆地上奥陶统五峰—龙马溪组厚度为 250～300m，五峰组—龙一$_1$层厚度约 30m 的优质页岩段是当前页岩气开发的主力层位[4]。该优质页岩段整体上属深水陆棚亚相的一套黑色硅质页岩、富有机质黑色碳质页岩，发育稳定、分布面积大，具有高孔隙度、高有机质丰度、高含气量等特征。孔隙度岩心实验分析值为 2.0%～6.8%，测井解释值为 3.6%～7.2%；有机质丰度岩心分析值为 3.0%～4.2%，测井解释结果为 2.7%～4.5%；含气量现场分析值为 2.0～3.5m³/t，测井解释结果为 2.9～7.4m³/t，整体上龙一$_1^1$和龙一$_1^3$小层最高。但优质储层内部非均质性仍然很强，龙一$_1^1$和龙一$_1^3$小层内部不同井段储层地质参数仍存在较大差异[5]。四川盆地两口评价井脉冲中子俘获测井监测显示，目前压裂规模下五峰组—龙一$_1$层压裂缝高约 12～13.4m，不足以完全改造开发目的层系；因此对目的层系开展储层分类评价，进一步提高优质储层钻遇率，有利于提高页岩气水平井产量。

总含气量由游离气和吸附气两部分组成，是页岩气评价和开发的核心参数，集成反映了有机碳含量、孔隙度、含气饱和度等储层特征关键参数[6]。为进一步区分优质页岩段储层品质，根据岩心观测和测井数据解释资料，将优质页岩段划分为 4 类储层：Ⅰ类储层富含有机质，笔石化石分布密度大、岩心面覆盖率超过 70%，密度相对较低，层理发育，总含气量（V_t）>4m³/t；Ⅱ类储层有机质含量相对较高，笔石化石分布密度约占岩心面的 50%～70%，总含气量（V_t）>$2m^3/t$；Ⅲ类储层有机质含量相对较低，笔石化石分布密度低于50%，总含气量（V_t）>$1m^3/t$；Ⅳ类储层有机质含量低，偶见笔石化石，总含气量（V_t）< $1m^3/t$。

采用上述储层分类标准，可进一步分析页岩气水平井钻遇不同类型储层比例，精确评价水平井钻井质量（图3），进一步厘清水平井钻遇优质储层比例与气井测试产量间关系，有助于优化水平井靶体位置、加强水平井随钻地质导向，从而提高页岩气水平井单井产量。

如图3所示，1井水平段长度1600m，压裂水平段长度1565m，压裂段数19段，加砂量2153t，测试产量$28.77\times10^4m^3/d$，单井EUR达$14925\times10^4m^3$；2井水平段长度1404m，压裂水平段长度868m，压裂段数12段，加砂量1257t，测试产量$3.84\times10^4m^3/d$，单井EUR为$2756\times10^4m^3$。通过水平井实际钻遇储层分析发现两口井钻遇Ⅰ类储层钻遇比例分别为50.84%和6.34%，表明两口井钻遇储层物质基础相差很大，从而两口井测试产量和EUR差别较大。

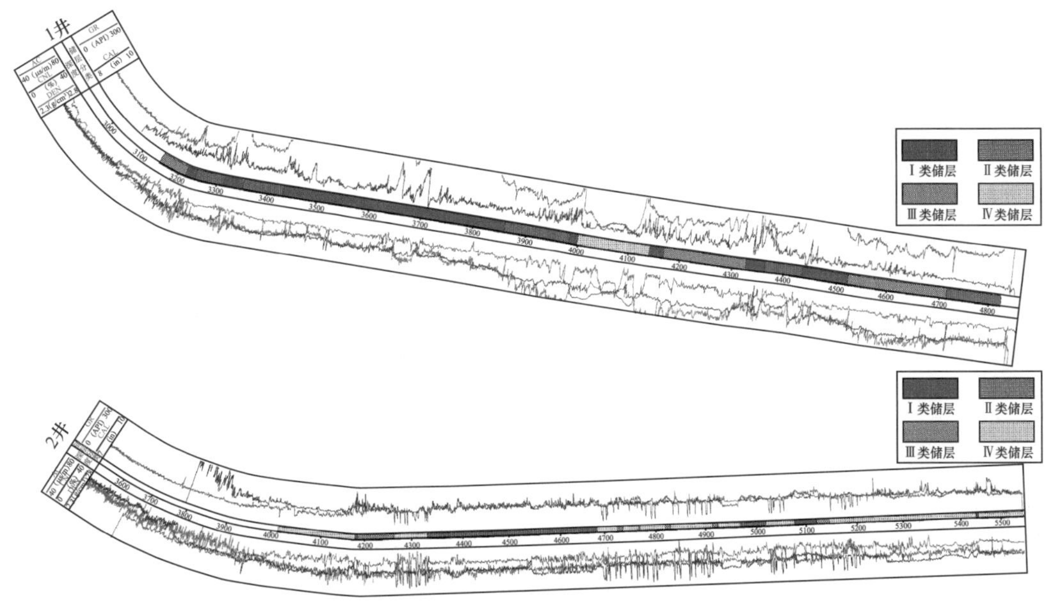

图3　页岩气水平井Ⅰ类钻遇储层率分析图

对页岩气示范区内已经完钻水平井开展实钻储层钻遇率分析（图3）。结果表明：Ⅰ类储层钻遇率与水平井测试产量间呈较好的相关性，不同区块水平井Ⅰ类储层钻遇比例相差较大；整体上长宁区块钻遇率普遍较高，威远区块钻遇率分布不均，昭通区块普遍偏低。Ⅰ类储层钻遇率与测试产量间关系结果表明：当Ⅰ类储层钻遇比例超过50%，水平井测试产量大于$15\times10^4m^3/d$的概率达92%（图4），同时结合气井测试产量与EUR间关系曲线（图2），当气井测试产量高于$15\times10^4m^3/d$，EUR在$8000\times10^4m^3$以上，按照目前单井综合投资5500万元，气价1.488元/m^3（含300元/10^3m^3补贴），可实现内部收益率约8%～10%，达到效益开发的要求。Barnett页岩气田水平井垂深1219～2590m，平均2286m，水平段长度304～1524m，平均1219m，水平井单井综合投资200万美元；Marcellus页岩气田水平井垂深610～3048m，平均2057m，水平段长度610～1828m，平均1128m，水平井单井综合投资300万～400万美元[7]。相比美国页岩气田水平井开发单井综合成本，国内页岩气示范区单井综合投资普遍偏高，进一步降低单井综合投资有利于提高页岩气开发效益、降低页岩气开发投资风险。

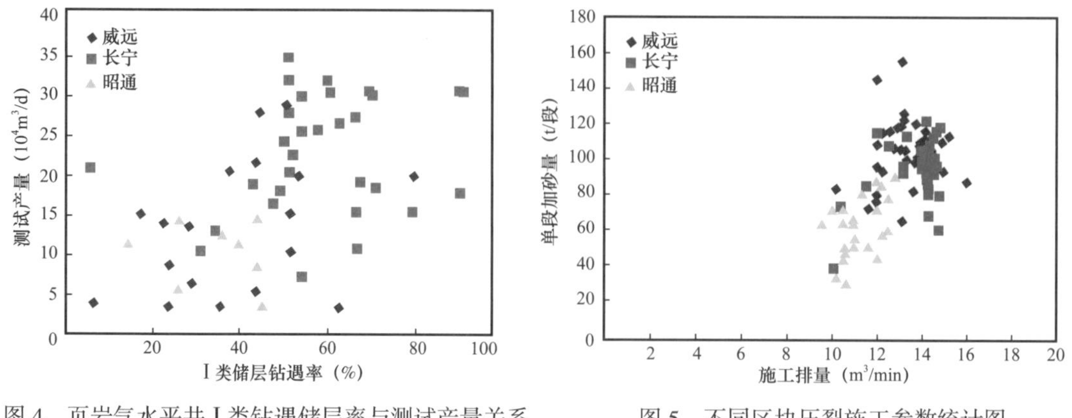

图 4　页岩气水平井Ⅰ类钻遇储层率与测试产量关系

图 5　不同区块压裂施工参数统计图

3　储层改造条件

天然裂缝、水平两向主应力非均质系数、岩石脆性等是影响体积压裂的关键因素[8, 9]。天然裂缝与人工裂缝相互沟通，能够较大程度地增加裂缝的复杂程度。因此同等压裂条件下天然裂缝发育地区改造效果更好。四川盆地内部长宁—威远地区主要发育的裂缝类型以层理缝和高角度充填缝为主（图 6）。高角度缝成因为成岩收缩，后期被方解石充填，偶见具有划痕的张开缝；裂缝分布密度低，天然裂缝欠发育。四川盆地周缘构造运动相对剧烈，受盆地边缘构造运动影响，主要发育高角度张开缝，常见缝面划痕；裂缝分布密度较大，天然裂缝较发育。整体上，昭通区块天然裂缝发育程度高于长宁、威远区块（表 1）。

图 6　四川盆地五峰组—龙一段露头和岩心观察裂缝发育类型图

（a）+（b）四川长宁双河剖面五峰组富有机质黑色页岩，见高角度层内收缩缝和页理构造；（c）昭通区块昭 104 井第 9 次取心段（龙一$_1$亚段）见高角度未充填构造缝；（d）昭通区块 YS111 井第 7 次取心段（龙一$_1$亚段）见强烈擦痕面；（e）昭通区块上 1 井第 11 次取心段（龙一$_1$亚段）见高角度充填构造缝；（f）昭通区块 YS108 井第 6 次取心段（龙一$_1$亚段）见强烈擦痕面

表 1　四川盆地部分取心井目的层段裂缝观察数据表

区块	井号	井深（m）	层位	裂缝数量（条）	裂缝长度（m）	裂缝密度（条 /10m）	裂缝类型
长宁	宁 203	2340～2348	五峰组—龙一$_1$亚段	7	1.79	0.6	高角度充填缝
	宁 210	2203～2245	五峰组—龙一$_1$亚段	5	0.47	1.2	高角度充填缝
威远	威 203	3178～3178	五峰组	2	0.14	0.3	高角度充填缝
	威 204	3519～3519	龙一$_1$亚段	3	0.22	0.4	高角度充填缝
昭通	昭 104	2039～2061	龙一$_1$亚段	8	1.45	3.6	高角度充填缝，大部分未充填
	YS111	2365～2397	五峰组—龙一$_1$亚段	15	1.17	4.6	高角度充填缝，大部分未充填

水平两向应力差是沟通天然裂缝与人工裂缝的主控因素之一，两向应力差较小，有利于裂缝转向、弯曲等，较易形成复杂裂缝，反之则较难形成复杂裂缝。水平两向主应力非均质系数（K_h）用于表征水平两向应力差，为最大水平主应力与最小水平主应力之差与最小水平主应力比值，是目前评价压裂改造能否形成复杂裂缝的主要参数。当 K_h 大于 0.2 时，水力裂缝趋向于单一缝；当 K_h 为 0～0.2 时，水力裂缝趋向于复杂裂缝或者网状裂缝[10, 11]。从水平两向主应力非均质系数分析，长宁区块有利于压裂改造沟通天然裂缝与人工裂缝（表 2）。

表 2　不同区块地应力场特征

区块	垂向地应力（MPa）	最大水平主应力（MPa）	最小水平主应力（MPa）	水平两向应力差（MPa）	水平两向应力非均质系数
长宁	62.3	45.4～62.1	34.9～47.7	10.5～14.4	0.19～0.23
威远	89.1	70.0～88.3	54.0～69.6	16.0～18.7	0.23～0.27
昭通	61.2	71.7～79.6	53.1～55.7	18.6～21.13	0.26～0.30

高岩石脆性是保持复杂裂缝导流能力的关键。岩石脆性指数越高，越易产生剪切破坏，形成剪切裂缝及粗糙的节理，并保持张开状态，同时保持一定的导流能力。矿物成分组成可用于评价岩石脆性指数（BI）。长宁区块五峰组—龙一$_1$亚段矿物成分中石英含量最高，平均含量占 40.25%；其次是方解石和白云石，平均含量分别为 13.07% 和 9.6%，含少量长石、黄铁矿；矿物脆性指数 52.54%。威远区块主力产层段石英含量占 34.64%，长石、方解石、白云石分别占 7.69%、12.06% 和 8.24%，矿物脆性指数 45.49%[12]；昭通区块主力产层段石英和长石含量占 47%，方解石和白云石分别占 13% 和 8%，黄铁矿含量约 1%，矿物脆性指数为 55.20%。黏土矿物含量总体较低，以伊 / 蒙混层和伊利石为主。黏土矿物含量介于 5.1%～58.2%，平均为 23.8%；黏土矿物均以伊利石（4%～95%，

平均 60.8%）和绿泥石（0～73%，平均 23.9%）为主，次为伊/蒙混层（0～92%，平均 15.3%），不含蒙皂石。以石英、白云石、斜长石等脆性矿物为主要组分，均不含高岭石、蒙皂石等膨胀矿物。依据北美页岩气开发经验，当脆性矿物指数高于 40% 可认为岩石是脆性的，脆性指数越高，压裂形成的裂缝网络越复杂。昭通区块脆性指数略高于威远和长宁区块，3 个区块整体上脆性指数均达到 40%，属脆性，适合体积压裂作业。

4　施工参数与压裂效果分析

4.1　施工参数分析

对于页岩气等非常规油气资源，采用大规模体积压裂是提高单井产量，获得高产商业油气流的技术关键。压裂施工设计参数包括：井身结构、完井方式、射孔及起裂方式、施工排量、压裂液及支撑剂性能、施工液量、加砂量等。研究表明，射孔及起裂方式、压裂液及支撑剂性能和施工液量、加砂量、施工排量对体积压裂裂缝成网和储层改造体积影响最大[11，13，14]。

表 3　相邻平台井组压裂施工参数对比表

井号		水平井段长（m）	压裂水平段长（m）	压裂段数（段）	加砂量（t）	施工排量（m^3/min）	返排率（%）	实际测试产量（10^4m^3/d）	EUR（10^4m^3）
平台1	1	1400	1026	13	1374	9.4～12.6	8.27	8.04	2800
	2	1200	1156	15	1726	11.0～12.0	9.65	21.02	9500
	3	1010	670	8	906	9.4～13.3	13.56	8.34	2300
	4	980	920	12	1287	12.0～12.5	18.70	13.40	5500
	5	1400	1350	18	1945	11.7～14.4	16.50	19.23	11500
	6	1350	1013	14	1388	9.1～14.4	20.79	10.40	9100
	7	1500	1358	18	2016	10.2～14.6	15.30	18.52	10600
平台2	1	1500	1680	23	2685	12.5～14.8	21.74	27.40	12800
	2	1800	1820	23	2655	11.3～14.7	24.46	15.43	12300
	3	1510	1318	18	1672.64	10.6～14.5	16.31	10.75	8000
	4	1000	1100	12	872.83	5.4～10.4	14.21	7.68	5239
	5	1000	1100	12	1029.9	8.2～11.5	13.51	7.72	11173
	6	1000	1023	8	312.37	5.3～10.1	23.78	5.55	3359

采用平台式“工厂化”作业模式的页岩气水平井，不同平台间受构造、沉积等作用的影响，储层仍然存在较大非均质性。同一平台内部或相邻平台之间，储层分布比较稳定。

选取两个完钻比例较高的相邻平台开展压裂施工参数分析，平台间隔仅 800～1000m，储层间差异相对较小，且采用拉链式作业，压裂施工中工艺上均采用：（1）滑溜水与活性胶液组合；（2）100 目石英砂与 40/70 目陶粒组合方式，微缝 + 支缝 + 主缝三级导流能力设计；（3）桥塞分段式加砂为主；（4）前置酸液作为前置液的一部分使用，提高较大范围和较远距离的处理效果，调整胶液注入时机和注入用量，保证裂缝高度延伸和平面上有效扩展，实时加入可降解暂堵剂实现缝内转向[9, 14, 15]。压裂液配方和支撑剂性能等压裂工艺设计均相同，能较好反映其他压裂施工参数对气井测试产量的影响（表 3）。

分析表明，压裂水平井单段加液量、压后返排率与气井测试产量间相关性不明显，单段加砂量和施工排量均与测试产量呈正相关；当加砂量大于 $100m^3$/ 段，施工排量大于 $12m^3/min$ 时，测试产量上升趋势明显（图 5）；表明储层条件基本相同情况下，压裂施工规模和施工排量对水平井测试产量影响明显（图 7）。

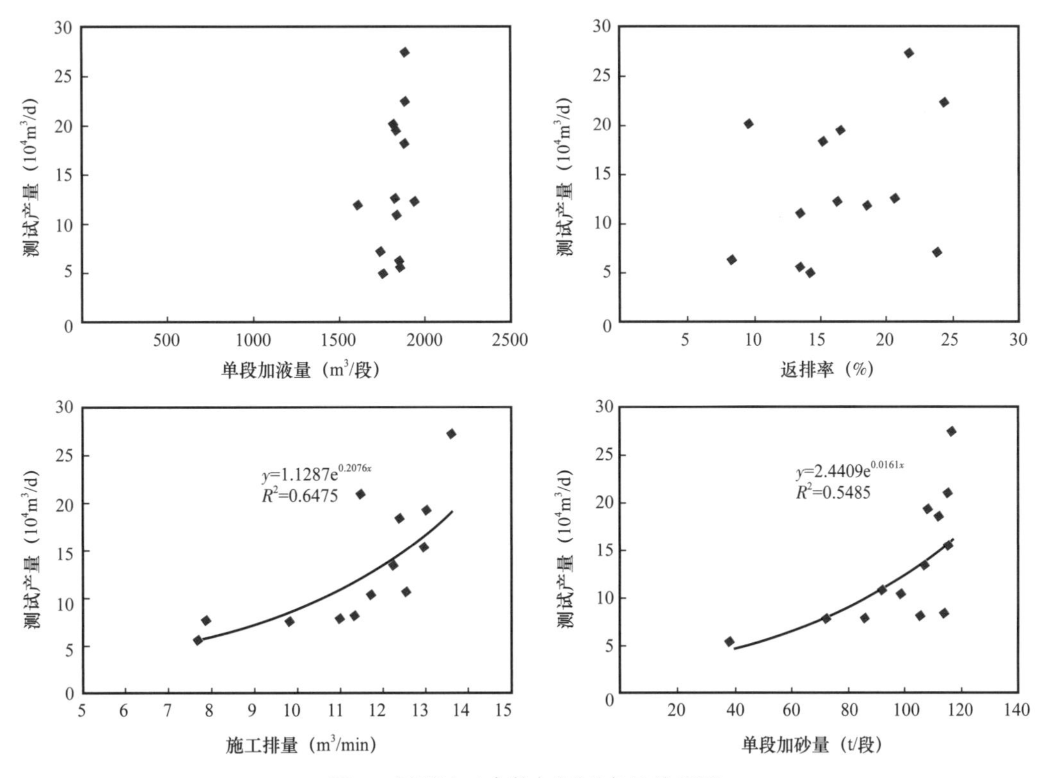

图 7　压裂施工参数与测试产量关系图

对 3 个区块平均压裂施工排量和单段加砂量分析（图 5），长宁、威远区块施工排量和加砂量相当；施工排量分布范围在 10.1～16.0m^3/min，平均 13.54m^3/min，单段加砂量分布范围 59.75～155.15t/ 段，平均 100.88t/ 段。昭通区块施工排量和单段加砂量均偏低，施工排量分布范围在 9.6～12.8m^3/min，平均 11.27m^3/min，单段加砂量分布范围 30.41～90.70t/ 段，平均 63.06t/ 段。国外典型页岩气田 Marcellus 气田 113.40～340.19t/ 段，施工排量 4.77～15.9m^3/min；Barnett 气田 36.29～453.59t/ 段，施工排量 7.95～15.9m^3/min[7]。与之相比，四川盆地 3 个页岩气示范区压裂加砂量均偏低，施工排量昭通区块偏低。

4.2 压裂效果分析

裂缝复杂指数（FCI）常由于描述水力压裂裂缝类型和形态的复杂性，反映体积压裂改造的效果。裂缝复杂指数是微地震监测的水力压裂缝网宽度（X_n）与长度（$2X_f$）的比值，即

$$FCI=X_n / 2X_f$$

FCI 值越大，表明产生的压裂裂缝越复杂、改造体积越大，改造效果越好。3 个区块均进行了微地震监测，将微地震监测获得的裂缝宽度和半长代入以上计算公式，得到 FCI 指数主要分布在 0.30～0.40，与国外页岩气田体积压裂后形成的人工裂缝复杂指数[16-18]相当，表明长宁、威远和昭通区块通过体积压裂均能够形成一定的复杂裂缝（图 8）。

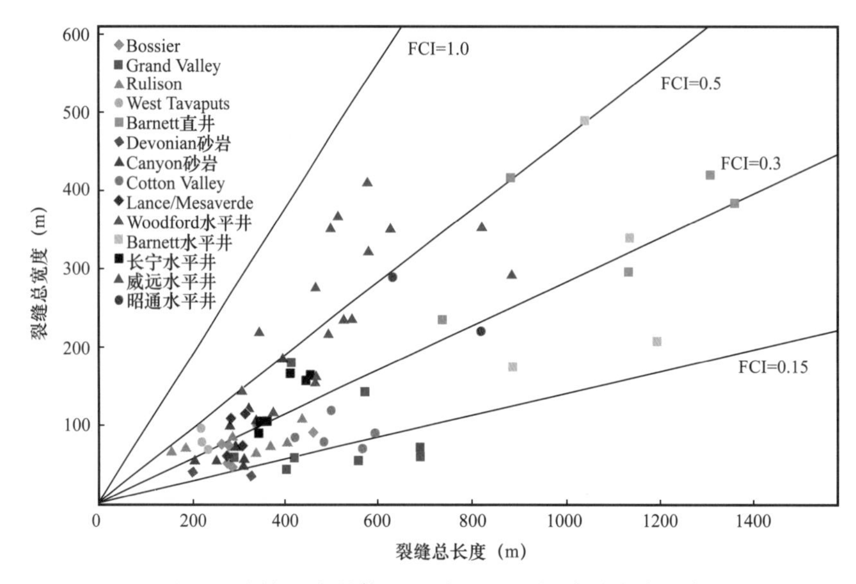

图 8 不同气田气井体积压裂后人工裂缝复杂指数分布

5 结论与建议

从水平井 I 类储层钻遇率、天然裂缝发育程度、主应力非均质系数、主力产层脆性指数、单段加砂量、平均压裂施工排量和裂缝复杂指数等 7 个方面（表 4），对四川盆地长宁、威远和昭通页岩气示范区水平井实施效果开展分析。

表 4 不同区块水平井产量影响因素统计表

区块	I 类储层钻遇率（%）	天然裂缝发育程度	主应力非均质系数（K_h）	主力产层脆性指数（BI）	单段加砂量（t/ 段）	平均压裂施工排量（m^3/min）	裂缝复杂指数（FCI）
长宁	60.78	低	0.19～0.23	52.54%	109.50	13.68	0.30
威远	41.61	低	0.23～0.27	45.49%	93.53	13.45	0.32
昭通	44.32	高	0.26～0.30	55.20%	67.19	11.27	0.36

（1）长宁区块水平井实际钻遇 I 类储层钻遇率最高达到 60.78%，完钻地质参数较好；区块内天然裂缝发育程度较低，区块水平主应力非均质性系数 0.19～0.23，脆性指数 52.54%，工程参数适合体积压裂施工；平均施工排量 13.68m^3/min，单段加砂量 109.50t/段，压裂施工参数相对合理；实际压裂裂缝复杂指数 0.30，表明体积压裂形成一定程度的复杂缝网。

（2）威远区块水平井 I 类储层钻遇率 41.61%，低于 50% 表明完钻地质参数较差；威远区块天然裂缝发育较低，主应力非均质系数 0.23～0.27，脆性指数 45.49%，工程参数较适合体积压裂施工；单段加砂量和平均施工排量分别为 93.53t/ 段和 13.45m^3/min，压裂施工参数相对合理；实际压裂裂缝复杂指数 0.32，表明体积压裂形成缝网较复杂；但水平井完钻地质条件较差、物质基础薄弱，制约了单井产量，进一步提高 I 类储层钻遇率有利于提高水平井单井产量。

（3）昭通区块 I 类储层钻遇率 44.32%，与威远区块相当，明显偏低，完钻地质参数较差；昭通区块天然裂缝相对发育，主应力非均质系数 0.26～0.30，脆性指数 55.20%，工程参数属优质、适合体积压裂施工；平均施工排量 11.27m^3/min，单段加砂量 67.19t/ 段，压裂施工参数较长宁和威远区块略低；实际压裂裂缝复杂指数 0.36，体积压裂能够形成一定程度的缝网。综合分析地质和工程参数及压裂实施效果，昭通区块虽然工程参数属优质，完钻地质条件和压裂施工参数制约了气井产量，进一步提高 I 类储层钻遇率和优化压裂施工参数是提高单井产量的主要技术方向。

参考文献

[1] 国家发展改革委，财政部，国土资源部，国家能源局 . 页岩气发展规划（2011-2015 年）[EB/OL].（2012-03-13）[2016-10-15]. http：//zfxxgk.nea.gov.cn/auto86/201203/P020120316383507834234.pdf.

[2] U.S. Energy Information Administration. Annual Energy Outlook 2016 with projections to 2040 [R/OL]. Washington DC：U.S. Energy Information Administration，Aug. 2016.（2016-09-15）[2016-10-15]. http：//www.eia.gov/outlooks/aeo/pdf/0383（2016）.pdf.

[3] 邹才能，杨智，朱如凯，等 . 中国非常规油气勘探开发与理论技术进展 [J]. 地质学报,2015,89(6)：979-1007.

[4] 杨洪志，张小涛，陈满，等 . 四川盆地长宁区块页岩气水平井地质目标关键技术参数优化 [J]. 天然气工业，2016，36（8）：60-65.

[5] 赵圣贤，杨跃明，张鉴，等 . 四川盆地下志留统龙马溪组页岩小层划分与储层精细对比 [J]. 天然气地球科学，2016，27（3）：470-487.

[6] 钟光海，谢冰，周肖 . 页岩气测井评价方法研究——以四川盆地蜀南地区为例 [J]. 岩性油气藏，2015，27（4）：96-102.

[7] Bruner KR，Smosna R. A comparative study of the Missippian Barnett shale，Fort Worth Basin，and Devonian Marcellus shale，Appalachian Basin [R/OL]. Albany，OR：U.S. Department of Energy，2011. http：//www.netl.doe.gov/File%20Library/Research/Oil-Gas/publications/brochures/DOE-NETL-2011-1478-Marcellus-Barnett.pdf.

[8] 王志刚 . 涪陵焦石坝地区页岩气水平井压裂改造实践与认识 [J]. 石油与天然气地质, 2014, 35 (3): 425–430.

[9] 尹丛彬，叶登胜，段国彬，等 . 四川盆地页岩气水平井分段压裂技术系列国产化研究及应用 [J]. 天然气工业，2014，34 (4)：67–71.

[10] 马旭，郝瑞芬，来轩昂，等 . 苏里格气田致密砂岩气藏水平井体积压裂矿场试验 [J]. 石油勘探与开发，2014，41 (5)：1–5.

[11] 鄢雪梅，王欣，张合文，等 . 页岩气藏压裂数值模拟敏感参数分析 [J]. 西南石油大学学报：自然科学版，2015，37 (6)：127–132.

[12] 张晨晨，王玉满，董大忠，等 . 四川盆地五峰组—龙马溪组页岩脆性评价与“甜点层”预测 [J]. 天然气工业，2016，39 (9)：51–60.

[13] 何建华，丁文龙，王哲，等 . 页岩储层体积压裂缝网形成的主控因素及评价方法 [J]. 地质科技情报，2015，34 (4)：108–118.

[14] 任勇，钱斌，张剑，等 . 长宁地区龙马溪组页岩气工厂化压裂实践与认识 [J]. 石油钻采工艺，2015，37 (4)：96–99.

[15] 刘旭礼 . 页岩气体积压裂压后试井分析与评价 [J]. 天然气工业，2016，36 (8)：66–72.

[16] Chong KK, Grieser B，Jaripatke O, et al. A completions roadmap to shale-play development：A review of successful approaches toward shale-play stimulation in the last two decades [C] //paper 130369-MS presented at International Oil and Gas Conference and Exhibition in China，8–10 June, Beijing, China. DOI：http：//dx.doi.org/10.2118/130369-MS.

[17] Cipolla C L, Warpinski N R，Mayerhofer M, et al.The relationship between fracture complexity，reservoir properties，and fracture-treatment design [C] //paper 115769-MS presented at the SPE Annual Technical Conference and Exhibition，21–24 Sept. 2010，Denver，Colorado, USA. DOI：http：//dx.doi.org/10.2118/115769-MS.

[18] 郭小哲，周长沙 . 页岩气储层压裂水平井三线性渗流模型研究 [J]. 西南石油大学学报 (自然科学版)，2016，38 (2)：86–94.

海相页岩有效产气储层特征——以四川盆地五峰组—龙马溪组页岩为例

雷丹凤，李熙喆，位云生，邱　振，卢　斌

（中国石油勘探开发研究院）

摘　要: 综合分析五峰组—龙马溪组页岩气储层特征，采用TOC值、孔隙度和含气量为含气性评价参数，脆性矿物含量为可压裂性评价参数来评价页岩气储层。以四川盆地SG区块为例，开展水平井段储层的精细评价与产量关系分析，建立该区块更为细化的储层分类标准，识别出有效产气储层。研究结果表明：五峰组—龙马溪组页岩气储层具有"四高"特征，即高TOC值（TOC≥2.0%）、高孔隙度（≥3.0%）、高含气量（≥$2.0m^3/t$）和高脆性矿物质量分数（≥40%），它们是页岩气储层评价的关键参数；SG区块页岩气储层类型可以细分为5类，其中Ⅰ类+Ⅱ类储层与单井产量关系最为密切，是该区块的有效产气储层；Ⅰ类+Ⅱ类储层关键参数特征为TOC≥4.0%、孔隙度≥5.0%、总含气量≥$4.5m^3/t$、脆性矿物质量分数≥60%，它们主要分布在龙一$_1^1$小层的下段和中段。建议可进一步优化该区块水平井钻井轨迹，将水平井靶体位置放在龙一$_1^1$小层中、下段，以提高有效产气储层钻遇厚度，达到提高单井产量并实现页岩气高效开发。

关键词: 页岩气；储层评价；储层类型；水平井靶体；四川盆地

国内海相页岩气资源丰富，可采资源量为（8～13）$\times10^{12}m^3$，并在四川盆地及周缘五峰组—龙马溪组海相页岩成功实现商业化开发[1-4]。截至2017年底，海相页岩气累计探明地质储量超过$1\times10^{12}m^3$，产量超过$90\times10^8m^3$[5]。针对五峰组—龙马溪组海相页岩，中国在四川盆地涪陵与川南威远、长宁、昭通形成四大页岩气商业化开发示范区，并在巫溪、荣昌、永川等地区相继取得勘探突破，页岩气勘探开发潜力巨大[1, 5-7]。

2016年国家能源局公布《页岩气发展规划（2016—2020年）》，并提出国内力争2020年页岩气产量实现$300\times10^8m^3$，2030年实现页岩气产量$800\times10^8m^3$到$1000\times10^8m^3$。为实现这些目标，需要不断增加钻井数量以扩大开采规模。同时，由于目前页岩气单井产量差异较大[6, 8]，提高单井产量则是实现页岩气大规模高效开发的关键所在。以川南威远—长宁页岩气开发区为例，在页岩气勘探开发初期，长宁区块与威远区块页岩气井平均测试日产气量分别为$10.9\times10^4m^3$和$11.6\times10^4m^3$；而后经过调整水平井靶体位置等一系列措施，平均测试日产气量分别提高至$23\times10^4m^3$和$17\times10^4m^3$[1, 6]。虽然影响单井产量的地质与工程因素很多，包括水平井钻遇储层品质、地层压力变化、压裂改造效果等[6, 8-10]，但水平井钻遇优质储层即有效产气储层累计厚度是单井高产的物质基础。因此，开展页岩气勘探开发目标层段精细评价，识别有效产气储层，可进一步优化水平井钻井轨迹，提高优质储层钻遇率，有利于增加单井产量，实现页岩气高效开发。以四川盆地SG区块五峰组—

龙马溪组页岩气为例，基于对评价井（直井）页岩储层开展分类评价，应用于水平井段（压裂段），结合水平井产量，识别出有效产气储层，提高水平井产量，以此指导该区块页岩气的高效开发。

1 地质概况

晚奥陶世—早志留世，四川盆地位于上扬子板块之上，受扬子板块与华夏板块碰撞作用的影响，其沉积环境由浅水的碳酸盐岩台地逐渐演化为碎屑陆棚环境[11-14]。在五峰组—龙马溪组沉积时期，四川盆地及周缘广泛沉积了一套页岩层系[3, 11, 15]，下段是以黑色富有机质页岩为主的深水陆棚相；上段是以灰色粉砂岩、页岩为主的浅水陆棚相[16]。奥陶纪末期受全球冈瓦纳冰川作用的影响，在五峰组顶部沉积了一段10～100cm的介壳灰岩或泥灰岩段，即观音桥段[11, 17, 18]。

近十年来，随着五峰组—龙马溪组页岩气勘探开发不断深入，该套页岩层系划分也在逐步细化[2, 5, 6, 19]（图1）。五峰组二分为五一段（页岩层段）和五二段（观音桥段）；龙马溪组总体上可以划分为两段：龙一段与龙二段。前者主要由灰黑色、黑色页岩、碳质页岩等组成，有机质质量分数一般高于1.0%，发育丰富的笔石化石；而后者主要为灰色、灰黑色页岩、粉砂质泥岩等组成，有机质含量一般偏低。中国石油根据其探区内地震、钻井岩心、测井及生产资料，结合野外露头资料，将龙一段细分为龙一$_1$和龙一$_2$亚段。龙一$_1$亚段主要为富有机质页岩段，TOC值一般高于2.0%。它被进一步细分为4个小层，分别为龙一$_1^1$、龙一$_1^2$、龙一$_1^3$和龙一$_1^4$[6, 19]。目前，龙一$_1^1$小层是威远、长宁区块页岩气主要开发的目的层[1, 6]。

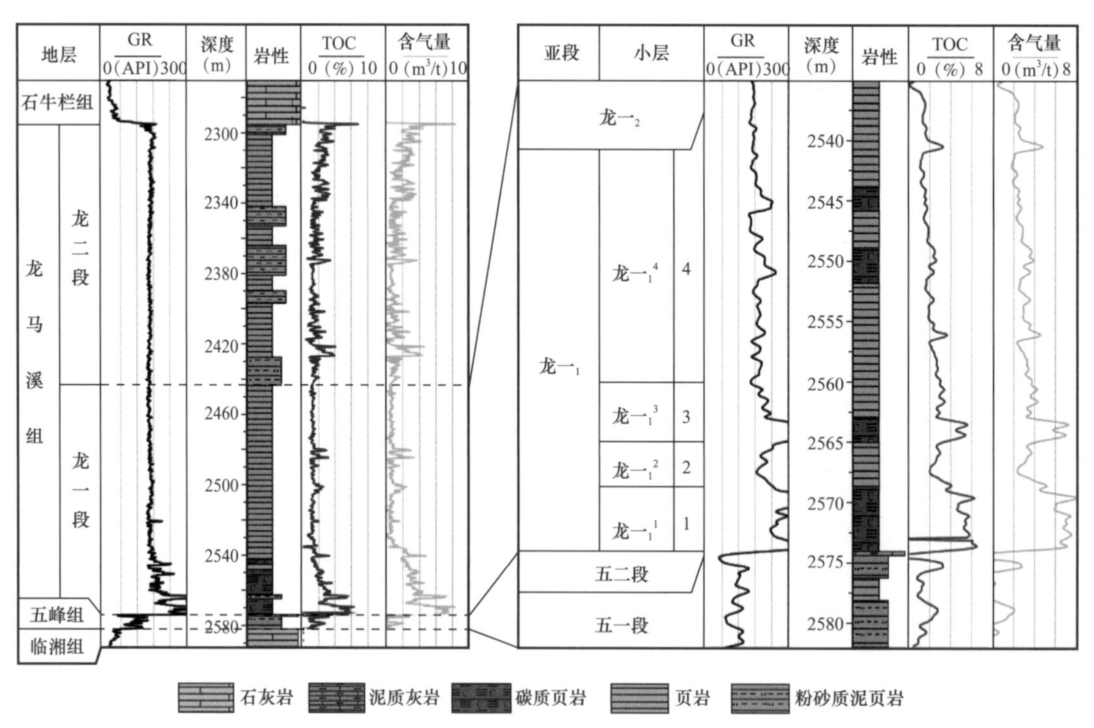

图1 四川盆地典型井五峰组—龙马溪组地层综合示意图

2 五峰组—龙马溪组页岩气储层特征与评价参数优选

页岩气储层品质（特征）是单井获得高产基础。美国最早实现页岩气商业化开发，诸多页岩气田的产气层段，如 Banett 页岩、Marcellus 页岩等，为具有一定连续厚度的高 TOC 值（>2.0%）的富有机质页岩层段[20, 21]。五峰组—龙马溪组页岩作为国内最早实现页岩气商业化开发的层系[3, 4, 22]，其页岩气储层特征是中国页岩气勘探开发过程中的研究热点之一。从早期的仅基于沉积相、有机质（干酪根类型、TOC 值、成熟度等）及物性（孔隙度、渗透率）等几个方面的研究[23–26]，逐步扩展到矿物组成（硅质、钙质、黏土等）、储集空间特征（孔隙类型、连通性等）、含气量（吸附气与游离气）、裂缝、脆性指数、压力系数等诸多方面的研究[27–38]，并逐步建立不同的页岩气储层评价参数体系。例如，文献［27］借鉴美国页岩气勘探经验，对四川盆地南方海相页岩进行储层评价，筛选出 TOC 值、无机矿物含量、物性、岩石力学性质和厚度 5 个指标作为储层评价关键参数，并给出了页岩储层的这些参数下限值；文献［28］采用 TOC 值、成熟度、脆性矿物含量、孔隙度、页岩厚度 5 个指标对页岩气储层进行分级评价，并可结合埋深优选出页岩气有利区；文献［32］通过对页岩储层破坏机理、力学特性和脆性评价方面的研究，综合开展页岩储层的脆性评价；文献［33］以上扬子古生界页岩气储层为例，用 TOC 值、吸附气含量、黏土含量、成熟度、孔隙度和可流动百分数 6 个指标进行储层评价；国土资源部 2014 年发布的《页岩气资源 / 储量计算与评价技术规范》[39] 明确提出页岩有效厚度、TOC 值、含气量、成熟度、脆性矿物含量、孔隙度等作为储层评价参数。尽管这些研究所选取的评价参数存在一些差异，但都对中国五峰组—龙马溪组页岩气的勘探评层选区及高效开发工作提供了有效支撑。

由于页岩储层相对致密，需要通过水平井分段压裂改造才能获得商业化开采，故上述这些地质特征可以归纳为两大类，即含气性评价参数与可压裂性评价参数[5]：（1）含气性评价参数，包括页岩有效厚度、干酪根类型、TOC 值、成熟度、孔隙度、渗透率、含气量等。这些参数与储量有关，决定了页岩气的储量丰度，是单井高产稳产的基础。（2）可压裂性评价参数，包括脆性矿物含量、黏土含量、岩石力学、天然裂缝等。这些参数与单井产量密切相关，决定了水平井压裂改造效果，是单井提产的关键。大量研究与页岩气勘探开发实践表明，我国五峰组—龙马溪组页岩气优质储层均具有“四高”特征[1, 5, 9, 22, 40, 41]，即高有机碳含量（TOC 值）、高孔隙度、高含气量与高脆性矿物含量（表 1）。

高有机碳含量（TOC 值）：五峰组—龙马溪组页岩有机质主要由浮游藻类、疑源类等非动物碎屑及笔石等动物碎屑组成[42–44]。优质储层段的 TOC 值普遍高于 2.0%，而主要产气层段一般高于 4%。例如，涪陵气田 JY1 井 TOC 值大于 2.0% 的层段作为甜点段，厚度达 38m，平均值高达 3.5%[40]，而主要产层 TOC 值普遍大于 4%[5]；威远—长宁气田优质储层段 TOC 值为 2.2%～5.3%[45]；昭通气田的优质储层段厚度达 35m，TOC 平均值为 3.5%～5.3%[41]。

高孔隙度：包括无机孔和有机质孔[46–48]，优质储层段的孔隙度普遍高于 3.0%，而主要产气层段一般高于 5.0%。例如，涪陵气田 JY1 井优质储层孔隙度为 2.8%～7.1%，平均值为 4.8%[40]，而主要产层的孔隙度普遍大于 5.0%[5]；威远—长宁气田优质储层的孔隙度一般为 3.3%～8.4%，昭通气田的孔隙度一般为 3.4%～7.4%[3]。

高含气量：页岩气的赋存状态包括吸附气和游离气两类。优质储层段的总含气量普遍高于 2.0m^3/t，而主要产气层段的一般高于 4.5m^3/t。例如，涪陵气田 JY1 井甜点段一般高于 2.0m^3/t，平均值为 3.0m^3/t[40]，而主要产层的平均含气量大于 6.0m^3/t[5]；威远—长宁气田优质储层段含气量为 2.3～6.5m^3/t[45]；昭通气田的优质储层段平均含气量为 2.3m^3/t，主要产层的含气量平均为 4.3m^3/t[49]。

高脆性矿物含量：主要由硅质、碳酸盐等脆性矿物组成。优质储层段的脆性矿物质量分数普遍高于 40%，而主要产气层段的一般高于 60%。例如，涪陵气田 JY1 井甜点段的脆性矿物以硅质为主，质量分数平均为 44%[40]，而主要产层的质量分数普遍大于 60%[5]；威远—长宁气田优质储层段脆性矿物主要由硅质、碳酸盐等矿物组成，其质量分数为 54%～74%[45]；昭通气田的优质储层段脆性矿物主要为硅质，其质量分数为 50%～80%，主要产层的含量一般高于 62%[49]。

这 4 个参数中，TOC 值、孔隙度和含气量为含气性评价参数，脆性矿物含量为可压裂性评价参数，它们常被用来作为页岩气储层评价的关键参数。

表 1　四川盆地及周缘典型页岩气田优质储层参数特征[3, 5, 40, 41, 49]

参数	TOC 值（%）	孔隙度（%）	含气量（m^3/t）	脆性矿物含量（%）
涪陵 *	平均 3.5	平均 4.8	平均 3.0	平均 44
威远—长宁	2.2～5.3	3.3～8.4	2.3～6.5	54～74
昭通	平均 3.5～5.3	3.4～7.4	平均 2.3	50～80

注 *：涪陵页岩气田数据主要来自 JY1 井。

3　五峰组—龙马溪组页岩气储层精细评价与有效产气储层识别

以四川盆地 SG 区块五峰组—龙马溪组页岩气为例，基于已有的 9 口探井 / 评价井（直井）岩心、测井、分析实验等资料，通过开展 TOC 值、孔隙度、含气量、脆性矿物含量等储层关键参数的精细评价，建立关键评价参数与测井曲线的关系，确定该区储层分类标准，以此进一步开展水平井段储层的精细评价，识别有效产气储层。

3.1　储层类型划分

四川盆地 SG 区块是我国五峰组—龙马溪组页岩气较早勘探地区之一，目前已钻探井 / 评价井数十口、水平井数百口以上。对其中 9 口评价井进行系统取心，目前已完成测井、各类储层实验等分析，为开展本区块页岩气储层评价提供了良好基础。储层类型划分是对储层品质优劣程度快速评价的有效手段，是开展储层评价的必要环节。近十年来，随着我国页岩气勘探开发的不断深入，对储层类型方面的认识，也从早期阶段仅区分页岩气储层与非储层[27]，逐步对页岩气储层类型进行详细划分。如《页岩气资源 / 储量计算与评价技术规范》[39] 对储层参数进行了由低到高的分类；中国石油也制定了页岩气储层划分标准，依据 TOC 值、孔隙度、含气量、脆性矿物含量 4 个关键参数划分出 Ⅰ 类、Ⅱ 类和Ⅲ类储层，其中 Ⅰ 类是最优质储层（TOC ≥ 3%、孔隙度≥ 5%、总含气量≥ 3m^3/t、脆性矿物质量分数≥55%），Ⅱ类、Ⅲ类依次变差（据《中国石油页岩气测井采集与评价技术管理规定》）。该储层类型划分标准被广泛用于SG区块页岩气勘探开发相关的研究之中。

SG区块取心段主要为五峰组—龙马溪组龙一$_1$亚段，厚度一般为44～54m。基于该区块不同评价井累计近800件/次样品实验分析数据（图2）表明，五峰组—龙马溪组龙一$_1$亚段页岩气储层4个关键参数相对偏高，具体如下：（1）TOC值变化较大，一般为1.5%～5.0%，平均值相对偏高，可达2.9%；（2）孔隙度一般为4.5%～7.5%，平均值为5.6%；（3）含气量一般为2.0～6.0m^3/t，平均值为3.8m^3/t；（4）脆性矿物主要为硅质，其含量一般为50%～78%，平均值为70%。随着SG区块水平生产井数量增多，现场大量生产数据表明：水平井段Ⅰ类储层（以TOC≥3%为标志）平均钻遇率可达77%～96%[1]，但在不同生产平台包括同一平台，水平井段压裂长度和Ⅰ类储层钻遇率均相近的不同单井之间的产量具有较大差异。例如，相邻两个生产平台的两口水平井SGH5-1和SGH6-4的压裂段长度分别为1427m和1425m，Ⅰ类储层钻遇累计厚度分别为1360m和1369m，钻遇率分别为95%和96%，采用相同压裂开采工艺，但它们之间的测试产量却存在着较大差异，分别为5.4×10^4m^3/d和22×10^4m^3/d。同时，涪陵页岩气田开发实践表明，该气田主要产气层具有较高TOC值（≥4%）、孔隙度（≥5%）、含气量（≥6.0m^3/t）和脆性矿物含量（≥60%）[5]特征，其储层品质明显高于SG区块所采用的当前储层分类方案中Ⅰ类储层。这些在一定程度上说明，Ⅰ类储层并不是都对产量具有直接贡献，可能仅是其中更优质部分储层具有直接贡献。为了进一步确定SG区块产量具有直接贡献的储层即有效产气储层，在已有的储层分类方案基础上，结合SG区块储层特征，开展了更加细化的储层分类。将页岩气储层划分为5种类型：分别为Ⅰ类、Ⅱ类、Ⅲ类、Ⅳ类和Ⅴ类，其品质依次变差，详见表2。

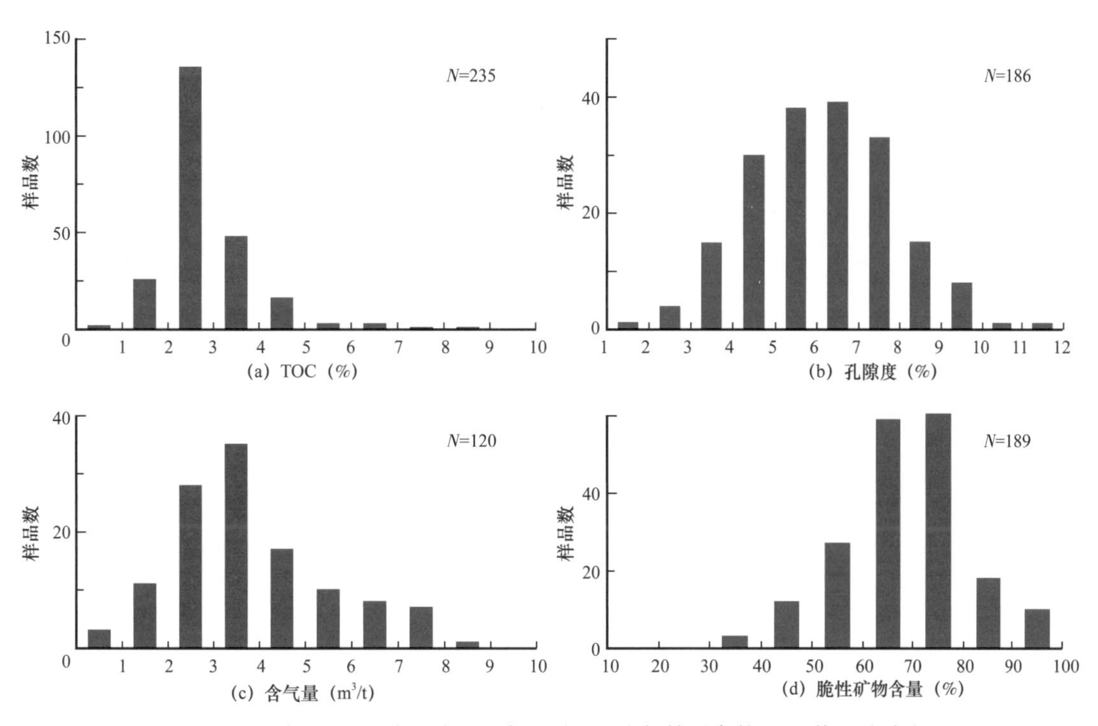

图2 四川盆地SG区块五峰组—龙马溪组页岩气储层参数TOC值、孔隙度、含气量与脆性矿物含量分布

表 2　四川盆地 SG 区块五峰组—龙马溪组页岩气储层分类划分表

类别		地质参数				测井参数			
		TOC 值（%）	孔隙度（%）	含气量（m^3/t）	脆性矿物（%）	GR（API）	DEN（g/cm^3）	AC（μs/m）	CNL（%）
储层	Ⅰ类	>5	>6	>6.0	≥ 60	>200	<2.52	>82	<18
	Ⅱ类	4～5	5～6	4.5～6.0	≥ 60	190～260	2.50～2.64	80～90	<18
	Ⅲ类	3～4	4～5	3.0～4.5	50～60	150～200	2.55～2.65	76～85	15～20
	Ⅳ类	2～3	3～4	2.0～3.0	40～50	140～180	2.58～2.67	73～82	>20
	Ⅴ类	1～2	2～3	1.0～2.0	30～40	130～160	2.64～2.70	70～75	>20
非储层Ⅵ类		<1	<2	<1.0	<30	<140	>2.68	<70	>20

其中，表 2 中测井参数的确定，主要是依据 SG 区块 9 口探井 / 评价井岩心实测的 4 个关键评价参数与不同类型测井数据开展大量统计分析的结果（图 3）。结果表明：（1）总有机碳（TOC）含量与自然伽马（GR）、密度（DEN）测井数据存在着较好的相关性，能够较好地区分出不同 TOC 值（图 3a）；（2）孔隙度高低可以通过声波（AC）、密度（DEN）测井数据交会图进行识别（图 3b）；（3）TOC 值与总含气量之间具有较好的线性相关性，可以通过 TOC 值计算出相应的含气量；（4）脆性矿物含量与中子测井（CNL）数据具有较好的线性相关，可以通过中子测井数据计算得到（图 3c）。

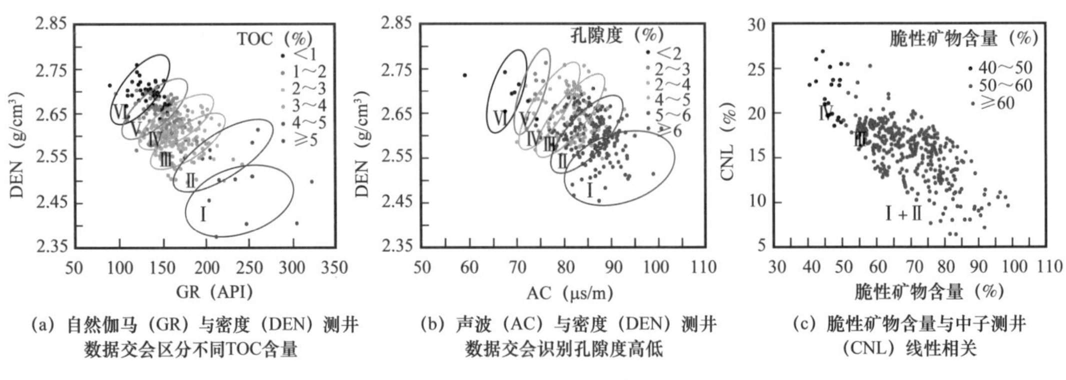

(a) 自然伽马（GR）与密度（DEN）测井数据交会区分不同TOC含量　(b) 声波（AC）与密度（DEN）测井数据交会识别孔隙度高低　(c) 脆性矿物含量与中子测井（CNL）线性相关

图 3　四川盆地 SG 区块储层地质参数（TOC 值、孔隙度和脆性矿物含量）与测井数据相关性分析

3.2　水平井压裂段储层精细评价与有效产气储层识别

基于上述储层分类，优选了 SG 区块已投产一年以上的 25 口水平井开展页岩气储层精细评价。评价对象主要是针对水平井段中的压裂段（图 4），精细刻画其储层类型分布，并分别统计Ⅰ类、Ⅱ类、Ⅲ类这 3 类页岩气储层（相当于早期方案中Ⅰ类优质储层）的累计厚度，以此分析它们与产量的关系。

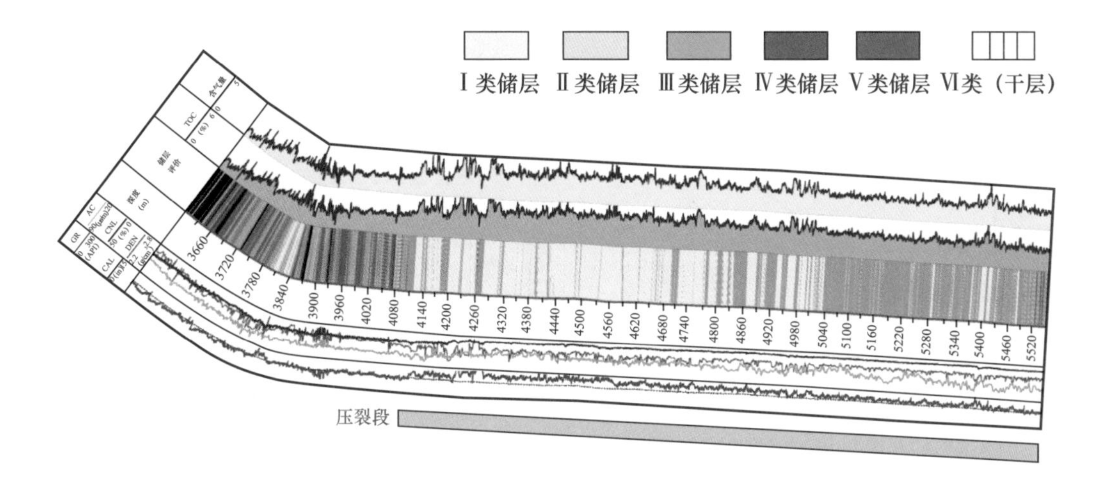

图4 四川盆地SG区块水平井压裂段储层精细评价综合示意

以该区块X平台6口水平井为例，它们的压裂段长度为919～1669m，测试产量变化相对较大，为（3.4～28.8）$\times10^4m^3/d$（表3）。通过对该平台6口水平井压裂段储层精细评价，它们的Ⅰ类与Ⅱ类储层的累计压裂长度与测试产量具有较好相关性（相关系数为0.85）（图5a），明显优于Ⅰ类、Ⅱ类、Ⅲ类这3类储层（相当于早期方案中Ⅰ类优质储层）的总压裂长度与测试产量的相关性（相关系数为0.52）（图5b）。尽管除了储层品质之外，影响水平井单井产量的地质与工程因素还很多，包括裂缝、地层压力变化、压裂改造效果等[6, 8, 9]。在X平台上这些水平井的压裂段长度具有较大差异，但它们的地质条件基本无变化，且采用相同的压裂开采工艺。故可以初步得出该区块X平台水平井段的Ⅰ类与Ⅱ类储层（以TOC≥4%为标志）压裂长度是影响其产量高低的关键性因素。

表3 四川盆地SG区块X平台6口水平井压裂段不同类型储层累计厚度与测试产量

井号	压裂长度（m）	Ⅰ类储层压裂累计长度（m）	Ⅱ类储层压裂累计长度（m）	Ⅲ类储层压裂累计长度（m）	Ⅰ+Ⅱ+Ⅲ类储层压裂累计长度（m）	Ⅰ+Ⅱ类储层压裂累计长度（m）	测试产量（$10^4m^3/d$）
SGH2-1	919	116	265	479	859	381	3.4
SGH2-2	1441	362	302	650	1314	664	5.3
SGH2-3	1232	52	44	510	607	96	3.6
SGH2-4	1565	1071	313	91	1475	1384	28.8
SGH2-5	1669	1087	182	356	1625	1269	20.0
SGH2-6	1446	16	219	1038	1274	235	6.4

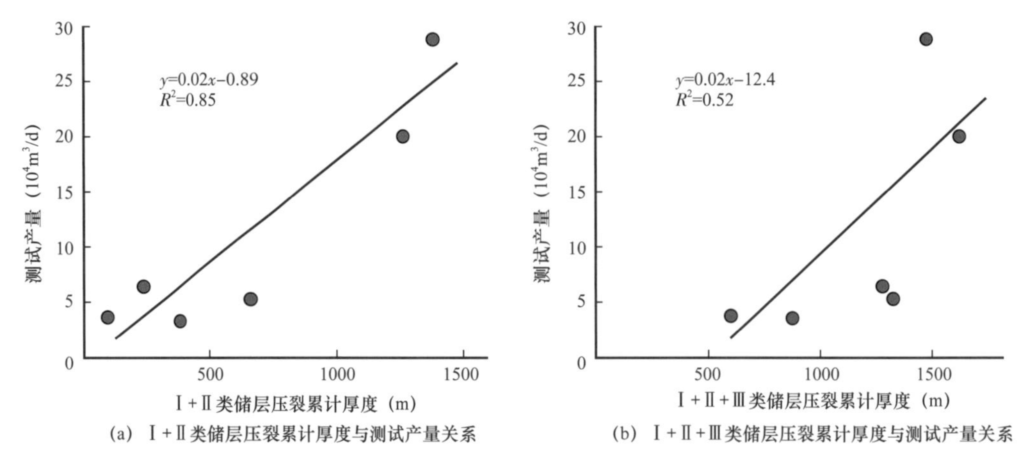

(a) Ⅰ+Ⅱ类储层压裂累计厚度与测试产量关系　(b) Ⅰ+Ⅱ+Ⅲ类储层压裂累计厚度与测试产量关系

图 5　四川盆地 SG 区块 X 平台 6 口水平井不同储层类型压裂累计长度与测试产量关系

为进一步揭示该区块水平井中Ⅰ类与Ⅱ类储层（以 TOC ≥ 4% 为标志）压裂长度与产量的关系，选择了来自不同生产平台的具有相同水平段压裂长度的两口生产井即 SGH5-1 和 SGH6-4 开展储层精细评价，并对储层类型进行针对性分析。它们的压裂段长度分别为 1427m 和 1425m，按照前人储层分类方案Ⅰ类储层（以 TOC ≥ 3% 为标志，相当于本文中Ⅰ类 + Ⅱ类 + Ⅲ类储层）钻遇累计厚度分别为 1360m 和 1369m，钻遇率分别为 95% 和 96%。如前所述，这两口水平井的地质条件与压裂开采工艺基本相同，但它们之间的测试产量相差较大，分别为 $5.4 \times 10^4 m^3/d$ 和 $22 \times 10^4 m^3/d$。然而，采用本文储层分类所评价的结果是，它们的压裂段Ⅰ类 + Ⅱ类储层（以 TOC ≥ 4% 为标志）累计厚度分别为 372m 和 1148m，分别与其测试产量（$5.4 \times 10^4 m^3/d$ 和 $22 \times 10^4 m^3/d$）具有非常好的对应关系。这进一步表明水平井段Ⅰ类 + Ⅱ类储层（以 TOC ≥ 4% 为标志）压裂长度与产量关系密切。

为减少其他地质因素、工程因素等方面对水平井产量的影响，优选出 SG 区块已投产一年以上的 25 口水平井，它们的地质条件差异不大，且无套管变形、落鱼、砂堵等工程事故。这些水平井压裂段储层精细评价结果如图 6 所示，它们的Ⅰ类 + Ⅱ类储层（以 TOC ≥ 4% 为标志）压裂长度与产量相关性明显好于Ⅰ类 + Ⅱ类 + Ⅲ类储层。因此，可以认为Ⅰ类 + Ⅱ类储层是 SG 区块页岩气开发的有效产气储层，其压裂长度是影响产量高低的关键性因素。

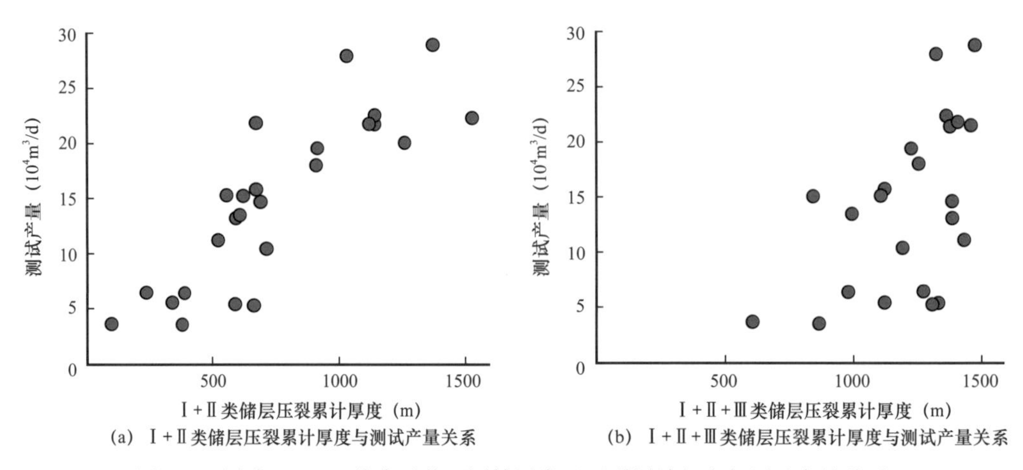

(a) Ⅰ+Ⅱ类储层压裂累计厚度与测试产量关系　(b) Ⅰ+Ⅱ+Ⅲ类储层压裂累计厚度与测试产量关系

图 6　四川盆地 SG 区块水平井不同储层类型压裂累计长度与测试产量关系

4 有效产气储层特征及开发建议

上述分析可知，四川盆地 SG 区块有效产气储层为Ⅰ类＋Ⅱ类储层，它们的关键储层参数特征为：TOC ≥ 4%、孔隙度≥ 5%、总含气量≥ 4.5m^3/t、脆性矿物质量分数≥ 60%。这些特征基本上与涪陵区块五峰组—龙马溪组页岩气主力产层一致[5]。以本文储层类型划分为基础，对该区块 9 口评价井（直井）五峰组—龙马溪组龙一$_1$亚段开展储层精细评价，并对这些井的不同储层类型发育程度进行统计分析。结果表明，Ⅰ类、Ⅱ类和Ⅲ类储层主要发育在龙一$_1^1$小层（图 7），Ⅳ类和Ⅴ类储层主要分布在龙一$_1^2$、龙一$_1^3$和龙一$_1^4$和五一段，五二段主要由泥灰岩等组成，是非储层段（图 1 和图 7）。SG 区块中页岩气有效产层的纵向分布稍不同于涪陵区块，后者的主要产气层为五峰组和龙马溪组底部（①②③小层，详见文献［5］）。

早期 SG 区块开发过程中水平井靶体距龙一$_1^1$小层相对较高，未把该小层作为主要开发目标。但在后期将水平井靶体调整到龙一$_1^1$小层内，开发效果明显提高。以川南威远—长宁区块为例，单井平均测试日产气量从（10.9～11.6）×10^4m^3逐步提高至（17～23）×10^4m^3[6]。这很大程度上是因为水平井靶体放在优质储层集中发育的龙一$_1^1$小层内，这样水平井能够钻遇更厚的优质储层，从而保证后期压裂改造的优质储层体积更大，最终获得更高的页岩气产量。

由图 7 可知，SG 区块龙一$_1^1$小层仍存在一定的非均质性，自下而上依次发育Ⅰ类、Ⅱ类和Ⅲ类储层，分别对应不同的电性特征，可进一步划分为下、中和上三段。Ⅰ类和Ⅱ类作为有效产气储层，分别发育在龙一$_1^1$小层的下段和中段。因此，建议进一步优化水平井钻井轨迹，将水平井靶体位置放在龙一$_1^1$小层中、下段，以提高有效产气储层的钻遇累计厚度，达到提高单井产量并实现页岩气高效开发。

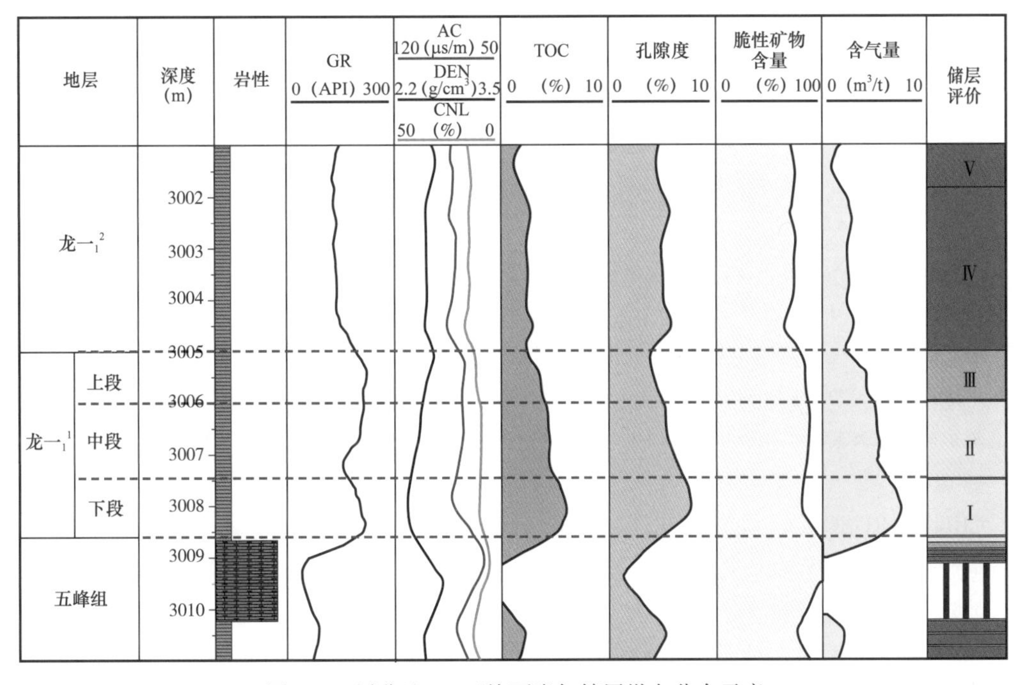

图 7 四川盆地 SG 区块页岩气储层纵向分布示意

5 结论

（1）综合前人资料分析，中国南方海相五峰组—龙马溪组页岩气优质储层均具有“四高”特征，即高有机碳含量（TOC 值）、高孔隙度、高含气量和高脆性矿物含量，前三者为含气性评价参数，后者为可压裂性评价参数，它们是开展页岩气储层评价的关键参数。

（2）基于四川盆地 SG 区块中已有测井、分析实验等资料，通过开展 TOC 值、孔隙度、含气量、脆性矿物质量分数等储层关键参数的精细评价，建立该区块更为细化的储层分类标准。以此进一步开展水平井段储层的精细评价，提出Ⅰ类 + Ⅱ类储层（以 *TOC* ≥ 4% 为标志）压裂长度是影响单井产量高低的关键性因素，这两类储层是该区块的有效产气储层。

（3）SG 区块页岩气的有效产气储层（Ⅰ类 + Ⅱ类储层）关键参数特征：TOC ≥ 4.0%、孔隙度≥ 5.0%、总含气量≥ $4.5m^3/t$、脆性矿物质量分数≥ 60%；它们主要分布在龙一$_1^1$小层的下段和中段。建议下一步将水平井靶体位置放在龙一$_1^1$小层的中、下段，以提高有效产气储层钻遇累计厚度，实现该区块页岩气的高效开发。

参考文献

［1］马新华，谢军 . 川南地区页岩气勘探开发进展及发展前景［J］. 石油勘探与开发，2018，45（1）：1–9.

［2］郭旭升，胡东风，魏志红，等 . 涪陵页岩气田的发现与勘探认识［J］. 中国石油勘探，2016，21（3）：24–37.

［3］邹才能，董大忠，王玉满，等 . 中国页岩气特征、挑战及前景（一）［J］. 石油勘探与开发，2015，42（6）：689–701.

［4］王志刚 . 涪陵页岩气勘探开发重大突破与启示［J］. 石油与天然气地质，2015，36（1）：1–6.

［5］马永生，蔡勋育，赵培荣 . 中国页岩气勘探开发理论认识与实践［J］. 石油勘探与开发，2018，45（4）：1–14.

［6］谢军 . 长宁—威远国家级页岩气示范区建设实践与成效［J］. 天然气工业，2018，38（2）：1–7.

［7］邱振，邹才能，李建忠，等 . 非常规油气资源评价进展与未来展望［J］. 天然气地球科学，2013，24（2）：238–246.

［8］贾成业，贾爱林，何东博，等 . 页岩气水平井产量影响因素分析［J］. 天然气工业，2017，37（4）：80–88.

［9］金之钧，胡宗全，高波，等 . 川东南地区五峰组—龙马溪组页岩气富集与高产控制因素［J］. 地学前缘，2016，23（1）：1–10.

［10］郭彤楼，张汉荣 . 四川盆地焦石坝页岩气田形成与富集高产模式［J］. 石油勘探与开发，2014，41（1）：28–36.

［11］CHEN X，RONG J Y，LI Y，et al. Facies patterns and geography of the Yangtze region，South China，through the Ordovician and Silurian transition［J］. Palaeogeogr Palaeoclimat Palaeoecol，2004，204：353–372

［12］ZOU Caineng，QIU Zhen，POULTON S W，et al. Ocean euxinia and climate change “double whammy” drove the Late Ordovician mass extinction［J］. Geology，2018，46（6）：535–538.

[13] SU Wenbo，WARREN D H，FRANK R E，et al.K–bentonite，black–shale and flysch successions at the Ordovician–Silurian transition，south China：Possible sedimentary responses to the accretion of Cathaysia to the Yangtze block and its implications for the evolution of Gondwana [J]. Gondwana Research，2009，15：111–130.
[14] YAN Detian，CHEN Daizhao，WANG Qingchen，et al. Large–scale climatic fluctuations in the latest Ordovician on the Yangtze block，south China [J]. Geology，2010，38（7）：599–602.
[15] 邱振，江增光，董大忠，等. 巫溪地区五峰组—龙马溪组页岩有机质沉积模式 [J]. 中国矿业大学学报，2017，46（5）：923–932.
[16] 刘树根，冉波，郭彤楼，等. 四川盆地及周缘下古生界富有机质黑色页岩：从优质烃源岩到页岩气产层 [M]. 北京：科学出版社，2014.
[17] RONG J Y. Ecostratigraphic evidence of regression and influence of glaciation of Late Ordovician in the Upper Yangtze area [J]. Stratigraphy Journal，1984，8：9–20.
[18] CHEN X，RONG J，FAN J，et al. The global boundary stratotype section and point（GSSP）for the base of the Hirnantian Stage（the uppermost of the Ordovician System）[J].Episodes，2006，29：183–196.
[19] 赵圣贤，杨跃明，张鉴，等. 四川盆地下志留统龙马溪组页岩小层划分与储层精细对比 [J]. 天然气地球科学，2016，27（3）：470–487.
[20] KINLEY T J，COOK I W，BREYER J A，et al. Hydrocarbon potential of the Barnett shale（Mississippian），Delaware basin，west Texas and Southeastern New Mexicao [J].AAPG Bulletin，2008，92（8）：967–991.
[21] 闫存章，董大忠，程克明，等. 页岩气地质与勘探开发实践丛书：北美地区页岩气勘探开发新进展 [M]. 北京：石油工业出版社，2009.
[22] 郭旭升. 南方海相页岩气“二元富集”规律：四川盆地及周缘龙马溪组页岩气勘探实践认识 [J]. 地质学报，2014，88（7）：1209–1218.
[23] 梁狄刚，郭彤楼，陈建平，等. 中国南方海相生烃成藏研究的若干新进展（一）：南方四套区域性海相烃源岩的分布 [J]. 海相油气地质，2008，13（2）：1–16.
[24] 梁狄刚，郭彤楼，边立曾，等. 中国南方海相生烃成藏研究的若干新进展（三）：南方四套区域性海相烃源岩的沉积相及发育的控制因素 [J]. 海相油气地质，2009，14（2）：1–19.
[25] 邹才能，董大忠，王社教，等. 中国页岩气形成机理、地质特征及资源潜力 [J]. 石油勘探与开发，2010，37（6）：641–653.
[26] 刘树根，马文辛，LUBA Jansa，等. 四川盆地东部地区下志留统龙马溪组页岩储层特征 [J]. 岩石学报，2011，27（8）：2239 – 2252.
[27] 蒋裕强，董大忠，漆麟，等. 页岩气储层的基本特征及其评价 [J]. 天然气工业，2010，30（10）：7–12.
[28] 李延钧，刘欢，刘家霞，等. 页岩气地质选区及资源潜力评价方法 [J]. 西南石油大学学报（自然科学版），2011，33（2）：28–34.
[29] 聂海宽，张金川. 页岩气储层类型和特征研究：以四川盆地及其周缘下古生界为例 [J]. 石油实验地质，2011，33（3）：219–225.
[30] 王玉满，董大忠，李建忠，等. 川南下志留统龙马溪组页岩气储层特征 [J]. 石油学报，2012，33（4）：551–561.

[31] 陈尚斌，朱炎铭，王红岩，等. 川南龙马溪组页岩气储层纳米孔隙结构特征及其成藏意义 [J]. 煤炭学报，2012，37（3）：438–444.
[32] 李庆辉，陈勉，金衍，等. 页岩气储层岩石力学特性及脆性评价 [J]. 石油钻探技术,2012,40（4）：17–22.
[33] 郭少斌，黄磊. 页岩气储层含气性影响因素及储层评价：以上扬子古生界页岩气储层为例 [J]. 石油实验地质，2013，35（6）：601–606.
[34] 张汉荣，王强，倪楷，等. 川东南五峰—龙马溪组页岩储层六性特征及主控因素分析 [J]. 石油实验地质，2016，38（3）：320–325.
[35] 李双建，袁玉松，孙炜，等. 四川盆地志留系页岩气超压形成与破坏机理及主控因素 [J]. 天然气地球科学，2016，27（5）：924–931.
[36] 姜振学，唐相路，李卓，等. 川东南地区龙马溪页岩孔隙结构全孔径表征及其对含气性的控制 [J]. 地学前缘，2016，23（2）：126–134.
[37] 赵金洲，许文俊，李勇明，等. 页岩气储层可压性评价新方法 [J]. 天然气地球科学,2015,26（6）：1165–1172.
[38] 赵金洲，沈骋，任岚，等. 页岩储层不同赋存状态气体含气量定量预测：以四川盆地焦石坝页岩气田为例 [J]. 天然气工业，2017，37（4）：27–33.
[39] 陈永武，王少波，韩征，等. DZ/T 0254—2014 页岩气资源 / 储量计算与评价技术规范 [S]. 北京：中国标准出版社，2014.
[40] 郭彤楼，刘若冰. 复杂构造区高演化程度海相页岩气勘探突破的启示 [J]. 天然气地球科学，2013，24（4）：643–651.
[41] 王鹏万，邹辰，李娴静，等. 昭通示范区页岩气富集高产的地质主控因素 [J]. 石油学报，2018，39（7）：744–753.
[42] 邱振，董大忠，卢斌，等. 中国南方五峰组—龙马溪组页岩中笔石与有机质富集关系探讨 [J]. 沉积学报，2016，34（6）：1011–1020.
[43] 腾格尔，申宝剑，俞凌杰，等. 四川盆地五峰组—龙马溪组页岩气形成与聚集机理 [J]. 石油勘探与开发，2017，44（1）：69–78.
[44] 邱振，邹才能，李熙喆，等. 论笔石对页岩气源储的贡献：以华南地区五峰组—龙马溪组笔石页岩为例 [J]. 天然气地球科学，2018，29（5）：606–615.
[45] 张鉴，王兰生，杨跃明，等. 四川盆地海相页岩气选区评价方法建立及应用 [J]. 天然气地球科学，2016，27（3）：433–441.
[46] LOUKS R G，REED R M，RUPPEL S C，et al. Spectrum of pore types and networks in mudrocks and a descriptive classification for matrix-related mudrock pores [J]. AAPG Bull，2012，96：1071–1098.
[47] LOUKS R G，REED R M，RUPPEL S C，et al Morphology，genesis，and distribution of nanometer-scale pores in siliceous mudstones of the Mississippian Barnett shale [J]. J Sediment Res，2009，79：848–865.
[48] CURTIS M E，Ambrose R J，SONDERGELD C H，et al. Microstructural investigation of gas shales in two and three dimensions using nanometer-scale resolution imaging [J]. AAPG Bull，2012，96：665–677.
[49] 王鹏万，李昌，张磊，等. 五峰组—龙马溪组储层特征及甜点层段评价：以昭通页岩气示范区 A 井为例 [J]. 煤炭学报，2017，42（11）：2925–2935.

页岩气井网井距优化

位云生，王军磊，齐亚东，金亦秋

（中国石油勘探开发研究院）

摘　要：基于页岩气“一井一藏”及“工厂化”作业的开发特点，一次性部署开发井是区块效益开发的关键，故井网井距优化对于提高页岩气采收率具有重要意义。以长宁国家级页岩气示范区为例，以单井动态分析结果为依据、以“多井平台”数值模拟为分析手段，建立以平衡基质接触面积、缝间干扰、井间干扰和裂缝—基质流入流出4种渗流关系为核心的井网井距优化方法，并论证井网井距优化流程：（1）通过干扰测试分析和施工参数类比，定性判断井距范围；（2）建立以支撑剂总体积为约束的裂缝参数优化模型，形成页岩气开发井距理论分析方法，定量评价以簇为单元的主裂缝长度、间距、条数、导流能力及裂缝穿透比，确定最优井距；（3）通过网格指数加密精细数值模拟，初步论证了龙马溪组一段采用“W”形的上下两层交错水平井部署的立体开发效果。结果表明：天然裂缝是影响井距优化的关键因素，长宁示范区天然裂缝不发育，现有压裂规模下采用300m井距、纵向采用两套水平井上下“W”形交错部署立体开发，页岩气采收率可提高15%以上。

关键词：页岩气；天然裂缝；井间干扰；水平井；井网井距；储量；采出程度

近十年，中国页岩气开发经历了国际合作评价、现场开发试验和初步规模开发3个阶段，完成了美国数十年才完成的原始积累过程[1]。现阶段的主要任务是如何将有效产量变为规模产量、将单井有效开发变为区块效益开发。

由于开发成本的限制，页岩气均采用平台化部署水平井、“工厂化”钻井和压裂及大规模连续作业方式，以实现经济开发。北美的开发经验表明，采用初期大井距后期加密的“滚动开发”方案，虽然可以降低开风险、保证第一批次气井的生产效果，但后期加密钻井压裂会导致老井与新井的“应力阴影”叠加，新井生产动态远低于预期，形成“1+1＜2”的开发效果[2, 3]。因此，页岩气井网井距必须一次性部署，以保证体积压裂对地层改造效果的最大化。要确保一次部署的合理性：若井网不合理、开发井距偏大，井间储层难以得到有效体积改造，剩余储量可能永远留在地下[4]；若开发井距偏小，压裂干扰风险加大，压力干扰也将加剧，严重影响开发效益[5]。以长宁页岩气示范区为例，以单井动态分析结果为依据、以“多井平台”数值模拟为分析手段，综合论证井网井距优化过程，建立适用于中国南方海相页岩气井网井距优化方法及流程。

1　井距优化方法与流程

页岩气井需要进行大型体积压裂改造后才能实现高产，体积压裂改造后，SRV（Stimulated Reservoir Volume）区内的储层结构复杂[6]，储层特征具有随机性，井距优化的影响因素较多，因此，基于类似“云分析”的论证思路，采用多种方法综合论证井距[7, 8]。

1.1 基于干扰测试的定性判断

干扰测试是判断井距是否合理的一种重要手段，但对于基质极其致密、靠体积压裂后复杂网络裂缝开发的页岩气来说，井间压力干扰对井距优化的影响不同[9]。为了研究页岩储层压力干扰对井距的影响，分析了几种干扰情况：（1）天然裂缝不发育，无人工裂缝连通，且页岩基质渗透率低至 10^{-5}mD，仅基质干扰，干扰程度弱，井距可进一步减小。（2）天然裂缝不发育，部分人工裂缝连通。人工裂缝连通后，短期内就会产生压力干扰，但由于基质渗透率低、各条主裂缝的流动范围是相对独立的，部分人工裂缝连通仍是局部的，井距仍可进一步优化。（3）天然裂缝发育，井间裂缝连通。天然裂缝发育的储层，有效渗流能力大大提高，甚至可达到常规中高渗气藏的级别，井间干扰是大面积的，会对平均单井 EUR 产生严重影响[2, 10]，因此，天然裂缝是影响井距优化的关键因素。

1.2 基于地质和施工参数类比的定性判断

最大限度地动用地质储量是气藏开发井距优化的主要目的。对于常规气藏来讲，有效储层连续性、连通性、渗透性等地质参数是定性判断开发井距是否合理的重要依据。页岩基质储层连续性较好，但连通性及渗透性极差，部分区块天然裂缝发育，主要通过长水平井、多簇射孔、大规模加砂压裂实现有效开发[11]，因此，形成的人工裂缝形态尺寸是决定开发井距的关键。天然气裂缝发育、两向水平应力差小，形成的裂缝网络复杂程度高，相同的液量和支撑剂量，改造范围小但改造程度高、泄流面积大；天然裂缝不发育、两向水平应力差大，形成的裂缝网络复杂程度低，即易形成主裂缝，相同的液量和支撑剂量，改造范围大但改造程度低、泄流面积小[12]。因此，天然裂缝、两向水平应力差等地质情况及钻井、压裂等施工参数是定性判断页岩气井开发井距的重要依据。

类比国内外已开发页岩气区块的天然裂缝发育情况、两向水平应力差、水平井长度、单段/簇压裂液量、支撑剂用量，可定性判断新区开发井距的合理程度。

1.3 基于产能评价模型的定量计算

目前国内外页岩气开发井距的论证主要采用定性判断、数值模拟方法[13]来大致确定，然后通过现场试验和测试分析进行调整，这种方法的优点是与地质条件直接结合，缺点是盲目性大、时间长、费用高。笔者提出在定性判断的基础上，采用理论方法进行井距优化。

1.3.1 数学模型

在 SRV 区内储层结构分析的基础上（图 1），提出主裂缝采用离散模型、SRV 区内采用分形模型、外围基质和天然微裂缝采用连续模型的思路，建立以簇为基本单元的（拟）稳态产能评价数学模型。

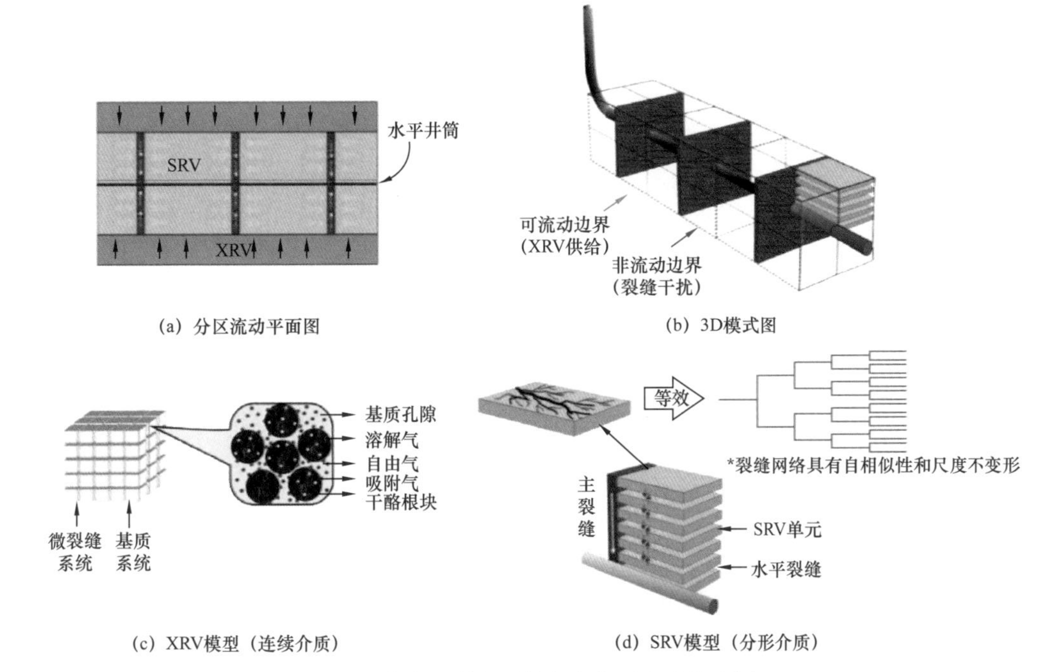

图 1　分段压裂水平井复合分区流动模型图

笔者采用的数学模型是已有成果的改进形式，这里只给出模型中的关键步骤，相关参量的具体表达形式可参考文献［14］。

1.3.1.1　XRV 区域内数学模型

在 XRV（External Reservoir Volume）区域内，流动空间假设满足双孔介质特征，可直接引用已有研究成果[14]，最终的无量纲压力表达形式满足：

$$\frac{\partial^2 \tilde{m}_{XRVD}}{\partial x_D^2} = \frac{s}{\eta_{XRVD}} \tilde{m}_{XRVD} \tag{1}$$

其中，m_{XRVD} 为 XRV 区气体拟压力；x_D 为 XRV 区内垂直于水平井筒方向的无因次流动距离；s 为 Laplace 空间转换因子；η_{XRVD} 为无因次扩散系数；上标“～”表示物理量在 Laplace 空间下的表达式。

1.3.1.2　SRV 区域内数学模型

在 SRV 区域内，渗透率和孔隙度空间分布规模满足幂律指数形式[15]，据质量守恒定律可以获得 SRV 连续性关系式：

$$\frac{\partial^{\gamma}}{\partial t^{1-\gamma}}\left[\frac{\partial}{\partial y}\left(\frac{K_{SRV} p_{SRV}^{\xi}}{Z_g \mu_g}\frac{\partial p_{SRV}^{\xi}}{\partial y}\right)\right]+\frac{K_{XRV}}{x_{HF}}\left(\frac{p_{XRV}}{Z_g \mu_g}\frac{\partial p_{XRV}}{\partial x}\right)_{x=x_{HF}}=\frac{\varphi_{SRV}^{\xi} p_{SRV} c_g}{Z_g}\frac{\partial p_{SRV}^{\xi}}{\partial t} \tag{2}$$

其中，γ 为分数阶参数；ξ 为流动区标识号；p_{SRV} 为 SRV 区域内压力，Pa；K_{SRV} 为 SRV 区内渗透率，m^2；K_{XRV} 为 XRV 区内渗透率，m^2；p_{SRV} 为 XRV 区域内压力，Pa；φ_{SRV} 为 SRV 内地层孔隙度；x_{HF} 为水力裂缝半长，m；Z_g 为气体偏差系数；μ_g 为气体黏度，Pa·s；t 为时间变量，s；c_g 为气体压缩系数，Pa^{-1}。

将 K_{SRV} 和 φ_{SRV} 定义代入上式，同时引入拟压力定义，可以获得拟压力控制方程：

$$\frac{\partial^{1-\gamma}}{\partial t^{1-\gamma}}\left[\frac{\partial m^2{}_{SRV}}{\partial y^2}+\frac{\partial m_{SRV}}{\partial y}\frac{d_f-d-\theta}{y}\right]=$$
$$\left(\frac{y}{L_{ref}}\right)^{\theta}\left[-\frac{K_{XRV}\varphi_{ref}}{x_{HF}K_{SRV}^{ref}}\left(\frac{\partial m_{XRV}}{\partial x}\right)_{x=x_{HF}}+\frac{\mu_{gi}c_{gt}\varphi_{ref}}{K_{SRV}^{ref}}\frac{\partial^{1-\gamma}}{\partial^{1-\gamma}}\left(\frac{\partial^{\gamma}m_{SRV}}{\partial t^{\gamma}}\right)\right] \tag{3}$$

其中，m_{SRV} 为 SRV 区气体拟压力，Pa；L_{ref} 为参考长度，m；d_f 为分形维数；d 为欧几里得维数；θ 为反常扩散系数；Z_{gi} 为原始压力下气体偏差系数；μ_{gi} 为原始压力下气体黏度，Pa·s；φ_{ref} 为参考孔隙度。

其中，K_{SRV}^{ref} 量纲为 $[L^2T^{1-\gamma}]$。利用无量纲定义处理式（3），则有

$$\frac{\partial^2\tilde{m}_{SRVD}^{\xi}}{\partial y_D^2}+\frac{\partial\tilde{m}_{SRVD}^{\xi}}{\partial y_D}\frac{d_f^{\xi}-d^{\xi}-\theta^{\xi}}{y_D}=y_D^{\theta}\left(\frac{s^{\gamma\xi}}{\eta_{SRVD}^{\xi}}+\frac{\lambda_{SRV,\xi}^{XRV}F_{SRV,\xi}^{XRV}}{s^{1-\gamma\xi}}\right)\tilde{m}_{SRVD}^{\xi} \tag{4}$$

其中，ξ 代表 R（右侧裂缝）和 L（左侧裂缝），式（4）的通解形式为

$$\tilde{m}_{SRVD}^{\xi}(y_D)=y_D^{\alpha_\xi}[C_1I_{n_\xi}(b_\xi y_D^{c_\xi})+C_2K_{n_\xi}(b_\xi y_D^{c_\xi})] \tag{5}$$

其中 I_n 和 K_n 为 n 阶 Bessel 函数（包括整数阶和分数阶）。利用 Fourier 有限余弦变换分别获得 SRV 区域对左右两侧裂缝的供给强度：

$$\left.\frac{\partial m_{SRVD}^{L}}{\partial y_D}\right|_{y_D=w_{HFD}/2}=\tilde{F}_{HF,R1}^{SRV}\tilde{m}_{HFD}^{R}-\tilde{F}_{HF,L1}^{SRV}\tilde{m}_{HFD}^{L} \tag{6}$$

$$\left.\frac{\partial m_{SRVD}^{R}}{\partial y_D}\right|_{y_D=w_{HFD}/2}=-\tilde{F}_{HF,R2}^{SRV}\tilde{m}_{HFD}^{R}+\tilde{F}_{HF,L2}^{SRV}\tilde{m}_{HFD}^{L} \tag{7}$$

其中，w_{HFD} 为无因次裂缝宽度；$\tilde{F}_{HF,R1}^{SRV}$ 为 R1 侧 SRV 区域对裂缝（HF）区域流量供给函数，其他相似符号意义一致。

1.3.1.3 裂缝区域内数学模型

利用物质守恒定律，分别在左右两条裂缝中建立连续性方程：

$$\frac{\partial}{\partial x}\left(\frac{k_{HF}\rho_g}{\mu_g}\frac{\partial p_{HF}^{\xi}}{\partial x}\right)+\frac{2}{w_{HF}}\left[\frac{\partial^{1-\gamma_\xi}}{\partial t^{1-\gamma_\xi}}\left(\frac{k_{SRV}\rho_g}{\mu_g}\frac{\partial p_{SRV}^{\xi}}{\partial y}\right)\right]_{y=\frac{w_{HF}}{2}}=\frac{\partial[\varphi_{HF}\rho_g(p_{HF}^{\xi})]}{\partial t} \tag{8}$$

其中，K_{HF} 为裂缝区域内渗透率，m^2；p_{HF} 为裂缝内压力，Pa；w_{HF} 为裂缝宽度，m；p_{SRV} 为 SRV 区域内压力，Pa；ρ_g 为气体密度，kg/m^3；φ_{HF} 为裂缝孔隙度。

将渗透率、孔隙度分布关系式代入式（8）形成控制方程：

$$\frac{\partial^2\tilde{m}_{HFD}^{\xi}}{\partial x_D^2}+2\frac{\lambda_{HF,\xi}^{SRV}}{s^{\gamma_\xi-1}}\left.\left(\frac{\partial\tilde{m}_{SRVD}^{\xi}}{\partial y_D}\right)\right|_{y_D=w_{fD}/2}=\frac{s}{\eta_{HFD}^{\xi}}\tilde{m}_{HFD}^{\xi} \tag{9}$$

推导获得左右两侧裂缝内的无因次拟压力分布为

$$\tilde{m}_{\mathrm{HFD}}^{\mathrm{L}}(x_{\mathrm{D}})=\tilde{q}_{\mathrm{HFD}}^{\mathrm{L}}AX_{\mathrm{SRV}}^{\mathrm{L}}(x_{\mathrm{D}})+\tilde{q}_{\mathrm{HFD}}^{\mathrm{R}}BX_{\mathrm{SRV}}^{\mathrm{L}}(x_{\mathrm{D}}) \tag{10}$$

$$\tilde{m}_{\mathrm{HFD}}^{\mathrm{R}}(x_{\mathrm{D}})=\tilde{q}_{\mathrm{HFD}}^{\mathrm{L}}BX_{\mathrm{SRV}}^{\mathrm{R}}(x_{\mathrm{D}})+\tilde{q}_{\mathrm{HFD}}^{\mathrm{R}}AX_{\mathrm{SRV}}^{\mathrm{R}}(x_{\mathrm{D}}) \tag{11}$$

其中，m_{HFD} 为裂缝区域内无因次气体拟压力；$\tilde{q}_{\mathrm{HFD}}^{\mathrm{L}}$ 为左侧 SRV 区域对裂缝区域的流量供给函数；$AX_{\mathrm{SRV}}^{\mathrm{R}},AX_{\mathrm{SRV}}^{\mathrm{L}},BX_{\mathrm{SRV}}^{\mathrm{R}},BX_{\mathrm{SRV}}^{\mathrm{L}}$ 为对应的压力分布函数。详细表达形式可参考文献[14]。

1.3.1.4 多裂缝模型

将多裂缝系统分解为一系列的单裂缝渗流单元，压降叠加原理广泛适用于线性偏微分数学方程中，依据压降叠加原理可以获得由 n_{f} 条裂缝引起的压力值，最终形成以各裂缝产量、井底压力为未知量的相同形式矩阵方程组：

$$\begin{bmatrix}\boldsymbol{A}-\boldsymbol{B}\\ \boldsymbol{B}^{\mathrm{T}}\quad 0\end{bmatrix}\cdot\begin{bmatrix}\boldsymbol{X}\\ s\tilde{m}_{wD}\end{bmatrix}=\begin{bmatrix}\boldsymbol{0}\\ 1\end{bmatrix} \tag{12}$$

其中，$\boldsymbol{A}$ 为系数矩阵，$\boldsymbol{B}$ 为系数为 1 的行列式，$\boldsymbol{B}^{\mathrm{T}}$ 为 $\boldsymbol{B}$ 的转至行列式，$\boldsymbol{X}$ 为未知量行列式，m_{wD} 为无因次井底压力，$\boldsymbol{0}$ 为系数为 0 的行列式。

利用 Stehfest 数值反演方法结合 Newton 迭代算法可快速求解拉普拉斯线性方程组式（12）。

1.3.2 优化方法

影响水平井产能的因素较多且不独立，各因素间相互干扰，是典型的非线性问题[16]。以式（12）数学模型为基础，建立以支撑剂总体积为约束的裂缝参数优化方法：

$$N_{\mathrm{prop}}=\frac{2K_{\mathrm{f}}}{K_{\mathrm{m}}}\frac{V_{\mathrm{prop}}}{V_{\mathrm{res}}}=\frac{x_{\mathrm{e}}}{y_{\mathrm{e}}}\sum_{m=1}^{n_{\mathrm{f}}}\left(\frac{2x_{\mathrm{fi}}^{m}}{x_{\mathrm{e}}}\right)^{2}\frac{K_{\mathrm{f}}w_{\mathrm{f}}^{m}}{K_{\mathrm{m}}\mathrm{x}_{\mathrm{f}}^{m}}=\sum_{m=1}^{n_{\mathrm{f}}}\left[\frac{(I_{\mathrm{x}}^{m})^{2}C_{\mathrm{fD}}^{m}}{\lambda}\right] \tag{13}$$

其中，K_{f} 为裂缝渗透率，m^2；K_{m} 为地层平均渗透率，m^2；x_{e} 为井距，m；y_{e} 为水平井长度方向控制范围，m；w_{f} 为裂缝宽度，m；I_{x} 为裂缝穿透比；N_{prop} 表示支撑数；V_{prop} 表示支撑剂体积，m^3；V_{res} 表示储层体积，m^3；λ 表示储层几何规模特征参数。

在给定支撑剂量条件下，裂缝长度和无量纲导流能力将会同时争夺支撑剂体积，当两者间达到某种平衡，生产井将达到较高的产能水平。图 2 反映了裂缝条数和间距对产能优化结果的影响：（1）图 2（a）反映了较大压裂段长（无因次穿透比 D_{fD}=0.75）下裂缝条数对气井产能的影响。对于给定的 N_{prop}，在整个无量纲导流能力 C_{fD} 变化范围内，条数较多系统产能均高于条数较少的裂缝系统，主要由于增加的裂缝接触面积提高产能幅度大于随之产生的裂缝干扰降低的产能幅度。两个系统在最优 C_{fDopt} 处产能达到最大值，随后随着 C_{fD} 增加产能减小。（2）图 2（b）反映了较小压裂段长（D_{fD}=0.4）下裂缝条数对气井产能的影响。当裂缝完全穿透地层时（I_{x}=1），裂缝条数对产能影响程度较小，裂缝条数较多情况产能略高。这意味着增加裂缝条数主要抵消了裂缝干扰的影响，而不能显著地提高气井产能。在低支撑数（N_{prop}=1）下，裂缝较多系统产能在整个 C_{fD} 变化范围内均高于裂缝较少的系统。在高支撑数（N_{prop}=10）下，裂缝数较多的产能较高的情况仅分布在高 C_{fD} 范围内（C_{fD}>50）。（3）图 2（c）反映了裂缝间距对产能影响（n_f=4）。在整个 C_{fD} 分布范围内，均匀分布模式（D_{fD}=0.75）下气井产能均高于非均匀布局。在任意固定 N_{prop} 数条件下，不同 D_{fD}（D_{fD}=0.5，0.75，0.9）下最高产能对应的最优 C_{fDopt} 值近似；在整个 C_{fD} 范围内，均匀布局模式高于非均匀布局模式，两种模式间差值随着 C_{fD} 增加而增加。

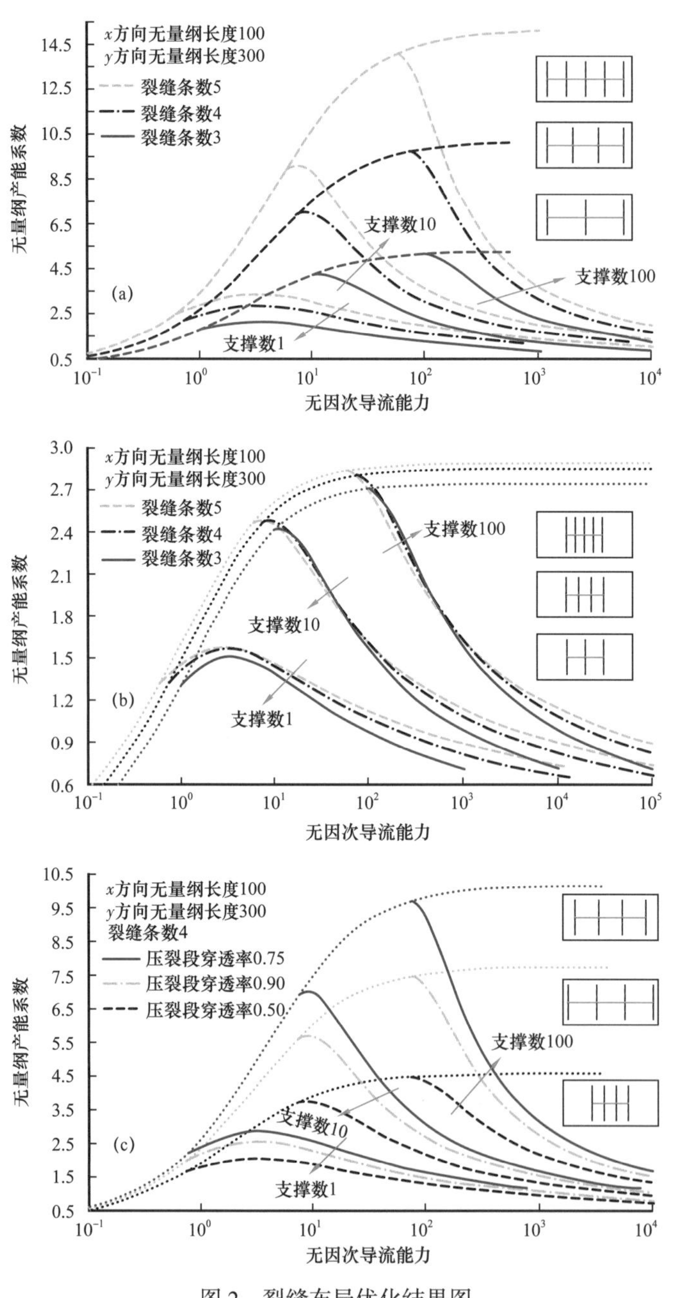

图 2　裂缝布局优化结果图

最终的优化结果通过增加裂缝条数和裂缝长度来增加裂缝系统与地层接触面积、调整裂缝有限导流能力以平衡裂缝内流入和流出关系、调整裂缝间距、裂缝与封闭边界相对位置降低裂缝相互干扰，以此达到最优产能水平。

1.3.3 优化流程

页岩气井通过压裂获取有效泄流面积和产量，压裂规模一定时，裂缝间距、裂缝条数、导流能力直接影响裂缝长度，从而影响开发井距；反过来讲，开发井距不同时，不同的裂缝参数组合获得的开发效果也不同，即裂缝参数、开发井距及裂缝—基质流入、流出动态是相互关联的（图 2）。

通过核心优化算法，调整平衡 4 种渗流关系实现气井（拟）稳态产能最大化，即平衡裂缝系统—基质接触面积、缝间干扰、井间干扰、裂缝—基质流入和流出动态，实现裂缝参数和井距的优化。具体流程如图 3 所示。

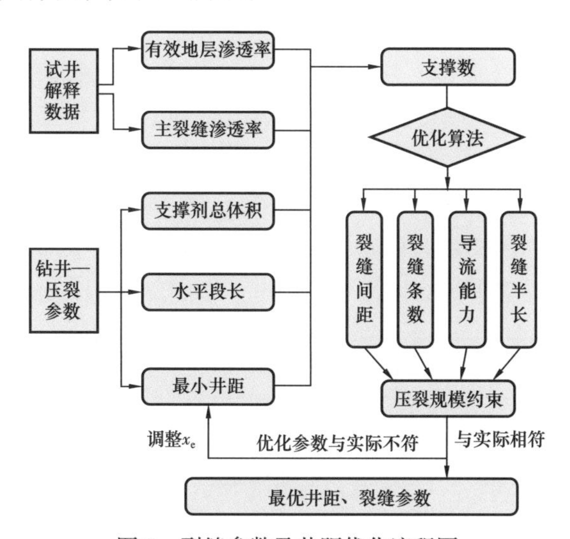

图 3 裂缝参数及井距优化流程图

2 井网部署论证

目前中国南方海相页岩气的开发（龙一$_1^1$小层）主要采用水平井方法，单井实现了效益开发，但区块储量采出程度不足 20%，要实现区块的高效科学开发，急需开展提高页岩气采收率研究；要保持页岩气规模上产和长期稳产又面临有效建产区域有限，提高页岩气采收率也是最现实的需求。

2.1 主裂缝形态分析

压裂模拟及大物模实验表明，射孔点的主裂缝纵向最大可延伸约 40m，远离射孔点，裂缝高度快速减小。同时，不同水平井靶体位置的开发效果表明，水平井筒附近层位压裂后储量动用程度高，远离井筒后大幅度降低。

综合分析认为，垂直于水平井筒的裂缝截面呈“星形”，这与人工裂缝最大延伸约 40m 和生产动态反映主体动用高度约 15m 是吻合的（图 4）。

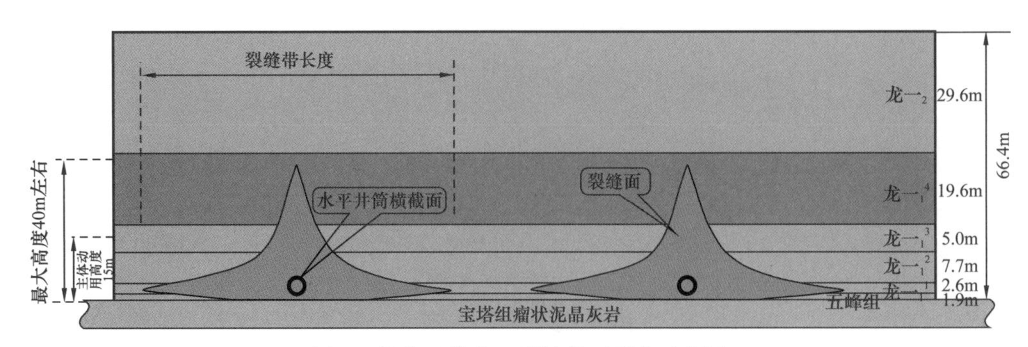

图 4　长宁区块人工裂缝截面形态示意图

2.2　纵向“W”形两层水平井交错部署论证

根据长宁区块各小层的储量分布比例，目前下部实际纵向上动用储量比例约 22%，且从下至上各层储量动用程度逐渐降低，特别是上部龙一$_1^3$和龙一$_1^4$小层动用程度很低，仍有多数可开发储量未有效动用。从储量基础上看，龙一$_1^4$小层含气量整体差于龙一$_1^1$、龙一$_1^2$、龙一$_1^3$、五一段 4 个层位，但平均仍在 $3m^3/t$ 以上，且厚度较大、游离气比例相对较高，具备开发动用的基础；从储层可压性上看，上部龙一$_1^4$小层石英含量与下部已开发层位基本相当，具备规模压裂条件。

在“星形”裂缝截面认识的基础上，采用“W”形的上下两层交错水平井部署对龙一段储层进行立体开发，有利于提高储量动用程度，空间配置关系如图 5 所示。

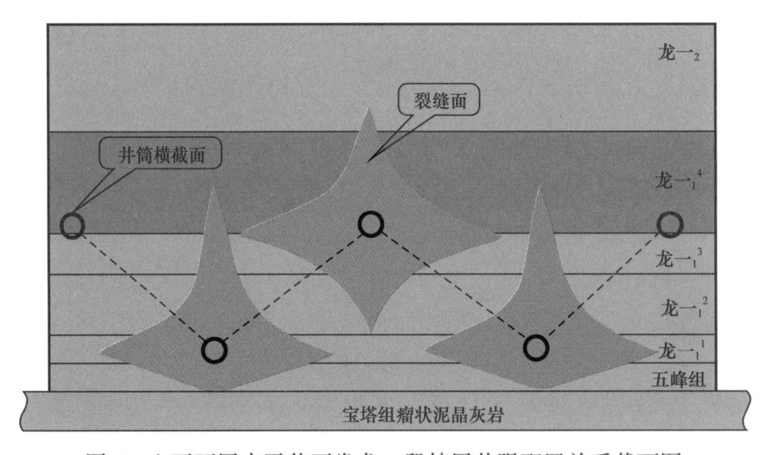

图 5　上下两层水平井开发龙一段储层井眼配置关系截面图

以南方海相页岩储层实际小层参数和裂缝分布形态认识为基础，建立储层和裂缝模型，通过裂缝网络指数加密处理和数值模拟，论证上下储层两层水平井“W”形交错部署的开发效果，并分两种情况进行了经济效益评价。

3　应用实例

以长宁 201 井区为例，目前完钻平台均采用一套水平井开发，井距以 400m 和 500m 为主，个别井距 300m，其他基本参数如表 1 所示。

表 1　长宁区块页岩储层原始参数表

参数	数值
目的层垂深（m）	2300
原始地层压力（MPa）	46
原始地层温度（K）	371
原始气体黏度（mPa·s）	0.023
原始气体偏因子	1.08

H2、H3、H6、H9 4 个平台 10 口井的干扰测试结果表明：若无跨井天然裂缝，300m、400m 和 500m 井距均未见到压力干扰。H6-4、H6-6、H9-6、H4-6、H12-4 井 5 口井压力恢复试井解释表明：主裂缝半长介于 36～64m，天然裂缝不发育的区域，300m 井距未见明显干扰；天然裂缝发育的区域，存在不同程度的井间干扰。成像测井（FMI）资料和岩心描述结果表明：长宁区块天然裂缝不发育（表 2）。因此，邻井部分人工裂缝沟通产生压裂干扰对井距优化影响较小。对比美国 4 大页岩气开发区块，长宁区块单段压裂液量和单簇加砂量与美国基本相当（表 3），但井距接近美国的两倍，因此，长宁区块开发井距存在优化空间。

表 2　长宁区块五峰组—龙一$_1^4$ 层天然裂缝发育情况表

井号	裂缝发育条数（条）		裂缝线密度（条 /m）	平均线密度（条 /m）
	高阻缝	高导缝		
宁 203	10		0.25	0.24
宁 209	7	1	0.19	
宁 210	8	2	0.25	

在以上定性判断的基础上，利用式（12）和图 3，计算长宁 H2 平台两口井目前的井距均为 500m，裂缝半长分别为 147.5m 和 122.1m，裂缝穿透率介于 0.9～0.95，合理井距分别介于 311～328m 和 257～271m，优化井距后，储量采出程度分别提高 7.9% 和 10.9%（表 4）。

表 3　长宁区块与美国 4 大页岩气区块井距对比表

区块	Barnett	Haynes-ville	Marcellus	Eagle Ford	长宁
水平段长度（m）	1219	1402	1128	1494	1496
天然裂缝情况	发育	不发育	较发育	发育	不发育
两向水平应力差（MPa）	1～2	8～14	6～9	7～11	10～13
单段压裂液量（m^3）	2720	1590	1590	2130	1910
单段支撑剂量（t）	129.7	162.3	181.2	112.6	97.8
单簇支撑剂量（t）	32.4	32.5	36.2	28.2	32.6
平均井距（m）	280	260	260	300	400～500

注：长宁数据截至 2017 年 6 月 30 日，美国 4 大页岩气区块数据截至 2015 年底。

表 4　长宁区块典型气井井距分析表

井名	水平段长（m）	EUR（10^8m^3）	采出程度（%）	优化井距（m）	优化后采出程度（%）
长宁 H2-7	1 500	0.97	14.0	311～328	21.9
长宁 H2-6	1 350	0.75	11.8	257～271	22.7

按照长宁区块五峰组—龙一段纵向储量及参数分布（表 5），模型宽度和长度分别为 1207.5m，2076.8m，开发井距为 301.7m，水品压裂段长为 1475m，人工裂缝“星形”展布，得到水平井的裂缝及控制参数见表 6。

表 5　长宁区块五峰组—龙一段各小层纵向储量及参数分布表

层位	有效厚度（m）	岩石密度（t/m^3）	总含气量（m^3/t）	储量丰度（$10^8m^3/km^2$）	储量比例（%）	等效基质孔隙度（%）	基质渗透率（$10^{-4}mD$）	裂缝孔隙度（%）	裂缝渗透率（mD）
龙一$_2$	29.6	2.62	2.27	1.68	21.9	1.83	0.16	3.5	800
龙一$_1^4$	19.6	2.61	5.67	2.76	35.9	4.58	0.65	3.5	800
龙一$_1^3$	5.0	2.58	7.74	0.95	12.4	6.21	1.51	3.5	800
龙一$_1^2$	7.7	2.60	7.12	1.36	17.7	5.75	1.19	3.5	800
龙一$_1^1$	2.6	2.57	9.1	0.58	7.6	7.33	2.69	3.5	800
五一	1.9	2.59	7.48	0.35	4.5	6.00	1.36	3.5	800

表 6　水平井、裂缝及控制参数表

模型宽度（m）	模型长度（m）	水平压裂段长（m）	井距（m）	簇间距（m）	裂缝半长（m）	簇数（个）	最小井底压力（MPa）	最小井口日产量（m^3）
1207.5	2076.8	1475	301.7	5	143	60	1.5	3000

通过网格指数剖分和数值模拟（图 6），模拟仅下部一套水平井，靶体位置位于龙一$_1^1$小层中部时，首年平均日产气量 $9\times10^4m^3$，单井 EUR$9897\times10^4m^3$，采收率为 23.8%。上下两层水平井同时开发，靶体位置分别位于龙一$_1^4$小层底部和龙一$_1^1$小层中部，模拟在相同压裂规模和压裂工艺下的首年平均日产量分别为 $5.5\times10^4m^3$ 和 $9\times10^4m^3$，单井 EUR 分别为 $7610\times10^4m^3$ 和 $9670\times10^4m^3$，采收率达到 41.5%。

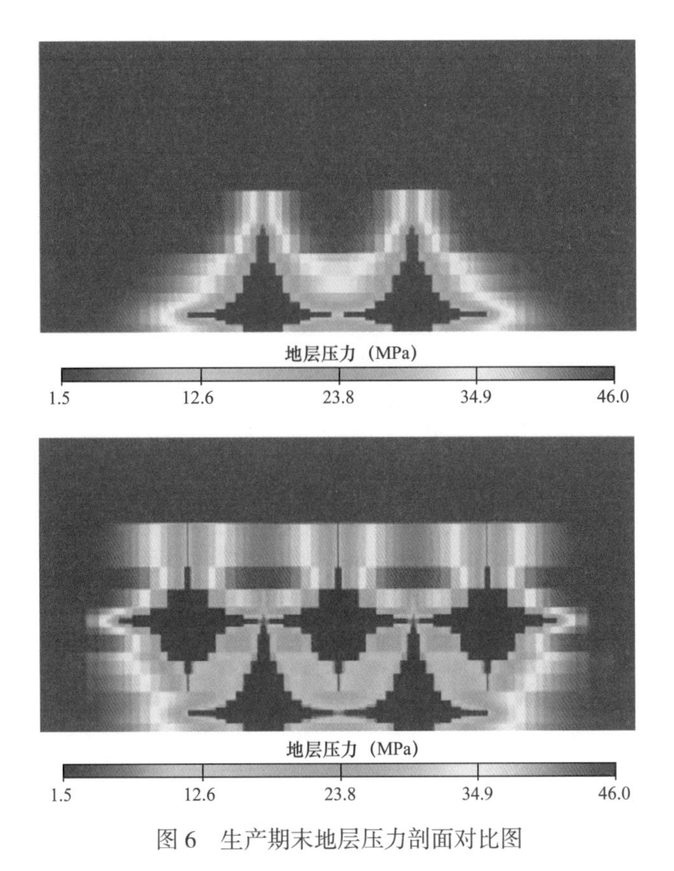

图 6　生产期末地层压力剖面对比图

下部一套水平井开发时，单井 EUR 9897×10^4m^3，单井综合投资 5500 万元，操作成本 300 元 /1000m^3 时，按 2018 年投产井实际补贴计算，评价内部收益率 12.56%（表 7）；若上下两层水平井采用同一井场、同一直井段工厂化钻井、压裂作业，在井场、钻井、地面、钻井液、压裂液、井管理等方面可节约投资 1500 万元，内部收益率也可达 13% 以上（表 7），可以实现高效开发，因此，长宁区块采用上下两层水平井"W"形交错部署立体开发，技术和经济上都是可行的。实施后，与下部一套水平井开发相比，可提高储量采出程度 17.7%，增加可采储量近 2×$10^{12}m^3$。若上部水平井单独部署，同样的单井综合投资，内部收益率低于 2%，当单井综合投资降至 4630 万元，或单井综合投资 5000 万元、操作成本降至 225 元 /1000m^3 时，上部井的内部收益率可达到 8%（表 8）。另一方面，上部井也可通过增大压裂规模，将单井首年平均日产量提高到 6.6×10^4m^3，单井 EUR 提高到 8700×10^4m^3（图 7），内部收益率也可达到 8%。

表 7　两种方案采出程度与收益对比表

方案	综合投资（万元）	EUR（10^8m^3）	采出程度（%）	内部收益率（%）
下部一层水平井	5500	9897	23.8	12.56
上下两层水平井"W"形部署	9500	17280	41.5	13.02

表 8　上下两层水平井“W”形部署时上部井投资和成本界限表

单井综合投资（万元）	操作成本（元 /1000m^3）	内部收益率（%）	
		下部井	上部井
5500	300	12.56	1.97
5000	300	19.38	5.08
4630	300		8
5000	225		8

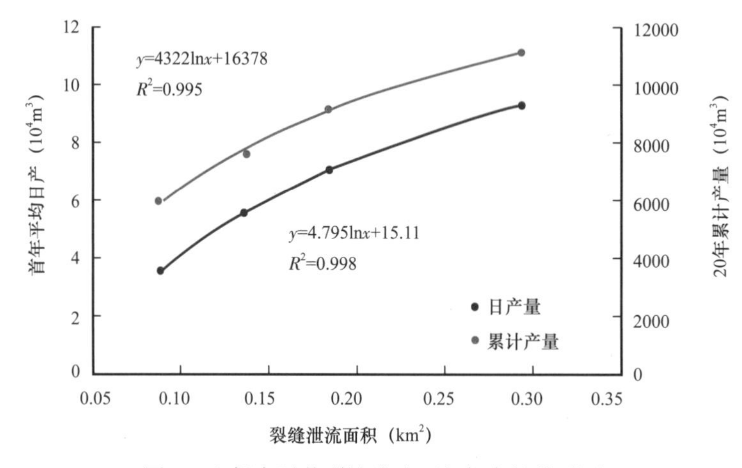

图 7　上部水平井裂缝泄流面积与产量关系图

4　结论与建议

页岩气井网井距优化是提高储量采出程度的重要技术手段，通过现场测试、理论分析及数值模拟研究，主要有如下几点认识与建议：

（1）天然裂缝是影响井距优化的关键因素。天然裂缝不发育区，基质和部分人工裂缝干扰对单井累计产量和井距优化的影响不大；反之，天然裂缝发育区，井间干扰对井距优化有较大影响，尽量避免发生。

（2）以簇为基本单元的（拟）稳态产能评价模型为基础、以支撑剂体积为约束的理论优化方法，实现了页岩气井距优化的定量优化。在干扰测试分析、地质和施工参数类比等定性判断井距的基础上，采用支撑剂体积为约束的理论迭代计算，可定量评价以簇为单元的主裂缝长度、间距、条数、导流能力及裂缝穿透比，从而可确定最优井距。建议长宁区块在目前的压裂规模下，主体采用 300m 井距，局部压裂不完善区，井距可进一步缩小。

（3）上下两层水平井“W”形交错部署是页岩气立体开发、提高储量纵向采出程度的有效手段。在 300m 井距基础上，采用上下两层水平井“W”形交错部署，采收率可提高 15% 以上，若上下两层水平井采用同一井场、同一直井段工厂化钻井、压裂作业，考虑在井场、钻井、地面、钻井液、压裂液、井管理等方面节约的投资，长宁区块采用上下两层水平井交错部署立体开发，技术和经济上都是可行的；若上部井网单独部署，需要进一步降低单井投资成本或增加压裂规模才能实现效益开发。

参考文献

[1] 国家能源局．页岩气发展规划（2016–2020年）[EB/OL]．

[2] Morales A，Zhang K，Gakhar K，et al. Advanced modeling of interwell fracturing interference：an Eagle Ford shale oil study–refracturing [C] //SPE Hydraulic Fracturing Technology Conference，9–11 February 2016，The Woodlands，Texas，USA.DOI：https：//doi.org/10.2118/179177–MS.

[3] Sardinha C M，Petr C，Lehmann J，et al.Determining interwellconnectivity and reservoir complexity through fracpressure hits and production interference analysis [C] //SPE/CSUR Unconventional Resources Conference–Canada，30 September–2 October 2014，Calgary，Alberta，Canada. DOI：https：//doi.org/10.2118/171628–MS.

[4] 贾爱林，位云生，金亦秋．中国海相页岩气开发评价关键技术进展[J]．石油勘探与开发，2016，43（6）：949–955.

[5] Yu W，Wu K，Zuo LH，Tan XS，Weijermars R. Physical models for inter–well interference in shale reservoirs：relative impacts of fracture hits and matrix permeability [C]．Unconventional Resources Technology Conference，1–3 August 2016，San Antonio，Texas，USA.DOI：URTEC–2457663–MS.

[6] 位云生，贾爱林，何东博，等．中国页岩气与致密气开发特征与开发技术异同[J]．天然气工业，2017，37（11）：43–52.

[7] 何东博，王丽娟，冀光，等．苏里格致密砂岩气田开发井距优化[J]．石油勘探与开发，2012，39（4）：458–464.

[8] 何东博，贾爱林，冀光，等．苏里格大型致密砂岩气田开发井型井网技术[J]．石油勘探与开发，2013，40（1）：79–89.

[9] Awada A，Santo M，Lougheed D，et al. Is that interference?A work flow for identifying and analyzing communication through hydraulic fractures in a multiwall pad [J] .SPE Journal，2016，21（5）：1–13.

[10] Yu W，Xu Y F，Weijermars R，et al. Impact of well interference on shale oil production performance：A numerical model for analyzing pressure response of fracture hits with complex geometries [C] // SPE Hydraulic Fracturing Technology Conference and Exhibition，24–26 January 2017，The Woodlands，Texas，USA.DOI：https：//doi.org/10.2118/184825–MS.

[11] 贾成业，贾爱林，何东博，等．页岩气水平井产量影响因素分析[J]．天然气工业，2017，37（4）：80–88.

[12] Suarez M，Pichon S. Completion and well spacing optimization for horizontal wells in pad development in the VacaMuertashale [C] //SPE Argentina Exploration and Production of Unconventional Resources Symposium，1–3 June 2016，Buenos Aires，Argentina. DOI：https：//doi.org/10.2118/180956–MS.

[13] Sahai V，Jachson G，Rai R R，et al. Optimal well spacing configurations for unconventional gas reservoirs [C] //SPE Americas Unconventional Resources Conference，5–7 June 2012，Pittsburgh，Pennsylvania USA. DOI：https：//doi.org/10.2118/155751–MS.

[14] Wei Y S, He D B, Wang J L. A coupled model for fractured shale reservoirs with characteristics of continuum media and fractal geometry [C] //SPE Asia Pacific Unconventional Resources Conference and Exhibition, 9–11 November 2015, Brisbane, Australia. DOI : https : //doi.org/10.2118/176843–MS.
[15] Wang W D, Shahvali M, Su Y L. A semi–analytical fractal model for production from tight oil reservoirs with hydraulically fractured horizontal wells [J]. Fuel, 2015, 158: 612–618.
[16] 王军磊，贾爱林，位云生，等.有限导流压裂水平井拟稳态产能计算及优化[J]. 中国石油大学学报(自然科学版)，2016，40(1)：100–107.

基于最优 SRV 的页岩气水平井压裂簇间距优化设计

任 岚[1]，林 然[1]，赵金洲[1]，吴雷泽[2]

（1. "油气藏地质及开发工程"国家重点实验室·西南石油大学；
2. 中国石化江汉油田分公司石油工程技术研究院）

摘 要：水平井分段压裂时的簇间距大小对页岩压裂改造具有重要影响，过小的簇间距设计将导致分簇主裂缝之间的改造区重叠，降低压裂改造效率；而过大的簇间距设计则会在主裂缝之间产生未改造区，影响储层的动用程度。目前的簇间距优化设计主要基于静态模式下进行且以储层潜在改造区为目标，没有发展与实际压裂过程相吻合并以动态改造体积（SRV）为目标的簇间距设计方法。为此，考虑裂缝扩展、压裂液滤失和应力干扰相互耦合的作用机制，建立了分簇裂缝扩展下的动态 SRV 计算模型，以最优 SRV 为体积压裂设计的目标，对簇间距进行优化设计。在国内涪陵页岩气藏开展了簇间距优化实例应用，验证了所提方法的可靠性，分析了影响最优簇间距的关键地质工程参数，并针对南、北不同区块不同地质特征，分别绘制了最优簇间距设计参考图版。结论认为，提出的簇间距优化设计方法对改善目前分簇射孔的盲目性、指导页岩气藏体积压裂优化设计具有重要意义。

关键词：页岩气；水平井；分段压裂；簇间距；储层；改造体积；数学模型；最优化方法

页岩气作为当今重要的非常规油气资源之一，在全球范围内受到了日益广泛的关注[1]。然而页岩气藏具有特低渗透[2, 3]、天然裂缝发育[4, 5]等特点，常规的压裂增产技术和开发方式已无法满足其经济开发的基本要求[6, 7]，需要借助新的增产工艺——水平井分段体积压裂技术，页岩气才能实现高效商业化开发[8-10]。近年来，得益于体积压裂技术的快速发展及推广应用，2014 年页岩气全球产量已突破 $2900\times10^8\mathrm{m}^3$，占天然气总产量超过 8%。

页岩气藏体积压裂技术通过水平井筒分段分簇设计，采用大排量、大液量、低砂比实施大规模水力压裂[11]。在逐段的压裂过程中，多条水力裂缝同时从不同的射孔簇点起裂延伸[12]，并激活和沟通附近储层中天然裂缝[13]，形成储层改造体积（SRV）[14]，从而提高页岩储层整体渗透性，改善页岩气的流动通道，提高单井产量[15]。水平井分簇射孔间距是体积压裂设计的关键参数，簇间距大小对页岩压裂改造具有重要影响，过小的簇间距设计将导致分簇主裂缝之间的改造区重叠，降低压裂改造效率；过大的簇间距设计将会在主裂缝之间产生未改造区，影响储层的动用程度。因此，合适的簇间距设计对提升储层缝网展布区域，提高压后效果具有重要意义。

国内外学者已针对页岩气水平井簇间距设计开展了相关研究，Cheng[16-18]和 Guo 等[19-21]分析了分簇裂缝延伸时，中间裂缝受到挤压而开度受限、延伸不足，导致加砂失败或缝网展布不足；以避免此类情况为目的，定性地进行了簇间距优化。Roussel 等[22-26]

考虑到多裂缝引发地应力变化，形成应力转向区域，将其视为储层潜在改造区，以此为最优化目标，定量地进行了簇间距的优化。以上簇间距设计均未以影响页岩压后产量的关键参数——SRV 为目标，然而，簇间距设计的最终目的在于最大化地提高压后产量，而根据页岩气压裂研究表明，压裂井产量与 SRV 呈现出显著的正相关关系[27, 28]。所以，直接以 SRV 为优化目标，对簇间距进行优化设计更为合理和有效。

要以 SRV 为目标优化簇间距，建立 SRV 的计算模型方法是实现簇间距优化设计的前提。目前 SRV 的评价方法主要有：微地震监测[29-32]、倾斜仪监测[33]、半解析方法[34, 35]等。其中，微地震和倾斜仪监测属于压裂实时评价，成本高昂，且不能用于压裂设计；半解析方法简单易用，但模型过于简化，未考虑主裂缝延伸情况，且模型中含有不确定的经验性参数。因此，以上方法难以实现对 SRV 的准确定量表征，基于 SRV 形成过程中的实际物理演化机制，提出了一种新的计算模型和方法，并以此 SRV 模型为基础，针对国内涪陵页岩气示范区内体积压裂，进行了簇间距优化设计，开展实例计算，研究分析了主要工程地质参数对最优簇间距的影响，并基于示范区内地质特征，绘制了最优簇间距设计图版。

1 SRV 计算模型

水平井分段分簇压裂过程中，多簇主裂缝同时延伸，产生诱导应力，进而改变其附近的地应力场分布；另一方面，压裂液沿着天然裂缝不断向储层中滤失，导致其附近的流体压力场不断升高。在地层应力场和流体压力场不断改变下，主裂缝附近区域内的天然裂缝发生失稳破坏，形成高导流区域，即为储层改造体积（SRV）。

结合裂缝延伸模型、裂缝诱导应力场模型、储层流体压力场模型和天然裂缝破坏准则，建立了一套新的 SRV 计算模型，能够对天然裂缝破坏区进行定量计算和表征。

1.1 裂缝延伸模型

针对水平井分段多簇压裂，建立页岩水平井分段多簇压裂的物理模型，并建立三维坐标系，其中 x 轴沿最小主应力方向，y 轴沿最大主应力方向，z 轴沿垂向应力方向。模型考虑单段压裂时多簇人工主裂缝同时从射孔簇点起裂，沿最大主应力方向延伸。

针对椭圆截面裂缝中的单相流动，Lamb[36]推导出了沿缝长方向上的缝内压力降落方程，并由 Nolte[37]进行了修正。即

$$q(y)=-\frac{\pi h_{\mathrm{f}}(y)W(y)^{3}}{64\mu}\frac{\mathrm{d}\left[p_{\mathrm{f}}(y)-\sigma_{\mathrm{n}}(y)\right]}{\mathrm{d}x} \tag{1}$$

其中，$q(y)$ 表示缝内 y 处位置流量，$\mathrm{m^3/min}$；μ 表示压裂液黏度，mPa·s；$h_{\mathrm{f}}(y)$ 表示 y 处裂缝缝高，m；$W(y)$ 表示延伸路径上 y 处的最大缝宽，m；$p_{\mathrm{f}}(y)$ 表示 y 处的流体压力，MPa；$\sigma_{\mathrm{n}}(y)$ 表示 y 处水力裂缝面受到的正应力，MPa；y 表示裂缝长度方向坐标值，m。

常规单缝压裂时，通常基于 England 等[38]推导的缝宽方程计算水力裂缝宽度：

$$W(y)=\frac{2\left(1-\nu^{2}\right)h_{\mathrm{f}}(y)\left[p_{\mathrm{f}}(y)-\sigma_{\mathrm{n}}(y)\right]}{E} \tag{2}$$

其中，E 表示岩石弹性模量，MPa；v 表示岩石泊松比。

但是，水平井分段压裂时，通常会采取多簇射孔策略，多条水力裂缝会从各个簇射孔点起裂并同时在地层中延伸。由于地层由弹性岩石构成，裂缝张开引起的诱导应力会使裂缝附近产生明显的应力阴影区域，造成较强的应力干扰现象[26]。因此，水平井分段分簇压裂时，基于经典弹性力学的缝宽方程将不再适用，故本模型将采用位移不连续方法（DDM）计算各条水力裂缝开度，计算方法后面阐述。

裂缝内任意位置点流量变化量等于该点裂缝体积变化量和滤失量之和，可得物质平衡方程[39]为

$$\frac{\pi}{64\mu}\frac{\partial}{\partial y}\left\{h_{\mathrm{f}}(y)W(y)^{3}\frac{\partial\left[p_{\mathrm{f}}(y)-\sigma_{\mathrm{n}}(y)\right]}{\partial y}\right\}=\frac{2h_{\mathrm{f}}(y)c_{\mathrm{L}}}{\sqrt{t-\tau(y)}}+\frac{\pi}{4}\frac{\partial\left[h_{\mathrm{f}}(y)W(y)\right]}{\partial t} \tag{3}$$

其中，$\tau(y)$ 表示裂缝中 y 点开始滤失的时间，min；c_{L} 表示压裂液滤失系数，$\mathrm{m/min^{0.5}}$；t 表示压裂施工时间，min。

裂缝高度由裂缝尖端应力强度因子和地层断裂韧性共同确定，裂缝尖端的应力强度因子由裂缝内压力、裂缝高度和地应力计算得到[40]：

$$K_{\mathrm{IC}}=\left[p_{\mathrm{f}}(y)-\sigma_{\mathrm{n}}(y)\right]\sqrt{\pi h_{\mathrm{f}}(y)/2} \tag{4}$$

其中，K_{Ic} 表示页岩断裂韧性，$\mathrm{Pa\cdot m^{0.5}}$。

求解条件为

$$W(y,t)\Big|_{|y|\geqslant L_{\mathrm{f}}(t)}=0 \tag{5}$$

$$-\frac{\pi h_{\mathrm{f}}(y)W(y)^{3}}{64\mu}\frac{\mathrm{d}\left[p_{\mathrm{f}}(y)-\sigma_{\mathrm{n}}(y)\right]}{\mathrm{d}y}\Bigg|_{y=0}=Q_{i} \tag{6}$$

$$Q_{\mathrm{pump}}=\sum_{i=1}^{M}Q_{i} \tag{7}$$

其中，Q_{i} 表示第 i 条水力裂缝内压裂液分流量，$\mathrm{m^3/min}$；L_{f} 表示水力裂缝半长，m；Q_{pump} 表示压裂施工泵注总排量，$\mathrm{m^3/min}$；M 表示水裂缝条数，即射孔簇数。

式（1）～式（4）共同构成关于 p、h_{f}、w_{f} 和 q 等 4 个未知变量的求解方程组，代入式（5）～式（7）边界条件，由于各个方程之间相互隐含求解参数，采用迭代法可对其进行耦合求解[41, 42]。

1.2 裂缝诱导应力场模型

水平井分段分簇压裂过程中，由于多条水力裂缝同时张开，导致地层产生弹性形变，从而产生诱导应力[43]。基于弹性力学理论模型，利用位移不连续方法（DDM）[44, 45]计算由水力裂缝产生的诱导应力场。

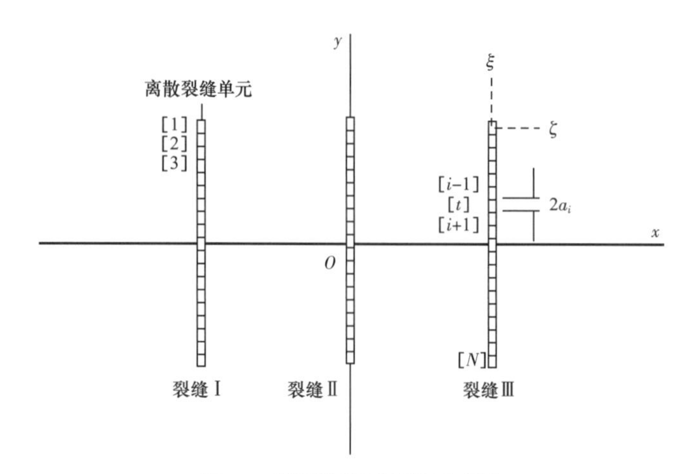

图 1　多裂缝单元离散示意图

建立如图 1 所示 $x—z$ 二维笛卡尔坐标系，将模型中的位移不连续边界，即水力裂缝离散为 N 段，每段长度 $2a_i$。分别以每段中心为原点建立该单元 $\xi—\zeta$ 局部坐标系，其中，ξ 沿离散裂缝单元切向方向，ζ 沿离散裂缝单元法向方向。

首先，建立离散裂缝 i 单元受到所有单元作用下的应力平衡方程组：

$$\left(\sigma_{\mathrm{t}}\right)_i=\sum_{j=1}^{N}\left(A_{\mathrm{tt}}\right)_{ij}\left(\hat{u}_{\mathrm{t}}\right)_j+\sum_{j=1}^{N}\left(A_{\mathrm{tn}}\right)_{ij}\left(\hat{u}_{\mathrm{n}}\right)_j \tag{8}$$

$$\left(\sigma_{\mathrm{n}}\right)_i=\sum_{j=1}^{N}\left(A_{\mathrm{nt}}\right)_{ij}\left(\hat{u}_{\mathrm{t}}\right)_j+\sum_{j=1}^{N}\left(A_{\mathrm{nn}}\right)_{ij}\left(\hat{u}_{\mathrm{n}}\right)_j \tag{9}$$

其中，$(\sigma_{\mathrm{t}})_i$、$(\sigma_{\mathrm{n}})_i$ 分别表示 i 单元在局部坐标系内所受切应力和正应力，MPa；$(\hat{u}_{\mathrm{t}})_j$、$(\hat{u}_{\mathrm{n}})_j$ 分别表示 j 单元在局部坐标系内的切向应变和法向法向，m；$(A_{\mathrm{tt}})_{ij}$、$(A_{\mathrm{nt}})_{ij}$、$(A_{\mathrm{tn}})_{ij}$、$(A_{\mathrm{nn}})_{ij}$ 分别表示 j 单元切向位移和法向位移不连续量分别在 i 单元上引起的切向应力分量和法相应力分量，i，j 取值 1～N；N 表示裂缝离散单元总数。

假设水力裂缝处于张开状态，且内部净压力（p_{net}）为缝内压力（p_{f}）与最小主应力（σ_{h}）之差，则任意 i 单元应力边界条件如下：

$$\left(\sigma_{\mathrm{t}}\right)_i=0 \tag{10}$$

$$\left(\sigma_{\mathrm{n}}\right)_i=-p_{\mathrm{net}}=-(p_{\mathrm{f}}-\sigma_{\mathrm{h}}) \tag{11}$$

其中，p_{net} 表示裂缝内净压力，MPa。

根据裂缝离散单元应力边界条件，联立式（8）、式（9）进行求解。由于共划分离散单元 N 个，故该方程组总共有 $2N$ 个线性方程，包含未知数$(\hat{u}_{\mathrm{t}})_i$和$(\hat{u}_{\mathrm{n}})_i$共 $2N$ 个，故方程组存在唯一解。其中，裂缝单元法向应变$(\hat{u}_{\mathrm{n}})_i$即为裂缝单元开度 w_{f}，将其带入裂缝延伸模型中，即可对裂缝几何参数与压力分布进行耦合求解。

求解得出$(\hat{u}_{\mathrm{t}})_i$和$(\hat{u}_{\mathrm{n}})_i$后，代入以下方程中进行求和，即可计算出坐标平面域内任一点的诱导应力分量：

$$\Delta\sigma_{xx}=\frac{G\hat{u}_{\mathrm{n}}}{2\pi(1-\nu)}\left[2nlF_3+\left(n^2-l^2\right)F_4+\zeta\left(lF_5+nF_6\right)\right] +\frac{G\hat{u}_{\mathrm{t}}}{2\pi(1-\nu)}\left[2n^2F_3-2nlF_4+\zeta\left(nF_5-lF_6\right)\right] \tag{12}$$

$$\Delta\sigma_{yy}=\frac{G\hat{u}_{\mathrm{n}}}{2\pi(1-\nu)}\left[2nlF_3+\left(n^2-l^2\right)F_4-\zeta\left(lF_5+nF_6\right)\right] -\frac{G\hat{u}_{\mathrm{t}}}{2\pi(1-\nu)}\left[2l^2F_3+2nlF_4+\zeta\left(nF_5-lF_6\right)\right] \tag{13}$$

$$\Delta\sigma_{xy}=\frac{G\hat{u}}{2\pi(1-\nu)}\zeta\left(lF_6-nF_5\right)+\frac{G\hat{u}}{2\pi(1-\nu)}\left[F_4+\zeta\left(lF_5+nF_6\right)\right] \tag{14}$$

$$\Delta\sigma_{zz}=\nu\left(\Delta\sigma_{xx}+\Delta\sigma_{yy}\right) \tag{15}$$

其中，$\Delta\sigma_{xx}$、$\Delta\sigma_{yy}$、$\Delta\sigma_{xy}$、$\Delta\sigma_{zz}$ 分别表示诱导应力分量，MPa；G 表示地层剪切模量，MPa；n 表示全局坐标 z 轴与离散单元局部坐标 ζ 轴夹角余弦值，无量纲；l 表示全局坐标 x 轴与离散单元局部坐标 ξ 轴夹角余弦值，无量纲；F_{k} 表示系数方程。

以上各式中的系数方程 F_1～F_6 和局部坐标转换方程参见不连续位移方法文献［44］。

由于原始地应力场和诱导应力场均为三维空间中的二阶张量场，其分量可以进行线性叠加。所以，根据式（12）～式（15）计算得到诱导应力后，可利用叠加原理计算当前地应力场，地层中任意点当前应力张量可表示为

$$\begin{bmatrix}\sigma_{xx} & \sigma_{xy} & \sigma_{xz}\\ \sigma_{yx} & \sigma_{yy} & \sigma_{yz}\\ \sigma_{zx} & \sigma_{zy} & \sigma_{zz}\end{bmatrix}=\begin{bmatrix}\sigma_{xx}^{(0)}+\Delta\sigma_{xx} & \sigma_{xy}^{(0)} & \sigma_{xz}^{(0)}+\Delta\sigma_{xz}\\ \sigma_{yx}^{(0)} & \sigma_{yy}^{(0)}+\Delta\sigma_{yy} & \sigma_{yz}^{(0)}\\ \sigma_{zx}^{(0)}+\Delta\sigma_{xz} & \sigma_{zy}^{(0)} & \sigma_{zz}^{(0)}+\Delta\sigma_{zz}\end{bmatrix} \tag{16}$$

其中，$\sigma_{xx}^{(0)}$、$\sigma_{yy}^{(0)}$、$\sigma_{zz}^{(0)}$、$\sigma_{xy}^{(0)}$、$\sigma_{xz}^{(0)}$、$\sigma_{yz}^{(0)}$ 分别表示初始地应力值分量，MPa；、σ_{xx}、σ_{yy}、σ_{zz}、σ_{xy}、σ_{xz}、σ_{yz} 分别表示当前地应力值分量，MPa。

1.3 储层流体压力场模型

页岩储层基质渗透率极低，天然裂缝发育，没有明显的底层与盖层，故将其在垂向上视作具有各向异性的双重介质巨厚储层。水力压裂过程中，压裂液会从人工裂缝壁面进入储层中，使储层内压力升高。于是，可将人工裂缝视为储层中的面源。

利用 Green 函数源函数方法，求解出定排量条件下，Laplace 域内的储层压力场增量随时间的变化：

$$\Delta\bar{p}(x,y,z,s)=\frac{2\mu h_{\mathrm{r}}}{\pi K_{\mathrm{m}}h_{\mathrm{rD}}s}\sum_{n=1}^{\infty}\frac{1}{n}\sin n\pi\frac{h_{\mathrm{f}}}{2h_{\mathrm{r}}}\sin n\pi\frac{z_{\mathrm{w}}}{h_{\mathrm{r}}}\sin n\pi\frac{z}{h_{\mathrm{r}}} \times\int_{-L_{\mathrm{f}}/L}^{+L_{\mathrm{f}}/L}\tilde{q}K_0\left[\sqrt{u+\frac{n^2\pi^2}{h_{\mathrm{rD}}^2}}\sqrt{\left(x_{\mathrm{D}}-x_{\mathrm{wD}}-\alpha\sqrt{K_{\mathrm{m}}/K_{\mathrm{mx}}}\right)^2+\left(y_{\mathrm{D}}-y_{\mathrm{wD}}\right)^2}\right]\mathrm{d}\alpha \tag{17}$$

其中，$\Delta\overline{p}$表示 Laplace 域内压力场，MPa；L 表示任意参考长度，一般取水平井长度，m；h_r 表示油气藏厚度，m；h_{rD} 表示无因次油气藏厚度，无量纲；K_{mx} 表示基质系统 x 方向上渗透率，mD；K_m 表示基质系统等效渗透率，mD；$\tilde{q}$ 表示裂缝壁面任意点单位面积流量（随缝长方向变化），m/min；Q 表示泵注排量，m^3/min；s 表示 Laplace 变量；α 表示裂缝无因次长度积分变量，无量纲；K_0 表示 0 阶 Bessel 函数；u 表示自定义函数；z_w 表示井底 z 坐标，m；x_D 表示无因次 x 坐标，无量纲；y_D 表示无因次 y 坐标，无量纲；x_{wD} 表示井底无因次 x 坐标，无量纲；y_{wD} 表示井底无因次 y 坐标，无量纲。以上各中间参数、无因次量和自定义函数 u 的表达式参见储层压力场解析方法文献［46］，任意时刻裂缝几何参数 L_f 和 h_f 可由裂缝延伸模型求解得出。

求解出拉普拉斯域中的储层压力场 $\Delta\overline{p}(x,y,z,s)$ 后，利用 Stehfest 数值反演[47，48]即可得到储层实际压力场 $\Delta p(x,y,z,t)$。

同理，由于储层压力场为三维标量场，可以进行线性叠加。所以，当同时进行多条裂缝延伸时，可利用叠加原理计算出任意时刻的储层压力场：

$$p(x,y,z,t)=p_0(x,y,z,t_0)+\sum_{i=1}^{M}\Delta p_i(x,y,z,t) \tag{18}$$

其中，p 表示当前储层压力场，即天然裂缝系统压力场，MPa；p_0 表示初始储层压力场，MPa；Δp 表示储层压力增量场，MPa；x、y、z 分别表示储层空间坐标，m；t_0 表示初始时间，s，t 表示施工时间，s。

1.4 天然裂缝破坏准则

页岩储层中天然裂缝较为发育，且方向性很强，表现出单轴方向性。在水力压裂过程中，流体压力和地层应力改变引发天然裂缝的张性或剪切破坏，从而形成储层改造体积（SRV），实现体积压裂。

目前对天然裂缝的破坏方程基本是根据 Warpinski 准则[49]得出的天然裂缝张性破坏和剪切破坏的平面二维破坏公式[50]。但是，此类公式只适用于垂直天然裂缝，不具有普适性。基于 Warpinski 准则，通过张量运算，推导出适用于天然裂缝在任意空间方位上的破坏判断准则。

设天然裂缝的单位法线向量为

$$\vec{n}=n_ie_i=\begin{bmatrix} n_x & n_y & n_z \end{bmatrix} \tag{19}$$

其中：

$$n_x=\sin\phi\cos\alpha$$
$$n_y=\sin\phi\sin\alpha$$
$$n_x=\cos\phi$$

其中，$\vec{n}$ 表示天然裂缝单位法线向量，无量纲；ni 表示天然裂缝单位法线向量分量，无量纲，$i\in\{x,y,z\}$；α 表示天然裂缝与水平主应力方向夹角（逼近角），（°）；ϕ 表示天然裂缝与水平面夹角（倾角），（°）。

此时，作用在天然裂缝面上的力为

$$\vec{F}=\bar{\bar{\sigma}}\vec{n}=\sigma_{ij}e_ie_jn_ke_k=\sigma_{ij}n_ke_i\delta_j^k=\sigma_{ij}n_je_i \tag{20}$$

其中，$\bar{\bar{\sigma}}$ 表示地应力二阶对称张量，MPa；σ_{ij} 表示应力张量分量；e_i、e_j、e_k 分别表示标准正交基矢量；$\vec{t}$ 表示天然裂缝壁面所受作用力，MPa；δ 表示 Kronecker 符号；i，j，k 分别表示坐标指标，取值 x、y、z。

将该作用分解到裂缝的法线方向上，即为裂缝壁面受到的正应力值：

$$\sigma_{\mathrm{n}}=\vec{F}\vec{n}=n_ke_k\sigma_{ij}n_je_i=n_k\sigma_{ij}n_j\delta_i^k=n_i\sigma_{ij}n_j \tag{21}$$

其中，σ_{n} 表示天然裂缝壁面所受正应力值，MPa。

根据力的合成原则，得到天然裂缝面上作用力沿裂缝壁面方向的切应力值：

$$\sigma_{\tau}=\sqrt{\vec{F}\vec{F}-\sigma_{\mathrm{n}}^2}=\sqrt{\sigma_{ij}n_je_i\sigma_{ij}n_je_i-\sigma_{\mathrm{n}}^2}=\sqrt{\sigma_{ij}n_j\sigma_{ij}n_j-\sigma_{\mathrm{n}}^2} \tag{22}$$

其中，σ_{τ} 表示天然裂缝壁面所受切应力值，MPa。

根据 Warpinski 准则，天然裂缝张性破坏判别式为

$$p_{\mathrm{nf}}>\sigma_{\mathrm{n}}+S_{\mathrm{t}} \tag{23}$$

其中，p_{nf} 表示天然裂缝内流体压力，等于当前储层压力（p），MPa；S_{t} 表示天然裂缝抗张强度，MPa。

天然裂缝剪切破坏判别式为

$$\sigma_{\tau}>\tau_0+K_{\mathrm{f}}\left(\sigma_{\mathrm{n}}-p_{\mathrm{nf}}\right) \tag{24}$$

其中，K_{f} 表示天然裂缝摩擦系数，无量纲；τ_0 表示天然裂缝内聚力，MPa。

通过式（21）和式（22）计算得到天然裂缝壁面的正应力和剪应力，再分别代入式（23）和式（24）破坏判断准则，即可进行天然裂缝破坏类型的判断。

1.5　SRV 计算流程

综合上述各方程，SRV 计算流程如图 2 所示。即：（1）输入储层地质参数和压裂施工参数，并建立相应的三维笛卡尔直角坐标系；（2）基于裂缝延伸模型与地层应力场模型中的缝宽方程，利用式（1）～式（9）耦合计算出裂缝扩展几何尺寸和空间位置；（3）基于地层应力场模型，利用式（12）～式（16）求解出当前地应力场；（4）基于储层压力场模型，利用式（17）、式（18）求解出当前储层压力场；（5）基于天然裂缝判断准则，利用式（21）～式（24）判断储层内任一点的天然裂缝是否发生破坏；（6）提取储层内所有破坏点的坐标数据及其破坏的类型，利用空间数值积分方法，分别计算张性破坏 SRV 和剪切破坏 SRV，并将两者的空间并集算作总体 SRV；（7）图像处理，输出 SRV 三维空间展布形态，进行下一时步的运算，直至程序结束。

2　实例计算与分析

大量现场数据表明，体积压裂效果与总体 SRV 呈现出显著的正相关性[27, 28]，故可基于以上 SRV 计算模型，以总体 SRV 最大化为优化目标，对射孔簇间距进行定量的最优化研究。

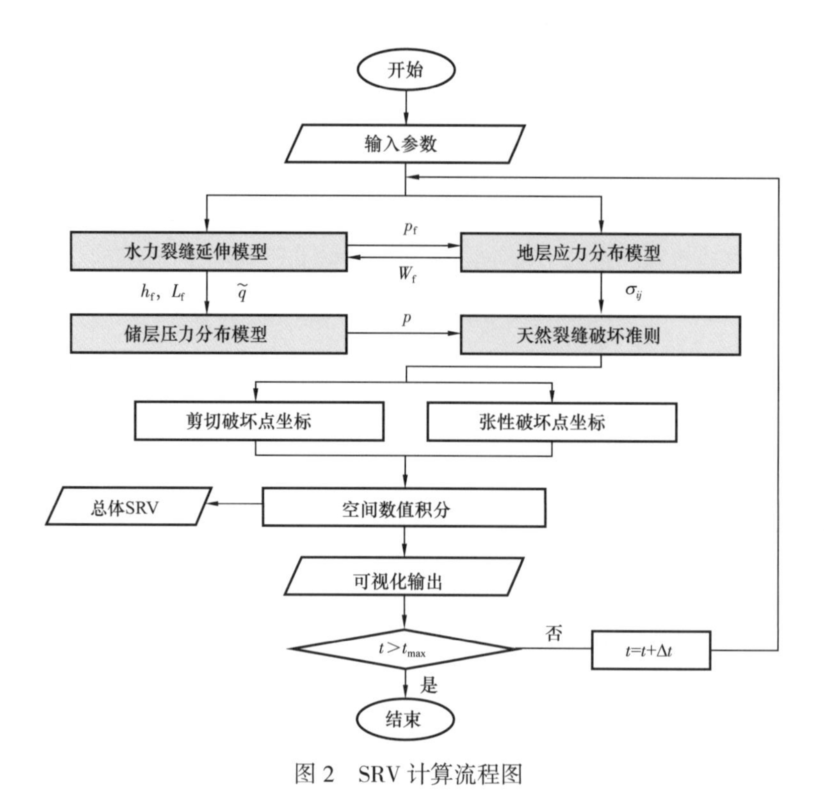

图 2 SRV 计算流程图

2.1 最优簇间距计算

为了验证上述 SRV 数学模型的可靠性与簇间距优化方法的可行性，该模型已在中国涪陵页岩气示范区内开展了实例应用。

涪陵页岩气示范区位于中国西南四川盆地内，是除北美以外最大的商业化页岩气田，其探明页岩气储量 $990\times10^8m^3$，主力产层为上奥陶统五峰组和下志留统龙马溪组海相页岩[51]。该气田借鉴了北美页岩气田（如 Barnett 页岩和 Eagle Ford 页岩）的成功开发经验，采用水平井工厂与分段分簇压裂增产相结合的开发模式，实现了示范区内页岩气的经济高效开采。

涪陵页岩气示范区主要分为南、北两区，两大区块由于埋深与构造差异，地应力分布与岩石力学参数有所不同。其中，北部储层埋深较浅，垂深约为 1500～3000m，地应力差约为 5MPa，杨氏模量约为 30GPa，泊松比约为 0.19；南部储层埋深较深，介于 3000～3500m，地应力差约为 8MPa，杨氏模量约为 25GPa，泊松比约为 0.21。

首先，以涪陵页岩气示范区北区体积压裂为例，进行簇间距优化实例计算。目标气藏地质参数与体积压裂施工具体参数如下：原始地层压力为 36.66MPa，基质渗透率（水平）为 0.0578mD，基质渗透率（垂向）为 0.00578mD，天然裂缝渗透率（水平）为 45mD，天然裂缝渗透率（垂向）为 4.5mD，孔隙度为 3.92%，地层压缩系数为 $4.00\times10^{-4}MPa^{-1}$，综合滤失系数为 $2.20\times10^{-3}m/min^{0.5}$，地层杨氏模量为 30GPa，地层泊松比为 0.19，最大水平应力为 54.86MPa，最小水平应力为 49.56MPa，垂向应力为 60.15MPa，天然裂缝黏聚力为 7MPa，天然裂缝摩擦系数为 0.3，天然裂缝抗张强度为 1MPa，天然裂缝逼近角为 28°，天然裂缝逼倾角为 88°，泵注排量为 $15.5m^3/min$，施工时间为 136min，压裂液黏度为

10mPa·s，射孔簇数为3簇。

基于本文SRV数学模型，编译相应计算程序，结合上表工程地质参数，设置不同簇间距（10m、20m、30m、40m），得到体积压裂后形成的总体SRV平面展布如图3所示。

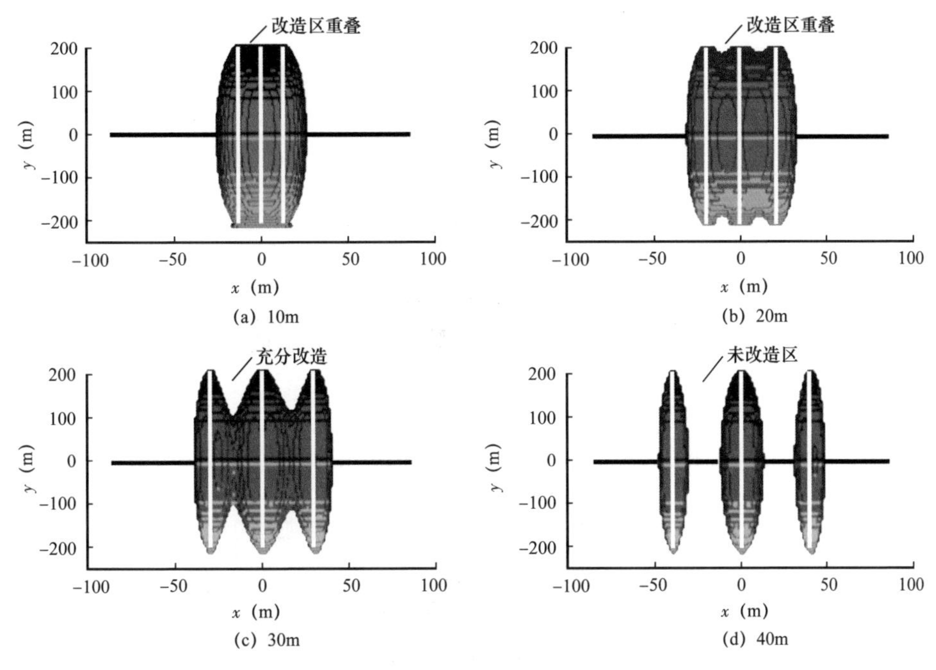

图3　不同簇间距下的总体SRV平面展布图

由图3可知，当簇间距较小时（10m、20m），各条主裂缝周围各自形成的SRV处于几乎完全叠加的状态，呈现出单一的规则椭球体形态；随着簇间距的增大（30m），各条主裂缝周围各自形成的SRV处于相互叠加与相互分离之间的临界状态——不存在过多叠加，也未彻底分离，呈现出不规则的椭球体形态，裂缝间区域内的天然裂缝被充分地激活，此时总体SRV达到全局最大值；随着簇间距的进一步增大（30m），各条主裂缝周围各自形成的SRV彻底相互分离，SRV转变成为3个不连续的规则椭球体形态，体积转而逐渐减小。

为求得最优簇间距，分别计算出簇间距10～50m范围内对应的SRV体积（图4）。

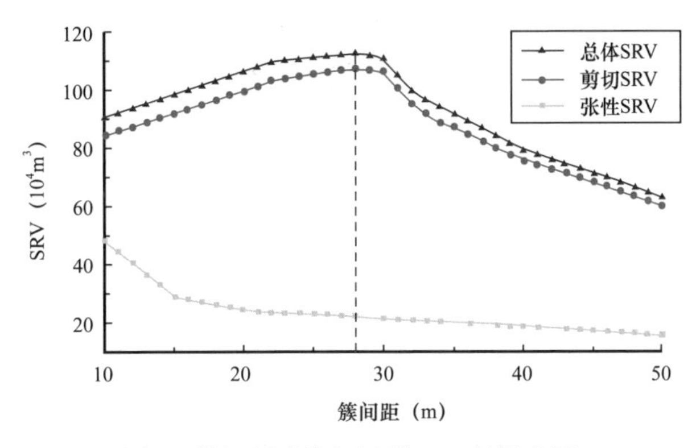

图4　簇间距对体积压裂SRV的影响图

由图4可知，随着簇间距的增大，体积压裂形成的张性SRV单调递减，而剪切破坏SRV与总体SRV则先增加后减小。当簇间距为28m时，总体SRV达到最大值（$112.5 \times 10^4 m^3$），故28m即为该特定工程地质条件下的最优簇间距。此时，储层压力、应力分布和总体SRV三维展布分别如图5至图7所示。

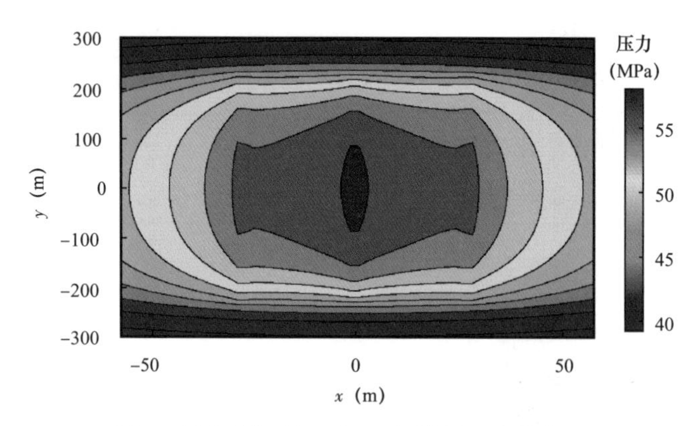

图5　最优簇间距下（28m）储层压力分布图

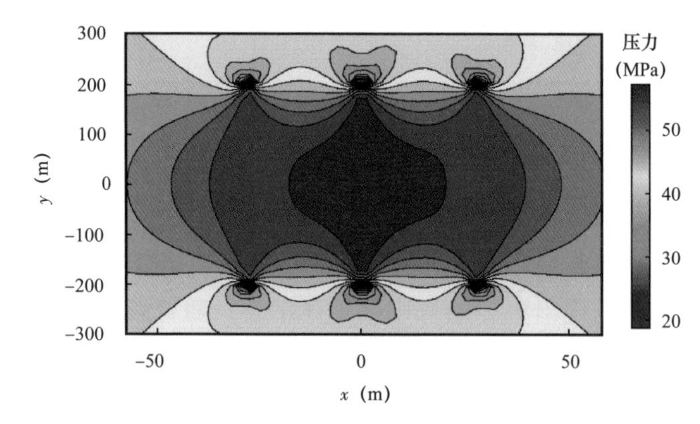

图6　最优簇间距下（28m）储层最小水平主应力分布图

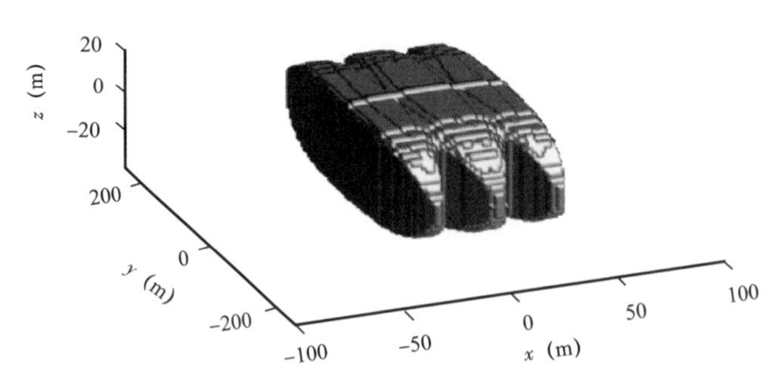

图7　最优簇间距下（28m）总体SRV三维展布图

2.2　敏感性因素分析

基于以上簇间距优化设计思路，研究涪陵页岩气示范区内地质与工程主要参数对最优簇间距的影响。分别针对地层水平主应力差、地层岩石杨氏模量、泊松比、压裂施工排量、压裂液总量和压裂液黏度进行敏感性分析。

图 8 表示出了不同地质参数（水平应力差、泊松比、杨氏模量）条件下对应的最优簇间距。随着水平应力差的增大，体积压裂 SRV 整体减小，最优簇间距也随之减小。原因在于：地应力差越大，天然裂缝越不容易被激活，SRV 沿主裂缝面垂直方向的扩展受限，椭球体变窄，体积随之缩小；此时需要更小的簇间距，使各条主裂缝相互靠拢，共同激活主缝之间区域内的天然裂缝，从而获得较大的 SRV。

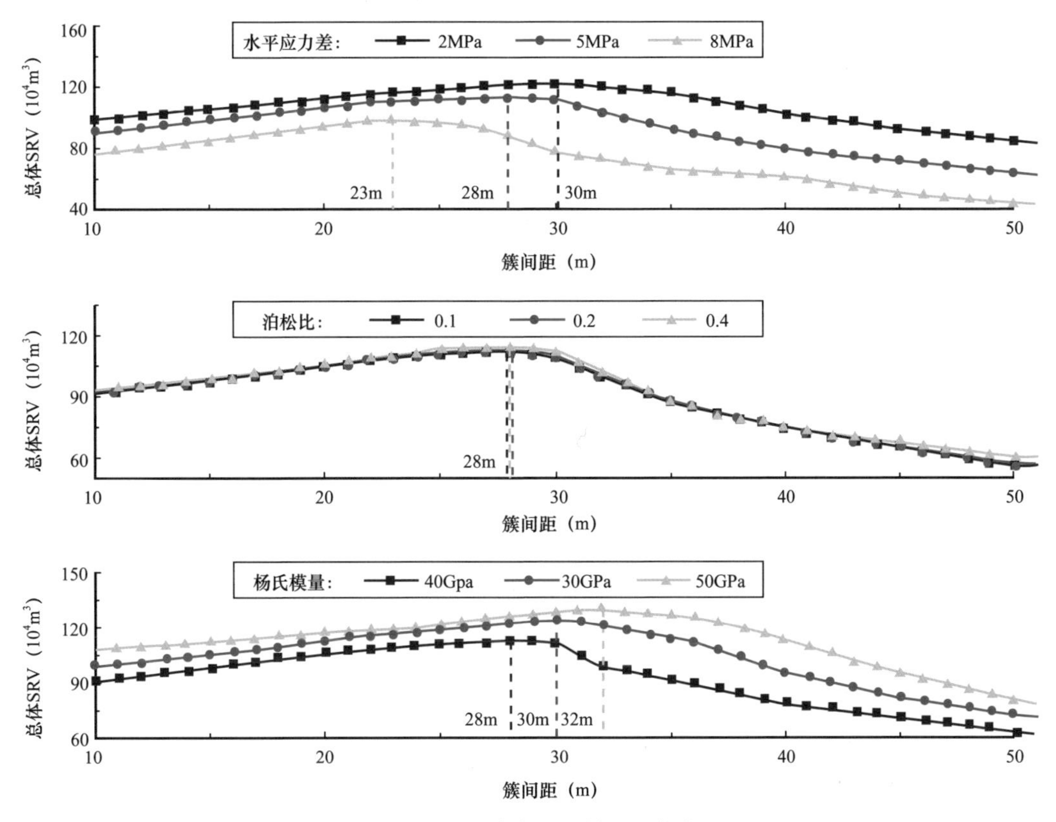

图 8　不同地质参数下的簇间距优化图

随着岩石泊松比的增大，体积压裂 SRV 整体变化不明显，最优簇间距几乎不变。原因在于：需要实施体积压裂的致密砂岩或页岩等岩石泊松比变化范围本身较小（0.1～0.4），而在此有限的变化范围内，并不能对总体 SRV 产生较大改变，也就几乎无法影响最优簇间距的大小。随着岩石杨氏模量的增大，体积压裂 SRV 整体增大，最优簇间距也随之增大。

原因在于：岩石杨氏模量越大，主裂缝延伸越长，缝内净压力越高，其应力影响区域也越大，有利于激活天然裂缝，并促使 SRV 沿三轴方向扩展，椭球体变大，体积随之增大；此时需要更大的簇间距，使各条主裂缝相互分开，充分利用应力影响区域，从而获得较大的 SRV。

图 9 表示出了不同工程参数（施工排量、压裂液总量、压裂液黏度）条件下对应的最优簇间距。随着压裂施工排量的增大，体积压裂 SRV 整体增大，最优簇间距也随之增大。原因在于：施工排量越大，主裂缝净压力越大，其应力影响区域也越大，有利于激活天然

裂缝，并促使 SRV 沿三轴方向扩展，椭球体变大，体积随之增大；此时需要更大的簇间距，使各条主裂缝相互分开，充分利用压力上升区域，从而获得较大的 SRV。

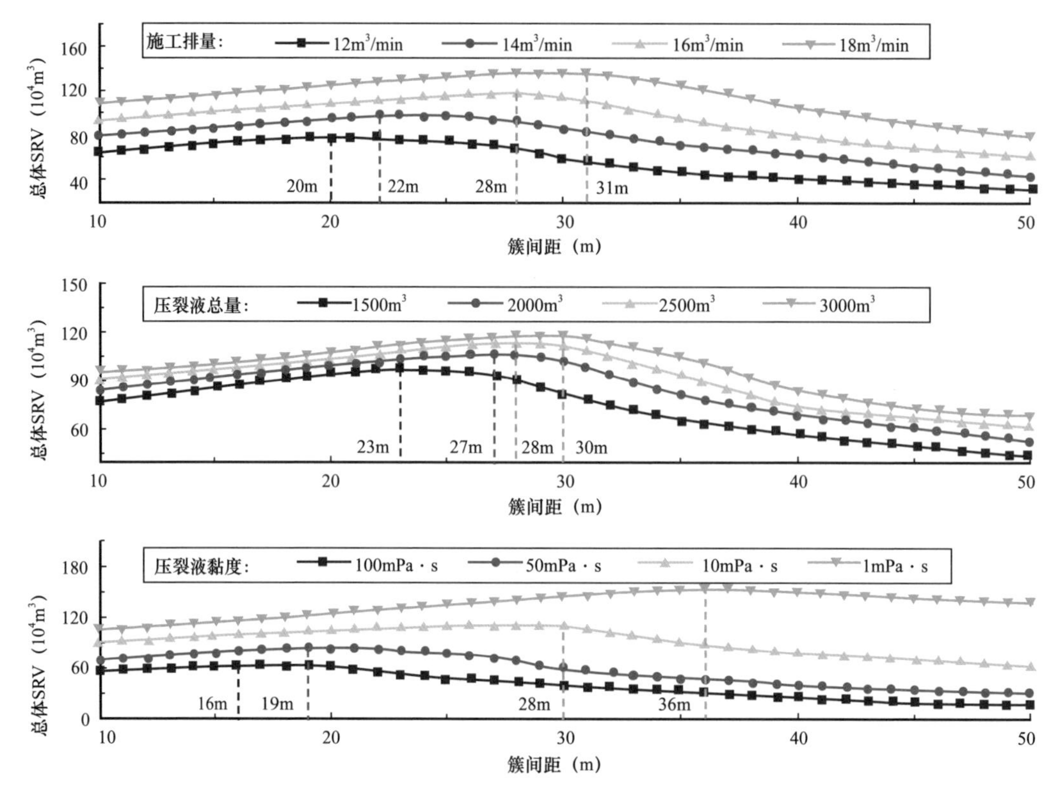

图 9　不同工程参数下的簇间距优化图

随着压裂液总量的增大，体积压裂 SRV 整体增大，最优簇间距也随之增大。原因在于：压裂液总量越大，主裂缝附近储层压力影响区域越大，天然裂缝激活区域越大，SRV 沿三轴方向扩展，椭球体变大，体积随之增大；此时需要更大的簇间距，使各条主裂缝相互分开，充分利用压力上升区域，从而获得较大的 SRV。

随着压裂液黏度的增大，体积压裂 SRV 整体减小，最优簇间距也随之减小。原因在于：压裂液黏度越大，其滤失量越小，并且高黏度流体在储层中的压力传播速度也越慢，最终导致主裂缝附近储层压力影响区域越小，天然裂缝激活区域越小，SRV 沿三轴方向扩展受限，椭球体变小，体积随之缩小；此时需要更小的簇间距，使各条主裂缝相互靠拢，共同激活主缝之间区域内的天然裂缝，从而获得较大的 SRV。

2.3　最优簇间距图版

目前，涪陵页岩气示范区内压裂施工通常采用黏度较低（10～20mPa·s）的滑溜水作为主要压裂液。因此，影响示范区内压裂最优射孔簇间距的因素主要包括泵注排量和压裂液总量两大工程因素。该区块水平井体积压裂设计方案中，泵注排量通常为 8～16m³/min，压裂液总量通常为 1000～3000m³。为此，分别基于该气藏南北两区块内的基本地质参数，通过 SRV 计算模型得到泵注排量与压裂液总量同时变化时对应的最优簇间距（图 10），并进行了双因素分析。

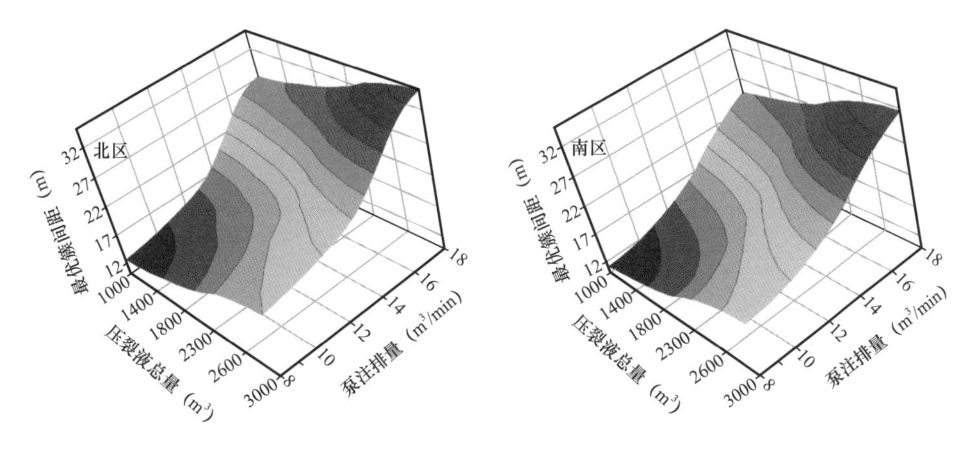

图 10　涪陵页岩气藏压裂参数与最优簇间距关系图

从图 10 可以看出，最优簇间距随着压裂液总量的增加而增大，也随着泵注排量的增加而增大。对于北区来说，最优簇间距范围为 12.5～35.0m；对于南区来说，最优簇间距范围为 11～32m，平均比北区小约 2m。此外，对于该区块目前压裂施工方案来说，泵注排量对最优簇间距的影响比压裂液总量的影响更大：当总液量由 $1000m^3$ 增加至 $3000m^3$ 时，最优簇间距将随之增加约 8m；当排量由 $8m^3/min$ 增加至 $18m^3/min$ 时，最优簇间距则增加约 15m。

基于上述研究，即可分别出针对涪陵页岩气示范区南、北两区块不同的地质特征，绘制出最优簇间距设计参考图版（图 11）。该图版可根据压裂设计中不同的泵注排量和压裂液总量，分布为南、北两区块内水平井完井射孔布簇提供最优簇间距的参考值，有助于在水力压裂过程中最大程度减小多簇主裂缝之间的干扰效应，提高压裂效率。

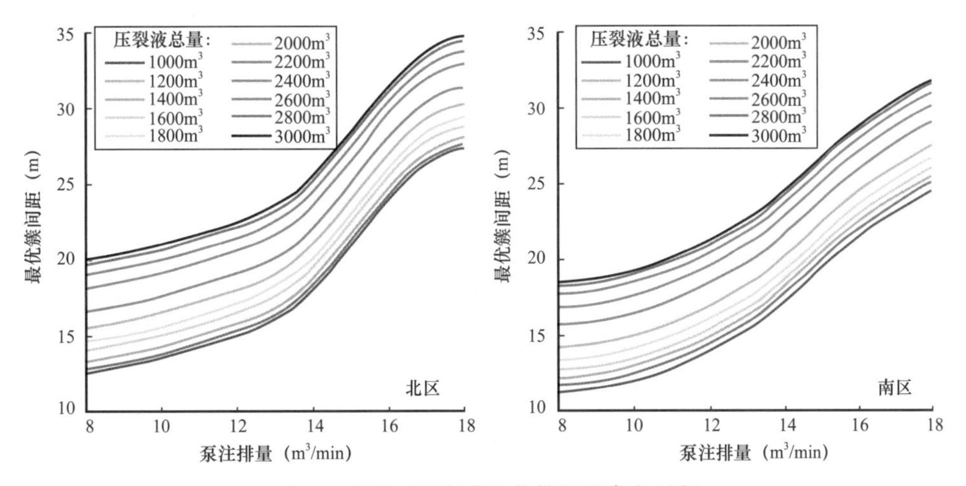

图 11　涪陵页岩气藏最优簇间距参考图版

3　结论

（1）基于裂缝延伸模型、裂缝诱导应力场模型、储层流体压力场模型和天然裂缝破坏准则，建立了一套新的 SRV 计算模型，并在该模型基础上提出了基于 SRV 最大化的水平井分段体积压裂簇间距优化方法。

（2）实例计算表明，涪陵页岩气示范区北区体积压裂设计中，存在最优簇间距约为28m；敏感性因素分析表明，随着地层水平应力差、压裂液黏度增大，最优簇间距减小；随着储层岩石杨氏模量、压裂施工排量、压裂液总量增大，最优簇间距增大；储层岩石泊松比对最优簇间距影响不明显；分别基于涪陵页岩气示范区南、北两区块的工程地质情况，绘制了最优簇间距设计图版。

（3）文中提出的水平井分段体积压裂SRV计算模型和簇间距优化设计方法对完善页岩体积压裂理论，指导页岩气藏体积压裂簇间距优化设计具有重要理论价值和矿场实际意义，能有效改善目前水平井分段射孔布簇的盲目性。

参考文献

[1] Boyer C, Clark B, Jochen V, et al.Shale gas : a global resource [J].Oilfield Review, 2011, 23 (3): 28–39.

[2] Luffel DL, Hopkins CW, Schettler PD Jr. Matrix permeability measurement of gas productive shales [C] // SPE Annual Technical Conference and Exhibition, 3–6 October 1993, Houston, Texas, USA.DOI : http : //dx.doi.org /10.2118/26633–MS.

[3] Sakhaee–Pour A, Bryant S.Gas permeability of shale [J].SPE Reservoir Evaluation & Engineering, 2012, 15 (4): 401–409.

[4] Cho Y, Ozkan E, Apaydin OG. Pressure–dependent natural–fracture permeability in shale and its effect on shale–gas well production [J]. SPE Reservoir Evaluation & Engineering, 2013, 16 (2): 216–228.

[5] Walton I, Mc lennan J. The role of natural fractures in shale gas production [C] //ISRM International Conference for Effective and Sustainable Hydraulic Fracturing, 20–22 May 2013, Brisbane, Australia. DOI : http : //dx.doi.org/10.5772/56404.

[6] 曾义金，陈作，卞晓冰.川东南深层页岩气分段压裂技术的突破与认识[J].天然气工业，2016，36（1）：61–67.

[7] 张廷山，赵国安，陈桂康，等.我国页岩气革命面临的问题及对策思考[J].西南石油大学学报（社会科学版），2016，18（2）：1–8.

[8] King GE. Thirty Years of gas shale fracturing : what have we learned? [C] // SPE Annual Technical Conference and Exhibition, 19–22 September 2010, Florence, Italy. DOI : http : //dx.doi.org/10.2118/133456–MS.

[9] Laffin M, Kariya M.Shale gas and hydraulic fracking [C] //20th World Petroleum Congress, 4–8 December 2011, Doha, Qatar.Doha, Qatar : World Petroleum Congress, 2011.

[10] Ren L, Zhao J Z Hu Y Q. Hydraulic fracture extending into network in shale : reviewing influence factors and their mechanism [J]. The Scientific World Journal, 2014, 2014: 847107.

[11] Seale RA, Donaldson J, Athans J. Multistage fracturing system : improving operational efficiency and production [C] // SPE Eastern Regional Meeting, 11–13 October 2006, Canton, Ohio, USA. DOI : http : //dx.doi.org/10.2118/104557–MS.

[12] Zhai Z Y，Fonseca E R，Azad A，et al.A New tool for multi-cluster and multi-well hydraulic fracture modeling [C] // SPE Hydraulic Fracturing Technology Conference，3–5 February 2015，The Woodlands，Texas，USA. DOI：http：//dx.doi.org/10.2118/173367–MS.

[13] 赵金洲，任岚，胡永全 . 页岩储层压裂缝成网延伸的受控因素分析 [J] . 西南石油大学学报（自然科学版），2013，35（1）：1–9.

[14] Fisher M K，Wright C A，Davidson B M，et al. Integrating fracture mapping technologies to optimize stimulations in the barnett shale [C] // SPE Annual Technical Conference and Exhibition，29 September–2 October 2002，San Antonio，Texas，USA. DOI：http：//dx.doi.org/10.2118/77441–MS.

[15] Mayerhofer M J，Lolon E，Warpinski N R，et al. What is stimulated reservoir volume? [J] . SPE Production & Operations，2010，25（1）：89–98.

[16] Cheng Y M. Boundary Element analysis of the stress distribution around multiple fractures：implications for the spacing of perforation clusters of hydraulically fractured horizontal wells [C] // SPE Eastern Regional Meeting，23–25 September 2009，Charleston，West Virginia，USA. DOI：http：//dx.doi.org/10.2118/125769–MS.

[17] Cheng Y M. Impacts of the number of perforation clusters and cluster spacing on production performance of horizontal shale–gas wells [J] . SPE Reservoir Evaluation & Engineering，2012，15（1）：31–40.

[18] Cheng Y M. Mechanical interaction of multiple fractures–exploring impacts of the selection of the spacing/number of perforation clusters on horizontal shale–gas wells [J] . SPE Journal，2012，17（4）：992–1001.

[19] 潘林华，张士诚，程礼军，等 . 水平井“多段分簇”压裂簇间干扰的数值模拟 [J] . 天然气工业，2014，34（1）：74–79.

[20] Guo J C，Lu Q L，Zhu H Y，et al. Perforating cluster space optimization method of horizontal well multi-stage fracturing in extremely thick unconventional gas reservoir [J] . Journal of Natural Gas Science and Engineering，2015，26：1648–1662.

[21] Lu Q L，Guo J C，Zhu H Y，et al. Cluster spacing optimization based on a multi–fracture simultaneous propagation model [C] // SPE/IATMI Asia Pacific Oil & Gas Conference and Exhibition，20–22 October 2015，Nusa Dua，Bali，Indonesia. DOI：http：//dx.doi.org/10.2118/176340–MS.

[22] Roussel NP & Sharma MM. Optimizing fracture spacing and sequencing in horizontal–well fracturing [J] . SPE Production & Operations，2011，26（2）：173–184.

[23] Roussel NP & Sharma MM. Strategies to minimize frac spacing and stimulate natural fractures in horizontal completions [C] // SPE Annual Technical Conference and Exhibition，30 October–2 November 2011，Denver，Colorado，USA. DOI：http：//dx.doi.org/10.2118/146104–MS.

[24] 尹建，郭建春，曾凡辉 . 水平井分段压裂射孔间距优化方法 [J] . 石油钻探技术，2012，40（5）：67–71.

[25] 曾凡辉，郭建春，刘恒，等 . 致密砂岩气藏水平井分段压裂优化设计与应用 [J] . 石油学报，2013，34（5）：959–968.

[26] LiuC，Liu H，Zhang Y P，et al. Optimal spacing of staged fracturing in horizontal shale–gas well [J] . Journal of Petroleum Science and Engineering，2015，132：86–93.

[27] Fisher M K, Heinze J R, Harris C D, et al. Optimizing horizontal completion techniques in the barnett shale using microseismic fracture mapping [C] // SPE Annual Technical Conference and Exhibition, 26–29 September 2004, Houston, Texas, USA. DOI : http : //dx.doi.org/10.2118/90051–MS.

[28] Zhao Y L, Zhang L H, Luo J X, et al. Performance of fractured horizontal well with stimulated reservoir volume in unconventional gas reservoir [J] . Journal of Hydrology, 2014, 512: 447–456.

[29] Denney D. Optimizing horizontal completions in the barnett shale with microseismic fracture mapping[J]. Journal of Petroleum Technology, 2005, 57 (3) : 41–43.

[30] Fisher M K, Wright C A, Davidson B M, et al. Integrating fracture mapping technologies to improve stimulations in the barnett shale [J] . SPE Production & Facilities, 2005, 20 (2) : 85–93.

[31] 刘尧文，廖如刚，张远，等 . 涪陵页岩气田井地联合微地震监测气藏实例及认识 [J] . 天然气工业，2016，36 (10) : 56–62.

[32] 巫芙蓉，闫媛媛，尹陈 . 页岩气微地震压裂实时监测技术——以四川盆地蜀南地区为例 [J] . 天然气工业，2016，36 (11) : 46–50.

[33] Astakhov D, Roadarmel W, Nanayakkara A. A new method of characterizing the stimulated reservoir volume using tiltmeter–based surface microdeformation measurements [C] // SPE Hydraulic Fracturing Technology Conference, 6–8 February 2012, The Woodlands, Texas, USA. DOI : http : //dx.doi.org/10.2118/151017–MS.

[34] Yu Guang, Aguilera R. 3D analytical modeling of hydraulic fracturing stimulated reservoir volume [C] // SPE Latin America and Caribbean Petroleum Engineering Conference, 16–18 April 2012, Mexico City, Mexico. DOI : http : //dx.doi.org/10.2118/153486–MS.

[35] Maulianda BT, Hareland G, Chen S. Geomechanical consideration in stimulated reservoir volume dimension models prediction during multi–stage hydraulic fractures in horizontal wells–glauconite tight formation in hoadley field [C] // Presented at the 48th USRock Mechanics/Geomechanics Symposium, 1–4 June 2014, Minneapolis, Minnesota.

[36] Lamb H.Hydrodynamics [M] .6th ed. Cambridge : Cambridge University Press, 1932.

[37] Nolte K G.Application of fracture design based on pressure analysis [J] .SPE Production Engineering, 1988, 3 (1) : 31–42.

[38] England A, Green A. Some two–dimensional punch and crack problems in classical elasticity [J] . Proceedings of the Cambridge Philosophical Society, 1963, 59 (2) : 489.

[39] Charlez P A. Rock mechanics : petroleum applications [M] . Paris : Editions Technip, 1997.

[40] Weng X W, Kresse O, Cohen C E, et al. Modeling of hydraulic–fracture–network propagation in a naturally fractured formation [J] . SPE Production & Operations, 2011, 26 (4) : 368–380.

[41] 张平，赵金洲，郭大立，等 . 水力压裂裂缝三维延伸数值模拟研究 [J]. 石油钻采工艺, 1997, 19(3): 53–59.

[42] 郭大立，纪禄军，赵金洲，等 . 煤层压裂裂缝三维延伸模拟及产量预测研究 [J] . 应用数学和力学，2001，22 (4) : 337–344.

[43] 李勇明，陈曦宇，赵金洲，等 . 水平井分段多簇压裂缝间干扰研究 [J] . 西南石油大学学报 (自然科学版)，2016，38 (1) : 76–83.

[44] Crouch S L. Solution of plane elasticity problems by the displacement discontinuity method. I. infinite body solution [J]. International Journal for Numerical Methods in Engineering, 1976, 10 (2): 301–343.

[45] Wu Kan, Olson J E. Investigation of the impact of fracture spacing and fluid properties for interfering simultaneously or sequentially generated hydraulic fractures [J]. SPE Production & Operations, 2013, 28 (4): 427–436.

[46] Ozkan E, Raghavan R. New solutions for well–test–analysis problems: part 1–analytical considerations[J]. SPEFormation Evaluation, 1991, 6 (3): 359–368.

[47] Stehfest H, Goethe–Univ J W, Germany F W. Remark on algorithm 368: numerical inversion of laplace transforms [J]. Communications of the ACM, 1970, 13 (10): 624.

[48] Stehfest H, Goethe–Univ J W, Germany F W. Algorithm 368: numerical inversion of laplace transforms[J]. Communications of the ACM, 1970, 13 (1): 47–49.

[49] Warpinski N R, Teufel L W. Influence of geologic discontinuities on hydraulic fracture propagation [J]. Journal of Petroleum Technology, 1987, 39 (2): 209–220.

[50] Ren L, Lin R, Zhao J Z, et al. Simultaneous hydraulic fracturing of ultra–low permeability sandstone reservoirs in China: Mechanism and its field test[J]. Journal of Central South University, 2015, 22(4): 1427–1436.

[51] Guo T L, Li J L, Lao M D, et al. Integrated geophysical technologies for unconventional reservoirs and case study within fuling shale gas field, Sichuan Basin, China [C] // Unconventional Resources Technology Conference, 20–22 July 2015, San Antonio, Texas, USA. DOI: http://dx.doi.org/10.15530/URTEC–2015–2152914.

页岩气生产数据联合分析方法研究与应用

王军磊[1]，位云生[1]，陈　鹏[2]，韩会平[2]

（1. 中油勘探开发研究院；2. 中油长庆油田公司）

摘　要: 生产数据分析方法集合了气藏工程和试井分析方法的优势，用以获取气藏动态信息、评价开发效果、预测气井产量，但受制于页岩气特殊的流动机理和开发方式，使用单一模型解释结果往往具有很大的不确定性。针对此问题，建立了以流态识别、解析分析和经验分析为核心的包含多种数据分析方法的综合模型，同时利用拟压力、拟时间处理气井压力、产量等原始生产数据，并结合系统分析原理分析拟生产数据，分析过程多种方法相互验证，不断提高解释结果的可信度，形成有效的页岩气生产数据联合分析方法。矿场实例具体演绎了整个分析流程，得到了合理的气藏动态信息，在此基础上预测气井产量，为后期开发方案调整提供了可靠的理论支持。

关键词: 页岩气；气井；拟变量；数据处理；数据分析；分析流程

页岩气是一种典型的非常规油气资源，具有较高的勘探成功率和较大的开发风险[1-2]。随着近些年来水平井钻完井、分段压裂和微地震监测技术的提高，页岩气正在成为一种重要的现实替代能源。目前中国页岩气的发展正处在起步阶段，四川长宁、威远地区已获得工业气流，证明了中国确实存在页岩气[3]。页岩气藏储量规模大小、单井产能高低等关键性问题决定着页岩气是否具有商业开采价值。

生产数据分析方法是一种重要的气藏动态描述手段，集合了试井分析和气藏工程的优点，通过处理和解释气井产量、压力等生产数据，获得储层渗透率、表皮系数、裂缝长度、导流能力、单井控制储量等参数，预测气井产量变化规律，从而为改进气藏开发效果、降低气藏开发风险提供有力的技术保证，具有实用、可靠、经济的特点。

页岩气非线性流动特性和气井不稳定生产使得压力 / 产量随时间的变化规律复杂[4,5]，长时间的非稳态流动期导致采用单一分析模型获得的解释结果不确定性高、可信度低[6]。针对页岩气的流动和开发特征，利用拟函数和叠加函数处理原始生产数据，联合多种分析模型分析生产数据，通过矿场应用演绎分析流程，获得了合理的评价结果，为后期开发方案调整提供理论依据，最终形成一套有效的页岩气生产数据联合分析方法。

1　页岩气生产数据联合分析方法

页岩气数据分析方法不同于常规方法，主要原因是其复杂的赋存及渗流机理[7-8]，包括气体从有机质颗粒或黏土表面的解吸、介质超低的渗透率、介质的应力敏感性、体积压裂形成复杂缝网结构的空间多尺度性以及气体的非达西流动效应等。

生产数据分析方法将气井生产过程中记录的产量、压力等信息进行分析处理，获取由这些信息反映出的地层特性。受制于页岩气特殊的渗流机理、复杂的开发方式和低质量、低分辨率的生产数据，目前仍没有一种完全成熟可靠的数据分析模型。联合多种分析模型分析气井数据，可以相互制约、验证，降低解释结果的不确定性，具体分析方法如图1所示。

通过步骤1～3筛选出合适的生产数据。生产数据分析方法使用的数据多，贯穿井的整个生命周期，但数据分辨率低、"噪声"大，数据源的质量决定了评价结果的可靠性，故在分析数据前需要评价、检查、剔除低质量的生产数据。对于一组原始的生产数据，需要对其进行诊断分析：（1）评价数据质量的可靠性，包括产量和压力数据、储层和流体参数、完井及增产措施等；（2）检查数据相关性，包括产量—压力、产量—时间和压力—时间的数据相关性检查；（3）初步诊断，主要是数据检查和整理，剔除错误数据。

利用步骤4～5中的数据分析模型分析气井产量、压力等数据，获得气藏、气井相关参数，这也是数据分析方法的核心部分。利用步骤6中的解析分析模型，通过调整参数拟合气井生产历史，以预测气井动态。在步骤7中，利用经验分析模型与解析分析模型进行相容性评价调整。

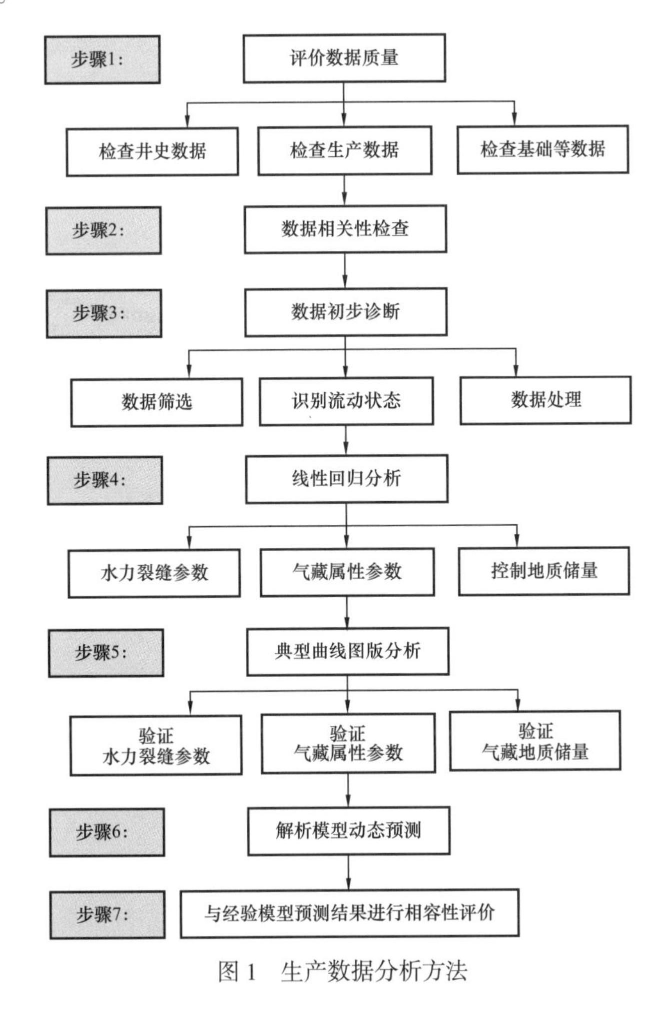

图1　生产数据分析方法

2 数据分析模型

数据分析模型包括流态识别模型、解析分析模型和经验分析模型。其中，流态识别模型是数据分析的基础。解析分析模型遵循系统分析原理，即正反问题：正问题，建立并求解一系列假定数学模型的标准信息（如理论公式、曲线图版）；反问题，利用实际系统反映出的信息（如产量、压力）与标准信息对比、拟合，确定出未知系统属于哪种已知系统。经验模型没有严格的理论推导，通过观察大量生产数据预测气井产量递减规律，预测结果不确定性大。

2.1 流态识别模型

流态指气体在地层中的流动特征，能够反映气井结构、气藏属性、缝网结构、气体物性等特征，流态识别是数据分析中最重要的步骤，贯穿整个分析流程。受气体高压物性、非达西流动、应力敏感和变产量生产等非线性影响，页岩气井原始生产数据变化规律复杂，将生产数据转换为拟压力、叠加拟时间或物质平衡拟时间后能清晰地反映出数据间的关联性。

使用各类诊断模型（如修正压力与物质平衡拟时间的双对数图版）识别气体流态（图 2）。不同的流态在诊断图版上有不同的曲线特征，总体上气体流态可分为两大部分：非稳态流动段和拟稳态流动段。其中，非稳定流动阶段生产数据主要受气井结构和地层属性影响，拟稳态阶段生产数据主要受地层非均质性和地层边界控制。

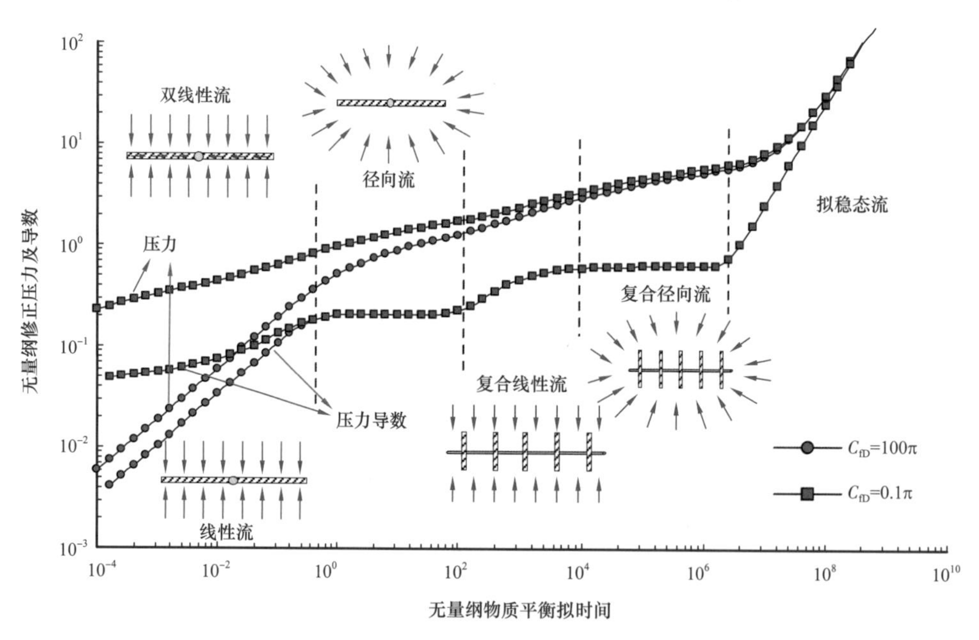

图 2 分段压裂水平井流态变化规律

2.1.1 线性回归分析

在正确识别流态的基础上利用解析分析模型定量分析生产数据。线性分析类似于常规试井分析，将生产数据处理为产量修正的拟压力差 $\Delta m^*/q_{gsc}$ 和物质平衡拟时间 t_{ca}^* 或叠加拟时间，利用拟变量间的变化规律，结合相应的诊断工具确定流态，根据不同流态对应的线性关系式获得裂缝及气藏属性参数。常见流态的线性关系及参数计算公式如表 1 所示，处理后的数据在半对数、双对数或平方根等诊断图版下将呈现线性关系。

表 1　常见流态的线性关系式

流动状态	线性流动关系	拟合参数
双线性流	$\dfrac{m(p_i)-m(p_w)}{q} vs. t_a^{1/4} \text{or} t_{a,BLS} = \sum_{j=1}^{n} \dfrac{q_j - q_{j-1}}{q_n}(t_{a,n}-t_{a,n-1})^{1/4}$	$\sqrt{K_f w_f} K^{1/4} = \dfrac{443.2T}{m_{BL} h(\phi \mu_{gi} c_{ti})^{1/4}}$
线性流	$\dfrac{m(p_i)-m(p_w)}{q} vs. t_a^{1/2} \text{or } t_{a,LS} = \sum_{j=1}^{n} \dfrac{q_j - q_{j-1}}{q_n}(t_{a,n}-t_{a,n-1})^{1/2}$	$\sqrt{K} x_f = \dfrac{40.93T}{m_L h(\phi \mu_{gi} c_{ti})^{1/2}}$
椭圆流	$\dfrac{m(p_i)-m(p_w)}{q} vs. \ln(A+B)$	$K = \dfrac{1422T}{m_E h}; x_f = \exp(b_E)$
拟径向流	$\dfrac{m(p_i)-m(p_w)}{q} vs. \lg(t_a) \text{or } t_{a,LS} = \sum_{j=1}^{n} \dfrac{q_j - q_{j-1}}{q_n} \lg(t_{a,n}/t_{a,n-1})$	$K = \dfrac{1637T}{m_R h}$
拟稳态流	$\dfrac{q}{m(p_i)-m(p_w)} vs. G_p \dfrac{m(p_i)-m(p_R)}{m(p_i)-m(p_w)}$	利用 x 轴截距计算储量 G_p

其中，L，r，x 为长度，m；h 为地层厚度，m；w 为裂缝宽度，m；x_f 为裂缝半长，m；ϕ 为地层孔隙度；t 为时间，h；t_a 为拟时间，h；p_R 为平均地层压力，MPa；p_i 为原始地层压力，MPa；p_w 为井底压力，MPa；m 为拟压力，MPa；K 为渗透率，mD；q 为产量，$10^4 m^3/d$；μ_{gi} 为原始地层压力下的气体黏度，mPa·s；c_{ti} 为原始地层压力下的综合压缩系数，MPa^{-1}；T 为地层温度，K；G_p 为地质储量，$10^8 m^3$；求和项为 Duhamel 叠加时间；m_{BL}、m_L、m_E、m_R 为对应流态的直线斜率；下标 BLS、LS、LS 分别表示双线性流、线性流、径向流；A、B 表示椭圆流常数。

2.1.2　典型图版分析

典型图版分析类似于现代试井解释方法，基于某一特定的数学模型得到无量纲典型曲线及相应诊断曲线，将处理后的生产数据与典型曲线拟合获得相关参数。与线性分析方法不同的是，典型曲线通常同时拟合多个流态，这样可以相互制约，降低分析结果的不确定性。

在 Fetkovich 等[9]、Blasingame 等[10]、Agrwal 等[9-11]学者基础上，Wattenbarger 等[12]建立线性流模型，Bello 等[13]将裂缝网络看作双重介质，提出双线性流模型，Brown 等[14]针对不同区域的流动能力，将体积压裂区（SRV）区内水力裂缝线性流、裂缝间地层线性流，裂缝端部 SRV 区外地层的线性流概括为三线性流模型。(多)线性流模型更接近于气体在 SRV 内的实际流动，能够有效分析页岩气井生产数据，同时近年来也有学者开始使用多重介质模型描述页岩气在多尺度流动空间中的流动规律。

2.2 经验分析模型

2.2.1 幂律指数方法

直接应用 Arps 递减评价页岩气藏储量会产生很大误差。页岩气在相当长一段时间内处于非稳态流动阶段，Ilk 等[15]以整个渗流期为研究对象，重新定义递减率 D，进而得到不同于传统双曲递减的幂律型指数递减模型。其中早期非稳态流动期主要受 $\tilde{D}_i t$ 影响，晚期拟稳态流动阶段受 $D_\infty t$ 影响。

$$q = \hat{q}_i \exp\left(-D_\infty t - \frac{D_1}{n} t^n\right) = \hat{q}_i \exp\left(-D_\infty t - \hat{D}_i t^n\right) \tag{1}$$

其中，D_1 为时间为 1 天时对应的递减常数，d^{-1}；D_∞ 为时间趋于无穷大时对应的递减常数，d^{-1}；$\hat{D}_i$ 为递减常数，定义为 D_1/n；n 为无量纲时间指数；$\hat{q}_i$ 为初始流量，$10^4 m^3/d$；t 为时间，d。

2.2.2 合成方法

根据递减指数 b 变化规律[16]，用解析模型分析非稳态数据得到地层渗透率和表皮系数，用经验模型分析拟稳态数据获得递减参数 b、D_{elf} 和 q_{elf}，预测产量。以线性流为例[12]，$t < t_{elf}$ 时的非稳态流和 $t > t_{elf}$ 时的拟稳态流对应的产量公式分别满足：

$$q_{elf} = \frac{1}{m_L \sqrt{t + b'}} \tag{2}$$

其中，m_L 为线性流直线斜率，$10^{-4} m^{-3} \cdot d^{0.5}$；$q_{elf}$ 为非稳态流量，$10^4 m^3/d$；b' 为直线截距，$10^{-4} m^{-3} \cdot d$。

$$q = \frac{q_{elf}}{\left[1 + bD_{elf}(t - t_{elf})\right]^{1/b}} \tag{3}$$

其中，b 为 Arps 无量纲递减指数；D_{elf} 为线性流结束时刻递减常数，d^{-1}；q 为拟稳态流量，$10^4 m^3/d$；t_{elf} 为线性流结束时间，d。

3 矿场实践

以某口水平井为例演绎生产数据分析方法。该区块中页岩储层致密，发育丰富的微米—纳米级孔隙，石英、长石等脆性矿物含量高，易于压裂改造。该井改造水平段长 1045m，改造段数为 10 段，自 2011 年 1 月投产，生产时间为 805 天，累计产气量为

$0.053\times10^8m^3$。原始地层压力为16.3MPa，地层温度为65℃，厚度为39.7m，孔隙度为5%，含气饱和度为65.25%，Langmuir体积为$3m^3/t$，Langmuir压力为2.8MPa。使用CH_4物性参数，原始气体压缩系数为$0.0246MPa^{-1}$，原始气体体积系数为0.00395。

3.1 生产数据筛选及处理

按步骤1～3筛选生产数据，按式（4）、式（5）处理数据。其中拟时间和物质平衡拟时间中涉及的平均地层压力通过地质储量迭代算法获得[17]。

$$m^*(p)=\frac{\mu_g(p_i)Z_g(p_i)}{p_i}\int_{pref}^{p}\frac{p}{\mu_g Z_g}dp \tag{4}$$

$$t_a^*=\mu_g(p_i)c_t^*(p_i)\int_0^t\frac{1}{\mu_g(p_{avg})c_t^*(p_{avg})}dt \tag{5}$$

$$t_{ca}^*=\frac{\mu_g(p_i)c_t^*(p_i)}{q_{gsc}}\int_0^t\frac{q_{gsc}(\tau)}{\mu_g(p_{avg})c_t^*(p_{avg})}d\tau \tag{6}$$

其中，c_t^*为考虑吸附气影响的综合气体压缩系数[18]：

$$c_t^*(p)=c_g(p)+c_d(p)=c_g(p)+\frac{p_{sc}V_L p_L T}{\phi T_{sc}Z_g(p_{sc})}\frac{Z_g(p)}{p(p+p_L)^2} \tag{7}$$

其中，c_t^*为综合压缩系数，MPa^{-1}；p_{avg}为原始地层压力，MPa；p_L为Langmuir压力，MPa；m^*（p）为气体拟压力，MPa；p_{sc}为标准压力，MPa；q_{gsc}为标况下流量，$10^4m^3/d$；T_{sc}为标准温度，K；t_a^*为气体拟时间，d；t_{ca}^*为气体物质平衡拟时间，d；V_L为Langmuir体积，m^3/t。

3.2 生产数据分析

在步骤3中，使用表1中的叠加时间$t^*_{a,\ LS}$识别流态，同时按步骤4对生产数据进行线性回归分析，分析结果见图3（a）。该井采用分段压裂技术，形成有效的缝网结构，提高了井筒附近地层的整体渗透率，但受制于气井短暂的生产历史，气体主要处于非稳态线性流阶段，线性流结束时间t_{elf}=403d，随后进入拟稳态生产阶段。利用探测边界移动规律式（8）可获得渗透率K，结合相应流态关系式获得裂缝半长x_f；利用流动物质平衡方程p/z^*-G_p和动态物质平衡方程$q_{gsc}/\Delta m^*-G_p/\Delta m^*$迭代计算获得控制储量$G$，如图3（b）所示。

$$K=\frac{46.3\phi\mu_g(p_i)c_g(p_i)q_{gsc}(t_{elf})L_f^2}{n_f^2G_p(t_{elf})} \tag{8}$$

其中，L_f为裂缝半长，m；n_f为裂缝条数。

由于主要流态是非稳态线性流和拟稳态流，按步骤5，将相关参数代入Wattenbarger无量纲定义（9）、（10），对参数进行微调（＜10%），与典型曲线拟合效果良好，如图4所示。

$$q_{\mathrm{Dd}}=\frac{1.866\mu_{\mathrm{g}}(p_{\mathrm{i}})B_{\mathrm{g}}(p_{\mathrm{i}})}{Khx_{\mathrm{f}}}\left(\frac{GB_{\mathrm{g}}(p_{\mathrm{i}})}{4x_{\mathrm{f}}\phi hS_{\mathrm{gi}}}\right)\frac{q_{\mathrm{gsc}}}{\Delta m^{*}} \tag{9}$$

$$t_{\mathrm{caDd}}=\frac{3.6\times10^{-3}K}{\phi\mu_{\mathrm{g}}(p_{\mathrm{i}})c_{\mathrm{g}}(p_{\mathrm{i}})}\left(\frac{4x_{\mathrm{f}}\phi hS_{\mathrm{gi}}}{GB_{\mathrm{g}}(p_{\mathrm{i}})}\right)^{2}t_{\mathrm{ca}}^{*} \tag{10}$$

其中，B_{g} 为气体体积系数；q_{Dd} 为无量纲产量；t_{caDd} 为无量纲物质平衡拟时间；S_{gi} 为原始含气饱和度；Δm^{*} 为拟生产压差，MPa。

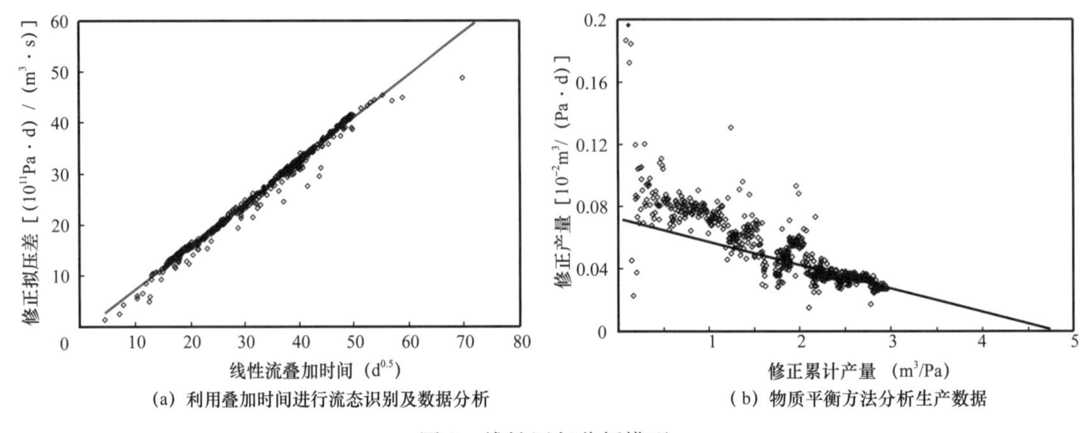

(a) 利用叠加时间进行流态识别及数据分析　(b) 物质平衡方法分析生产数据

图 3　线性回归分析模型

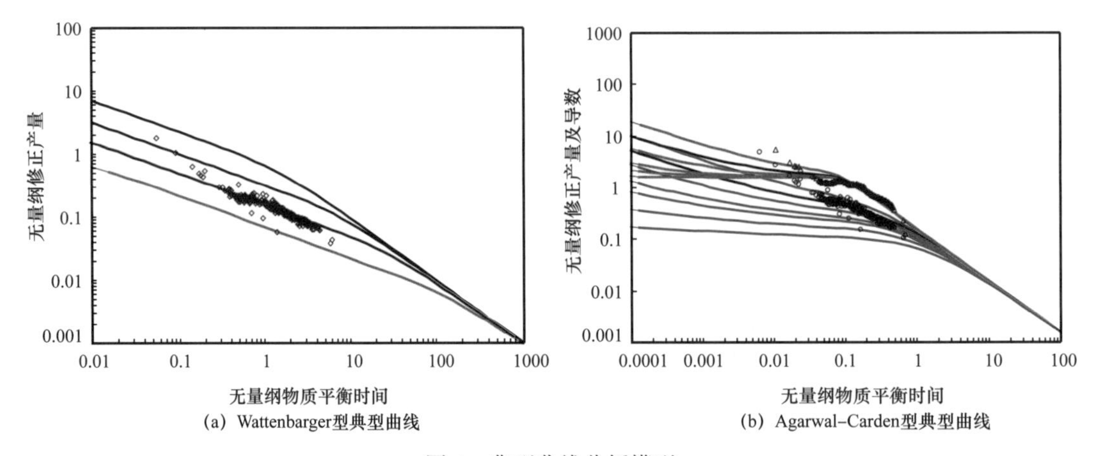

(a) Wattenbarger型典型曲线　(b) Agarwal-Carden型典型曲线

图 4　典型曲线分析模型

同理使用 Agarwal-Gardenr 图版，对应的大裂缝穿透比（$x_{\mathrm{f}}/r_{\mathrm{e}}$=0.5）拟合结果也预示着长时间的线性流动期。每种分析模型都有多个解释结果，不同模型间相互调整、验证，可以降低解释结果的不确定性，最终评价结果如表 2 所示。

以上分析结果仅仅考虑了气体高压物性、解吸和气井变产量生产的影响，重新定义拟压力和拟时间函数[4]能够进一步研究气体非达西流动和介质应力敏感性对数据分析的影响。

表 2　多种分析模型解释结果

数据分析模型	渗透率（mD）	裂缝半长（m）	地质储量（10^8m^3）	吸附气储量（10^8m^3）	自由气储量（10^8m^3）
线性分析	0.0017	42.83	0.392	0.184	0.208
物质平衡	—	—	0.478	0.225	0.253
Wattenbarger 图版	0.0011	44.16	0.415	0.195	0.220
Agarwal-Garden 图版	0.0023	48.25	0.423	0.198	0.225

3.3　气井产量预测

基于解析分析模型，利用物质平衡方程联立拟稳态气井产能方程进行页岩气产能预测[19]，20 年内的气井产量变化规律如图 5（a）所示。气井保持 4.82MPa 恒压生产，产量随时间递减，且递减率逐渐减小，直到后期保持相对稳定，最终采收率为 25.1%。气藏可采储量集中在生产早期，主要来自孔隙中储存的自由气，生产后期主要来自吸附态的解吸气。由于拟稳态阶段生产数据过少，经验分析模型不确定性大，使用 b=0，0.1，0.5，1，2 五组参数预测气井产量（图 5b），对比解析预测模型（表 3），发现经验指数型递减（b=0）较为接近，提高了解析预测结果的可信度。

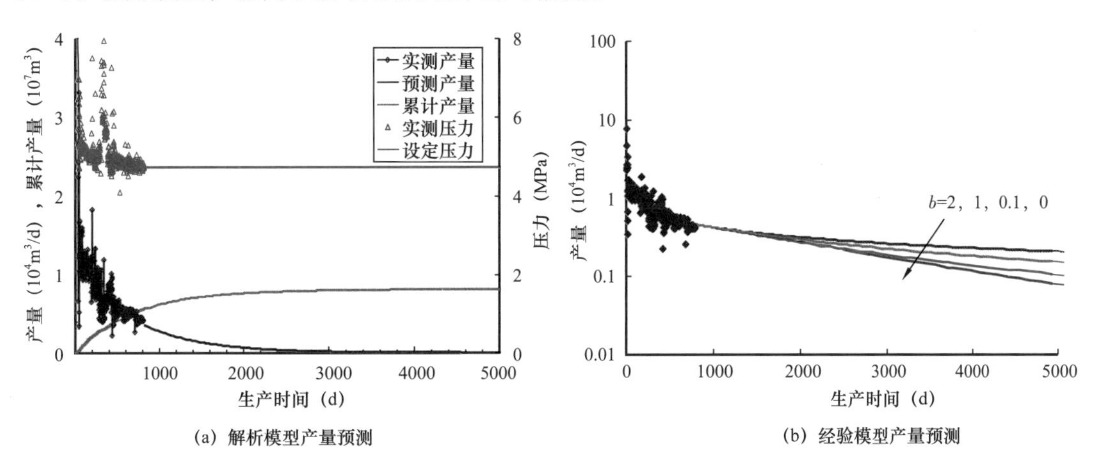

(a) 解析模型产量预测　(b) 经验模型产量预测

图 5　气井生产规律预测

表 3　不同分析模型的预测结果（20 年内）

解析模型		可采储量（10^8m^3）	采收率（%）
解析模型		0.082	19.8
经验模型	b=0	0.089	22.7
	b=0.1	0.094	24.5
	b=0.5	0.124	31.6
	b=1	0.162	41.3
	b=2	0.223	56.8

气藏弹性采收率仅取决于气井流压，而预测期内的气藏采收率取决于气井流压和气体渗流阻力。提高气藏渗透率、增大气井与地层接触面积能够有效减小渗流阻力，但需要大型的体积压裂，而在合理配产基础上，降低气井压力能够充分发挥气体弹性能量，增加气体流动势能和气体解吸量（图 6），从而提高页岩气藏采收率（表 4）。

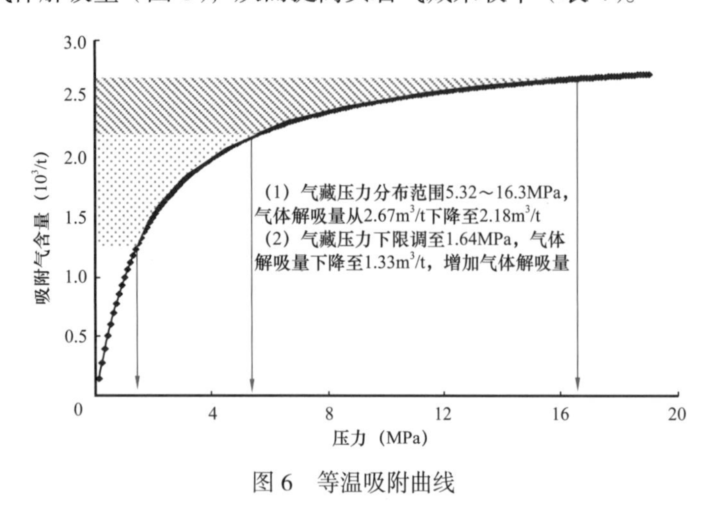

图 6　等温吸附曲线

表 4　基于解析模型的预测结果

气井压力（MPa）	最终可采储量（%）	预测期采收率（%）	预测期可采比例（%）
6.17	20	15.5	77.5
3.05	30	25.1	83.6
1.64	40	42.8	85.6
0.78	60	47.1	78.5
0.35	70	50.5	72.1
0.17	80	56.7	70.8

4　结论

（1）页岩气生产数据分析方法主要包括数据处理方法及数据分析模型。拟变量数据处理方法可以消除页岩气非线性流动的影响，是线性回归、典型曲线拟合等解析法的数据分析来源；幂律指数、解析合成等经验法直接拟合原始生产数据，预测气井产量变化规律。

（2）解析分析模型可以解释地层渗透率、气藏地质储量等参数、预测气井产量，解释结果可靠性强，但模型受制因素多、数据处理过程繁复。经验分析模型方便灵活，不需要数据处理，但缺乏理论依据。集合不同分析模型的优点联合分析生产数据，不同模型间相互调整、验证，可以降低评价结果的不确定性，提高预测结果的可靠性。

参考文献

[1] 赵文智，董大忠，李建忠，等.中国页岩气资源潜力及其在天然气未来发展中的地位[J].中国工程科学，2012，14(7)：46-52.

[2] 张东晓，杨婷云.页岩气开发综述[J].石油学报，2013，34(4)：792-801.

[3] 王兰生，廖仕孟，陈更生，等.中国页岩气勘探开发面临的问题与对策[J].天然气工业，2011，31(12)：1-4.

[4] Nobakht M，Clarkson C R，Kaviani D.New and improved methods for performing rate transient analysis of shale gas reservoirs[J].SPE Reservoir Evaluation & Engineering，2012，7(2)：335-486.

[5] Cipplla C L，Lolon E P，Erdle J C，et al.Reservoir modeling in shale gas reservoirs[J].SPE Reservoir Evaluation & Engineering，2010，8(3)638-653.

[6] 白玉湖，杨皓，陈桂华，等.页岩气产量递减曲线的不确定性分析方法[J].石油钻探技术，2013，41(4)：97-100.

[7] Civan F.Effective correlation of apparent gas permeability in tight porous media[J].Transport in Porous Media，2010，8(2)：375-384.

[8] Wang F P，Reed R M.Pore networks and fluid flow in gas shales[C].SPE Annual Technical Conference and Exhibition，4-7 October 2009，New Orleans，Louisiana，USA.DOI：https：//doi.org/10.2118/124253-MS.

[9] Fetkovich M L.Decline curve analysis using type curves[J].Journal of Canadian Petroleum Technology，1980，32(6)：1065-1077.

[10] Blasingame T A，McCray T L，Lee W J.Decline curve analysis for variable pressure drop/variable flowrate systems[C]//SPE Gas Technology Symposium，22-24 January 1991，Houston，Texas，USA.DOI：https：//doi.org/10.2118/21513-MS.

[11] Agarwal R G，Gardner D C，Kleinsteiber S W，et al.Analyzing well produciton data using combined-type-curve and decline-curve analysis concepts[J].SPE Reservoir Evaluation & Engineering，1999，2(5)：478-486.

[12] Wattenbarger R A，Ei-Banbi A H，Villegas M E，et al. Production analysis of linear flow into fractured tight gas wells[C]//SPE Rocky Mountain Regional/Low-Permeability Reservoirs Symposium，5-8 April 1998，Denver，Colorado，USA. DOI：https：//doi.org/10.2118/39931-MS.

[13] Bello R O，Wattenbarger R A.Multi-stage hydraulically fractured shale gas rate transient analysis[C]//North Africa Technical Conference and Exhibition，14-17 February 2010，Cairo，Egypt.DOI：https：//doi.org/10.2118/126754-MS.

[14] Brown M，Ozkan E，Raghavan R，et al.Practical solutions for pressure-transient responses of fractured horizontal wells in unconventional shale reservoirs[J].SPE Reservoir Evaluation & Engineering，2011，12(4)：663-676.

[15] Ilk D，Rushing J A，Perego A D，et al.Exponential vs.hyperbolic decline in tight gas sands-understanding the origin and implications for reserve estimates using Arps'decline curves[C]//SPE Annual Technical Conference and Exhibition，21-24 September 2008，Denver，Colorado，USA.DOI：https：//doi.org/10.2118/116731-MS.

[16] Kupchenko C L, Mattar L.Tight gas production performance using decline curves [C] //CIPC/SPE Gas Technology Symposium 2008 Joint Conference, 16–19 June 2008, Calgary, Alberta, Canada.DOI : https : //doi.org/10.2118/114991–MS.

[17] John L, Wattenbarger R A.Gas reservoir engineering [M] .Texas : SPE Inc, 1996.

[18] Ertekin T, King G R, Schewerer F C.Dynamic gas slippage : a unique dual–mechanism approach to the flow of gas in tight formations [J] .SPE Formation & Evaluation, 1986, 1 (6) : 43–52.

[19] 徐兵祥，李相方，Haghighi M，等．页岩气产量数据分析方法及产能预测[J]．中国石油大学学报：自然科学版，2013，37(3)：119–125.